Ensuring Secure and Ethical STM Research in the AI Era

Hewa Majeed Zangana
Duhok Polytechnic University, Iraq

Marwan Omar
Illinois Institute of Technology, USA

IGI Global
Scientific Publishing
Publishing Tomorrow's Research Today

Vice President of Editorial   Melissa Wagner
Director of Acquisitions   Mikaela Felty
Director of Book Development   Jocelynn Hessler
Production Manager   Mike Brehm
Cover Design   Phillip Shickler

Published in the United States of America by
       IGI Global Scientific Publishing
       701 East Chocolate Avenue
       Hershey, PA, 17033, USA
       Tel: 717-533-8845
       Fax: 717-533-7115
       Website: https://www.igi-global.com E-mail: cust@igi-global.com

Library of Congress Cataloging-in-Publication Data

Names: Zangana, Hewa Majeed, 1987- editor | Omar, Marwan, 1982- editor
Title: Ensuring secure and ethical STM research in the AI era / edited by
    Hewa Majeed Zangana, Marwan Omar.
Description: Hershey, PA : IGI Global Scientific Publishing, [2025] |
    Includes bibliographical references and index. | Summary: "This book
    aims to critically examine and address the complex challenges and
    opportunities that arise from the integration of artificial intelligence
    into science, technology, and medicine (STM) research"-- Provided by
    publisher.
Identifiers: LCCN 2025021277 (print) | LCCN 2025021278 (ebook) | ISBN
    9798337342528 h/c | ISBN 9798337342535 s/c | ISBN 9798337342542 ebook
Subjects: LCSH: Science--Research--Moral and ethical aspects |
    Science--Research--Technological innovations | Science--Research--Data
    processing | Artificial intelligence--Scientific applications |
    Artificial intelligence--Moral and ethical aspects | Artificial
    intelligence--Security measures
Classification: LCC Q180.55.M67 E57 2026 (print) | LCC Q180.55.M67
    (ebook)
LC record available at https://lccn.loc.gov/2025021277
LC ebook record available at https://lccn.loc.gov/2025021278

British Cataloguing in Publication Data
A Cataloguing in Publication record for this book is available from the British Library.

All work contributed to this book is new, previously-unpublished material.
The views expressed in this book are those of the authors, but not necessarily of the publisher.
This book contains information sourced from authentic and highly regarded references, with reasonable efforts made to ensure the reliability of the data and information presented. The authors, editors, and publisher believe the information in this book to be accurate and true as of the date of publication. Every effort has been made to trace and credit the copyright holders of all materials included. However, the authors, editors, and publisher cannot assume responsibility for the validity of all materials or the consequences of their use. Should any copyright material be found unacknowledged, please inform the publisher so that corrections may be made in future reprints.

Table of Contents

Detailed Table of Contents

Chapter 1

Hewa Majeed Zangana, Duhok Polytechnic University, Iraq
Marwan Omar, Illinois Institute of Technology, USA
Jamal N. Al-Karaki, Zayed University, UAE

The integration of Artificial Intelligence (AI) into scientific, technical, and medical (STM) research has unlocked transformative possibilities, enhancing discovery, analysis, and innovation at unprecedented scales. However, this rapid advancement also raises critical concerns regarding data privacy and security. This chapter explores the emerging standards, frameworks, and best practices designed to protect sensitive information within AI-driven STM research environments. It examines the ethical, legal, and technical challenges posed by AI systems, including data ownership, informed consent, bias mitigation, and cybersecurity risks. Furthermore, the chapter highlights international regulations, such as GDPR and HIPAA, and discusses the role of transparency, accountability, and secure data governance in maintaining the integrity and trustworthiness of AI-powered research. By offering a comprehensive overview, this chapter aims to equip researchers, policymakers, and technologists with the knowledge needed to navigate the evolving landscape of data protection in the AI era.

Chapter 2

Hewa Majeed Zangana, Duhok Polytechnic University, Iraq
Senny Luckyardi, Universitas Komputer Indonesia, Indonesia
Firas Mahmood Mustafa, Duhok Polytechnic University, Iraq
Shuai Li, University of Oulu, Finland

The integration of Artificial Intelligence (AI) into agricultural research has introduced transformative innovations that promise increased productivity, sustainability, and precision in farming practices. However, the rapid advancement and deployment of AI systems in this domain also bring significant ethical and security concerns. Issues such as data privacy, algorithmic bias, equitable access, environmental impacts, and transparency present complex challenges that must be carefully navigated. This chapter explores the critical ethical considerations and security risks associated with AI applications in agricultural research, while also highlighting the immense opportunities for advancing scientific understanding, improving food security, and promoting sustainable practices. By examining current trends, case studies, and regulatory frameworks, the chapter provides a comprehensive overview of how agricultural AI can be ethically and securely developed and deployed, ensuring that innovation serves both humanity and the environment responsibly.

Chapter 3

Dilshad Ahmad Mhia-alddin, University of Mosul, Iraq
Akram Mahmoud Hussein, University of Mosul, Iraq

The integration of Artificial Intelligence (AI) into Scientific, Technical, and Medical (STM) research has transformed traditional workflows, enabling accelerated discovery and innovation. However, the expiry of contracts governing AI-driven STM projects introduces complex legal challenges, potential risks, and evolving regulatory demands. Issues such as intellectual property ownership, data governance, confidentiality, and the continuity of AI systems post-contract termination require careful consideration. This chapter explores these multifaceted legal concerns, examines real-world case studies, identifies risks emerging from insufficient contract management, and proposes regulatory frameworks to mitigate negative outcomes. Through a critical analysis, it underscores the necessity for adaptive legal strategies that uphold the integrity, security, and ethical standards of STM research in the AI era.

Chapter 4

Noble Worlanyo Antwi, Illinois Institute of Technology, USA

This chapter investigates the critical role of threat detection in securing multi-cloud environments, a rapidly evolving area as organizations adopt platforms such as Amazon Web Services (AWS), Microsoft Azure, and Google Cloud Platform (GCP). It analyzes traditional security mechanisms, including firewalls and intrusion detection systems, highlighting their limitations in cloud-native infrastructures. The chapter explores advanced practices such as Artificial Intelligence (AI)-driven analytics, Machine Learning (ML), User and Entity Behavior Analytics (UEBA), Zero Trust Architecture (ZTA), and Extended Detection and Response (XDR). A vendor-neutral threat detection architecture is proposed for centralized monitoring and automated incident response. Ethical, legal, and compliance considerations are also discussed, aligning security practices with standards like GDPR and HIPAA. The chapter concludes with recommendations for holistic, intelligence-driven security and identifies future research opportunities in cross-cloud frameworks and explainable AI.

In the contemporary era of digital transformation, English remains the lingua franca of scientific discourse, serving as a gateway to global recognition and academic collaboration. Simultaneously, artificial intelligence (AI) tools—ranging from grammar checkers to large language models—are reshaping the landscape of scientific writing. This chapter critically explores the intersection of English language proficiency and AI-powered writing assistance in science, technology, and medical (STM) research. It discusses the ethical implications of AI-generated content, authorship attribution, and linguistic biases, while considering the academic pressures faced by non-native English-speaking researchers. By examining both opportunities and challenges, this chapter emphasizes the need for equitable language practices and responsible AI use to maintain the integrity and inclusivity of scientific communication.

The integration of Artificial Intelligence (AI) into scientific, technological, and medical(STM) research raises critical questions about balancing open science and responsible science. Open science promotes transparency, accessibility, and knowledge sharing to foster collaboration, innovation, and reproducibility. However, as sensitive data and proprietary technologies become central to research, security and intellectual property protection are significant concerns. This chapter explores the tension between these objectives and provides a framework for balancing transparency with security in AI-driven research. It discusses the ethical, legal, and technical challenges faced by researchers and institutions in adopting practices that promote both open access and secure data management. The chapter offers insights into reconciling open science initiatives with responsible science practices that prioritize privacy, confidentiality, and safety, highlighting the importance of policies and strategies for secure and ethical research practices.

Chapter 7

Ben Kennedy, Capitol Technology University, USA
Atif Mohammad, Capitol Technology University, USA
Matthew Wyandt, Capitol Technology University, USA

The Synthetic Cognitive Augmentation Network Using Experts (SCANUE) significantly advances the foundational Synthetic Cognitive Augmentation Network from a conceptual model into a sophisticated experimental platform designed to meet complex cognitive needs. SCANUE's specialized agents depend critically on high-quality, domain-specific datasets that accurately reflect intricate cognitive tasks such as nuanced planning, emotional inference, and conflict resolution, closely replicating human prefrontal cortex functions (Tate et al., 2024b). To address limitations inherent in generic data sources, SCANUE integrates several public dialogue corpora, including Taskmaster-1, MultiWOZ, DailyDialog, GoEmotions, and SocialIQa, into a unified framework (Byrne et al., 2019; Budzianowski et al., 2018; Li & Ding, 2017; Demszky et al., 2020; Sap et al., 2019), supplemented with AI-generated dialogues. This modular and biologically inspired system, enhanced by adaptive learning and Human-in-the-Loop methodologies, effectively aligns with user intentions to advance cognitive augmentation.

Chapter 8

Rebet Keith Jones, Capitol Technology University, USA

The rapid integration of artificial intelligence (AI) and machine learning (ML) into biomedical and health research has the potential to transform patient care, diagnosis, and treatment outcomes. However, as these technologies evolve, concerns surrounding algorithmic bias and fairness have emerged. In the context of healthcare, biased algorithms can exacerbate disparities in health outcomes, leading to inequality in care and undermining trust in AI-driven systems. This chapter explores the ethical implications of algorithmic bias in biomedical research, focusing on the factors contributing to bias in datasets, model design, and decision-making processes. Additionally, it examines various strategies and frameworks aimed at promoting fairness and equity in AI applications. Through a multidisciplinary lens, the chapter presents a critical analysis of how algorithmic fairness can be achieved, with particular emphasis on practical solutions and regulatory considerations to safeguard both the integrity of research and the well-being of diverse patient populations

Preface

INTRODUCTION

The accelerating integration of Artificial Intelligence (AI) into scientific, technical, and medical (STM) research is reshaping how knowledge is created, analyzed, and shared. From predictive modeling in healthcare to automated data interpretation in environmental science, AI technologies are not merely tools—they are becoming co-creators in research. Yet, with this transformative potential comes a complex array of new responsibilities. Researchers, institutions, and policymakers must now contend with challenges that range from data privacy and system security to algorithmic fairness and equitable access.

This reference volume was conceived in response to a growing recognition: that AI's impact on STM research is not just technical—it is deeply ethical, legal, and societal. As editors, we have gathered a range of expert voices to explore the most pressing issues at this intersection. Each chapter offers a distinct lens, grounded in practical experience and scholarly rigor, while collectively contributing to a broader understanding of responsible AI deployment. Whether examining contractual pitfalls in AI-based research projects or questioning linguistic biases in AI-assisted scientific writing, the volume addresses both foundational and emerging topics.

We believe this book fills a critical gap in the current discourse by offering a structured, multidisciplinary view of AI's influence on scientific practice. Rather than treat privacy, ethics, and governance as ancillary concerns, these chapters place them at the center of inquiry. In doing so, this volume serves not only as a technical reference but also as a guide for reflection, helping readers navigate the evolving obligations and opportunities that AI introduces into the research landscape.

The concluding chapters, including those on synthetic cognitive augmentation and algorithmic bias in biomedical research, point toward a future in which AI does far more than automate tasks—it shapes knowledge itself. These contributions challenge us to ask fundamental questions about trust, accountability, and inclusivity in AI-powered systems. They highlight that fairness in design, transparency in function, and inclusiveness in outcomes must be considered integral to any innovation strategy, not optional enhancements.

Moreover, chapters on cloud security, responsible open science, and AI's role in linguistic equity remind us that technological advancement must be tempered by social and institutional foresight. These discussions emphasize the importance of infrastructure—both technical and policy-based—that can sustain long-term innovation while safeguarding core scientific values. By providing frameworks and case studies across disciplines, the book equips readers to make informed, ethically grounded decisions in complex and evolving contexts.

Ultimately, our goal is to offer a resource that supports the thoughtful development and deployment of AI within STM research. As editors, we thank the authors for their intellectual contributions and commitment to responsible innovation. We invite readers—from researchers and technologists to institutional leaders and policymakers—to use this volume as a springboard for deeper engagement with the urgent questions posed by AI. The path forward requires not only technical excellence but also ethical clarity, legal preparedness, and a shared vision of science that serves both progress and the public good.

CHAPTER OVERVIEWS

Chapter 1: Data Privacy and Security Standards in AI-Powered Scientific Research

This chapter explores the ethical, legal, and technical dimensions of data privacy and security in AI-enabled scientific, technical, and medical (STM) research. It surveys key regulatory frameworks such as GDPR and HIPAA, and addresses challenges including data ownership, informed consent, and cybersecurity risks. Emphasizing transparency, accountability, and secure data governance, the chapter provides researchers and policymakers with best practices for protecting sensitive information in AI-intensive research environments.

Chapter 2: Ethical and Secure AI Applications in Agricultural Research: Challenges and Opportunities

Focusing on the agricultural domain, this chapter examines the dual potential and peril of AI in transforming farming and food systems. It analyzes ethical concerns such as algorithmic bias, environmental impact, and equitable access, alongside data security and regulatory issues. Through case studies and a review of current trends, the chapter outlines how AI can be ethically harnessed to promote sustainability, food security, and inclusive innovation in agricultural research.

Chapter 3: Contract Expiry in AI-Driven STM Research Legal Challenges, Risks, and Regulatory Considerations

This chapter delves into the legal complexities that arise when contracts governing AI-based STM research projects expire. It explores issues such as intellectual property rights, data continuity, and system shutdowns, emphasizing the importance of proactive contract management. Drawing on real-world examples and emerging regulatory trends, the chapter proposes adaptive legal frameworks to mitigate risks and uphold research integrity in the post-contract landscape.

Chapter 4: Threat Detection in Multi-Cloud Environments

As organizations increasingly adopt multi-cloud infrastructures, this chapter addresses the evolving landscape of threat detection and cybersecurity. It critiques the limitations of traditional tools and introduces advanced AI- and ML-driven strategies such as User and Entity Behavior Analytics (UEBA) and Zero Trust Architecture (ZTA). A vendor-neutral detection model is proposed, along with discussions on regulatory alignment and future research opportunities in explainable AI and cross-cloud threat intelligence.

Chapter 5: The Role of English Language and AI in Scientific Writing: Ethical and Academic Implications

This chapter investigates the intersection of AI tools and English as the dominant language in STM communication. It explores the academic and ethical implications of AI-generated content, particularly for non-native English speakers facing linguistic and institutional pressures. By analyzing authorship, bias, and inclusivity, the chapter calls for equitable language practices and responsible AI use to maintain fairness and integrity in global scientific discourse.

Chapter 6: Open Science vs. Responsible Science – Balancing Transparency and Security

This chapter examines the tension between the principles of open science—transparency, collaboration, and accessibility—and the growing need for data security and intellectual property protection. It outlines a strategic framework for balancing these objectives in AI-driven research, addressing legal, ethical, and institutional challenges. The chapter offers practical guidance for harmonizing openness with responsibility, ensuring both scientific progress and secure knowledge stewardship.

Chapter 7: Beyond Intelligence: The Synthetic Cognitive Augmentation Network Using Experts

Detailing the development of the Synthetic Cognitive Augmentation Network Using Experts (SCANUE), this chapter presents an experimental AI platform designed to emulate complex human cognitive functions. It discusses the integration of specialized public and synthetic datasets, as well as adaptive and human-in-the-loop methodologies. By closely modeling tasks like emotional inference and conflict resolution, SCANUE advances the frontier of biologically inspired AI for cognitive augmentation.

Chapter 8: Algorithmic Bias and Fairness in Biomedical and Health Research

Focusing on the critical issue of fairness in AI-powered healthcare, this chapter explores how algorithmic bias can influence clinical outcomes and exacerbate health disparities. It analyzes the sources of bias in data and model design and reviews existing frameworks for promoting equity in biomedical AI applications. Through a multidisciplinary lens, the chapter presents strategies for ensuring ethical, inclusive, and trustworthy AI in health research and practice.

Chapter 1
Data Privacy and Security Standards in AI-Powered Scientific Research

Hewa Majeed Zangana
https://orcid.org/0000-0001-7909-254X
Duhok Polytechnic University, Iraq

Marwan Omar
Illinois Institute of Technology, USA

Jamal N. Al-Karaki
https://orcid.org/0009-0000-7833-3970
Zayed University, UAE

ABSTRACT

The integration of Artificial Intelligence (AI) into scientific, technical, and medical (STM) research has unlocked transformative possibilities, enhancing discovery, analysis, and innovation at unprecedented scales. However, this rapid advancement also raises critical concerns regarding data privacy and security. This chapter explores the emerging standards, frameworks, and best practices designed to protect sensitive information within AI-driven STM research environments. It examines the ethical, legal, and technical challenges posed by AI systems, including data ownership, informed consent, bias mitigation, and cybersecurity risks. Furthermore, the chapter highlights international regulations, such as GDPR and HIPAA, and discusses the role of transparency, accountability, and secure data governance in maintaining the integrity and trustworthiness of AI-powered research. By offering a comprehensive overview, this chapter aims to equip researchers, policymakers, and technologists with the knowledge needed to navigate the evolving landscape of

DOI: 10.4018/979-8-3373-4252-8.ch001

data protection in the AI era.

1. INTRODUCTION

Artificial Intelligence (AI) is revolutionizing the landscape of scientific, technical, and medical (STM) research by enabling unprecedented levels of data analysis, automation, and innovation. From accelerating drug discovery to predicting environmental changes, AI systems are now central to many groundbreaking advances. However, this transformation also introduces profound challenges related to data privacy and security, particularly given the sensitivity and scale of the datasets involved (Alhitmi, Mardiah, Al-Sulaiti, & Abbas, 2024; Brightwood & Jame, 2024).

In AI-powered research ecosystems, vast amounts of personal, proprietary, and often confidential information are collected, processed, and analyzed. Ensuring the security and privacy of this data is not just a technical necessity but also a legal and ethical imperative. Breaches or mishandling of sensitive data can undermine public trust, violate regulatory requirements, and hinder scientific progress (Gawankar, Nair, Pawar, Vhatkar, & Chavan, 2024; Almeida & Barr, 2025).

Recent scholarly discussions have highlighted how AI's unique capabilities—such as deep learning and predictive analytics—complicate traditional approaches to privacy protection. Unlike classical systems, AI models often derive insights from patterns hidden within enormous and heterogeneous datasets, making the identification, classification, and safeguarding of sensitive data a nontrivial task (Abolaji & Akinwande, 2024; Akhtar & Rawol, 2024).

Moreover, the dynamic nature of AI training and inference presents risks such as data leakage, adversarial attacks, and model inversion, where attackers may reconstruct sensitive information from trained models (Awad, Babu, Barka, & Shuaib, 2024; Arya, Sharma, Devi, & Padmanaban, 2024). These risks are particularly critical in sectors like healthcare, where personal health information must be protected under strict regulatory regimes like HIPAA and GDPR (Arefin, 2024; Dhinakaran, Raja, Jasmine, Kumar, & Ramani, 2025).

Several innovative frameworks and solutions are emerging to address these challenges. Privacy-preserving techniques such as differential privacy, federated learning, and secure multiparty computation are being increasingly incorporated into AI workflows to minimize risks while maintaining model performance (Kumar, Lokeshwari, & Shanmugam, 2024; Arya, Sharma, Devi, Padmanaban, & Kumar, 2023). Similarly, researchers advocate for AI systems that are transparent, accountable, and designed with privacy by design principles from their inception (Abbas & Qazi, 2024; Ismail & Aloshi, 2025).

The integration of encryption methods, secure cloud infrastructures, and AI-driven threat detection systems also plays a pivotal role in safeguarding STM research environments (Gopireddy, 2021; Mohammed, Omar, & Nguyen, 2018). In this regard, advances in AI-based cybersecurity mechanisms, including anomaly detection and zero-day attack mitigation, are proving to be game-changers (Jones & Omar, 2024; Mbah & Evelyn, 2024).

However, the evolving complexity of data ecosystems—spanning cloud platforms, Internet of Things (IoT) devices, and even AI-powered biomedical sensors—demands a multidisciplinary approach that incorporates legal, technological, and ethical perspectives (Anidjar, Packin, & Panezi, 2023; Bhamidipaty et al., 2025). Without such a holistic view, security measures may lag behind the rapid pace of AI innovation, exposing critical research infrastructures to vulnerabilities (Farea, Alhazmi, Samet, & Guzel, 2024; Huff et al., 2023).

Further compounding these issues is the globalization of STM research, which often involves cross-border data flows and collaborations. This raises complex questions about jurisdiction, compliance, and harmonization of privacy standards (Gupta, Amarnani, Soanki, & Kishore, 2025; Gemiharto & Masrina, 2024). For example, privacy expectations and regulations can vary dramatically between regions, making it essential for researchers and institutions to adopt adaptable, globally informed data governance strategies (Daraf & Badi, 2023; Gholami & Omar, 2024).

In addition, emerging discussions around synthetic data and data anonymization techniques suggest promising avenues to balance innovation with privacy (Gholami & Omar, 2023). Yet, these methods are not without their own ethical dilemmas, such as ensuring that synthetic datasets do not inadvertently introduce biases or inaccuracies into AI models.

In response to these growing challenges, scholars and practitioners are advocating for a new generation of data privacy and security standards specifically tailored to AI-powered research (Omar & Zangana, 2025; Jones, Omar, Mohammed, Nobles, & Dawson, 2023). These standards emphasize adaptive risk management, dynamic compliance frameworks, and the integration of ethical considerations into every phase of AI system development and deployment.

Thus, as STM research continues to be transformed by AI technologies, it becomes imperative to not only advance our technical defenses but also cultivate a research culture rooted in privacy awareness, ethical responsibility, and resilient security architectures (Nguyen, Mohammed, Omar, & Banisakher, 2018; Hamza & Omar, 2013).

The following sections of this chapter will delve deeper into:

- The key challenges and vulnerabilities in AI-powered scientific research.
- Current regulatory landscapes and compliance requirements.

- Cutting-edge privacy-enhancing technologies (PETs) and secure data management frameworks.
- Practical strategies for researchers to implement secure and ethical AI applications in STM domains.

By providing an integrative overview, this chapter aims to offer researchers, technologists, and policymakers a roadmap for navigating the complexities of data privacy and security in AI-enhanced scientific inquiry.

To better illustrate the growing concerns around AI-driven research, Figure 1 presents a flowchart summarizing the main data privacy challenges emerging in AI-powered scientific research.

Figure 1. Flowchart of data privacy challenges in AI-powered research

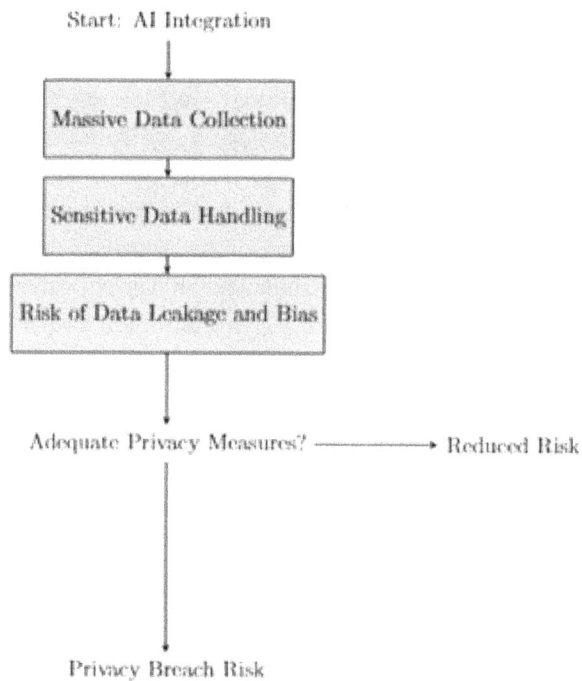

2. UNDERSTANDING DATA TYPES IN SCIENTIFIC RESEARCH

Scientific research heavily depends on the accurate collection, classification, and analysis of data. Different types of data serve distinct roles and purposes, which underscores the importance of understanding these categories to ensure reliable research outcomes, particularly as artificial intelligence (AI) and emerging technologies become increasingly embedded in scientific practice (Abbas & Qazi, 2024; Gholami & Omar, 2023).

Figure 2 presents a high-level visualization of the primary types of data encountered in scientific research, highlighting the relative importance of each in the context of AI integration.

Figure 2. Types of data in scientific research

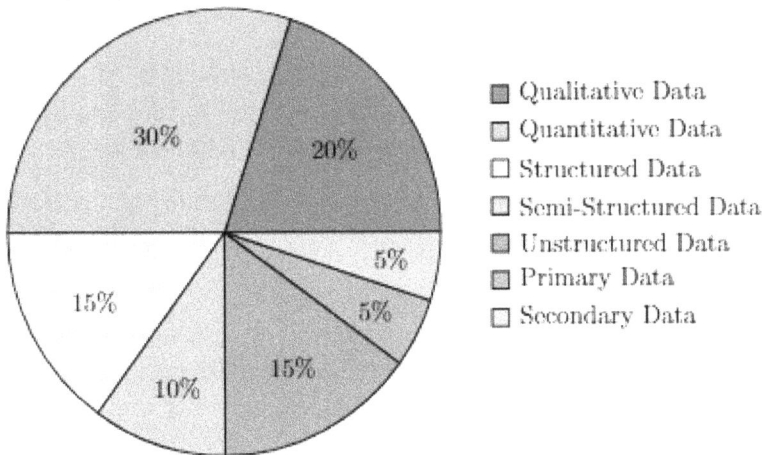

2.1. Qualitative and Quantitative Data

At the broadest level, scientific data is categorized into **qualitative** and **quantitative** types.

- **Qualitative data** describes qualities or characteristics that are often non-numeric, focusing on attributes, categories, and descriptions. It is crucial in fields like sociology, psychology, and digital communication systems, where the interpretation of human behavior and interaction is key (Gemiharto & Masrina, 2024; Brightwood & Jame, 2024).

- **Quantitative data**, on the other hand, involves measurable quantities and numeric information. It forms the backbone of experimental sciences, healthcare studies, and cybersecurity analytics (Awad, Babu, Barka, & Shuaib, 2024; Arya, Sharma, Devi, & Padmanaban, 2024).

As research domains evolve—especially with the integration of AI-powered technologies—both types are becoming increasingly intertwined. In healthcare, for example, AI sensors collect quantitative biometric data, while patient feedback might represent qualitative insights (Bhamidipaty et al., 2025).

2.2. Structured, Semi-Structured, and Unstructured Data

Understanding data structures is essential for leveraging AI tools and big data analytics:

- **Structured data** is highly organized and easily searchable, fitting neatly into databases and spreadsheets. For instance, hospital records and genomic sequencing outputs are structured data critical to AI-driven healthcare (Daraf & Badi, 2023; Arefin, 2024).
- **Semi-structured data** does not conform rigidly to tabular formats but contains organizational properties, like metadata in digital communication (Gemiharto & Masrina, 2024).
- **Unstructured data**, such as images, videos, and text, is more challenging to analyze but holds immense potential. AI advances in areas like cybersecurity and education increasingly rely on mining insights from unstructured datasets (Mbah & Evelyn, 2024; Ismail & Aloshi, 2025).

The explosion of semi-structured and unstructured data, particularly in IoT environments and AI-driven healthcare, demands advanced AI models capable of nuanced interpretation (Farea, Alhazmi, Samet, & Guzel, 2024; Arya et al., 2023).

2.3. Primary and Secondary Data

- **Primary data** refers to information collected firsthand by researchers. This type of data is essential for developing new AI-driven cybersecurity mechanisms and personalized medicine applications (Akhtar & Rawol, 2024; Dhinakaran, Raja, Jasmine, Kumar, & Ramani, 2025).
- **Secondary data** involves the use of existing datasets. In domains like AI-powered marketing and finance, researchers heavily depend on secondary

data to analyze trends and enhance privacy strategies (Alhitmi, Mardiah, Al-Sulaiti, & Abbas, 2024; Gupta, Amarnani, Soanki, & Kishore, 2025).

Researchers must assess the validity, reliability, and privacy implications of secondary datasets, particularly given heightened data privacy concerns in education and business sectors (Ismail & Aloshi, 2025; Brightwood & Jame, 2024).

2.4. Big Data and Real-Time Data

In contemporary scientific research, **big data** and **real-time data** streams have become crucial:

- **Big Data** consists of massive datasets characterized by volume, velocity, and variety. For instance, healthcare organizations are using big data combined with AI to improve diagnostics and patient privacy (Bhamidipaty et al., 2025; Almeida & Barr, 2025).
- **Real-time data**, often seen in cloud environments, enables instantaneous decision-making—a critical feature in cybersecurity for detecting zero-day threats (Jones & Omar, 2024; Gopireddy, 2021).

AI-powered systems must handle these data types while safeguarding against breaches and ensuring regulatory compliance (Huff et al., 2023; Gawankar, Nair, Pawar, Vhatkar, & Chavan, 2024).

2.5. Sensitive and Non-Sensitive Data

Data sensitivity classification is increasingly vital in AI applications:

- **Sensitive data** includes personally identifiable information (PII), protected health information (PHI), and confidential corporate records. Protecting such data is crucial for maintaining trust and compliance with global standards (Almeida & Barr, 2025; Anidjar, Packin, & Panezi, 2023).
- **Non-sensitive data** can be freely distributed without harming individuals or organizations.

AI solutions like privacy-preserving encoding mechanisms and differential privacy methods are being actively developed to enhance protection of sensitive data (Farea et al., 2024; Kumar, Lokeshwari, & Shanmugam, 2024).

2.6. Synthetic and Augmented Data

Given the limitations in accessing high-quality, privacy-compliant data, **synthetic data**—artificially generated datasets—has emerged as a crucial asset.

Gholami and Omar (2023, 2024) highlight that synthetic data can improve model training efficiency while mitigating privacy risks. Moreover, it fosters safer innovation across domains such as finance, healthcare, and education.

2.7. Data Privacy and Ethical Concerns

Finally, the understanding of data types must be grounded in strong ethical principles. Across all sectors—from AI in IoT devices (Awad et al., 2024) to AI-driven healthcare (Dhinakaran et al., 2025)—ethical data management, privacy preservation, and transparency are non-negotiable imperatives.

Emerging frameworks like the AI-powered threat detection systems and AI-powered cybersecurity strategies provide promising pathways toward secure, privacy-compliant innovation (Mbah & Evelyn, 2024; Mohammed, Omar, & Nguyen, 2018).

Moreover, researchers must remain vigilant about the policy implications of using varied data types, particularly as regulations evolve globally in response to AI's disruptive potential (Omar & Zangana, 2025; Nguyen, Mohammed, Omar, & Banisakher, 2018).

3. THREAT LANDSCAPE IN AI-POWERED RESEARCH

The rapid integration of Artificial Intelligence (AI) into research environments—across healthcare, business, security, education, and beyond—has transformed traditional processes into intelligent, automated systems. However, this transformation has simultaneously expanded the threat landscape, introducing complex challenges concerning data privacy, cybersecurity, ethical governance, and regulatory compliance.

3.1. Data Privacy and Security Vulnerabilities

AI-powered research often involves handling sensitive personal, medical, financial, and operational data. The deployment of AI in cloud environments enhances computational capacity but simultaneously introduces vulnerabilities such as data breaches and unauthorized access (Gopireddy, 2021; Hamza & Omar, 2013). In sectors like healthcare, AI systems used for diagnosis and management significantly heighten concerns regarding patient data confidentiality (Arefin, 2024; Gawankar et al., 2024; Bhamidipaty et al., 2025).

Moreover, AI-driven marketing and financial platforms are increasingly at risk of data exploitation, where massive datasets become lucrative targets for cyber-criminals (Alhitmi et al., 2024; Brightwood & Jame, 2024). The complexity grows when research organizations rely on third-party cloud vendors without stringent data protection frameworks, leading to vulnerabilities in the data supply chain (Nguyen, Mohammed, Omar, & Banisakher, 2018).

3.2. AI-Augmented Cybersecurity Threats

While AI itself can bolster cybersecurity, its misuse has accelerated the so-phistication of cyber threats. Attackers now exploit machine learning models for adversarial attacks, data poisoning, and model inversion (Akhtar & Rawol, 2024; Jones & Omar, 2024). Furthermore, AI-based systems for IoT security introduce new threat vectors, especially when inadequate attention is paid to the lifecycle security of deployed devices (Arya, Sharma, Devi, & Padmanaban, 2024; Arya, Sharma, Devi, Padmanaban, & Kumar, 2023; Awad et al., 2024).

Healthcare research labs, biotechnology firms, and academic institutions are prime targets for cyberespionage campaigns that seek to disrupt critical operations or exfiltrate sensitive information (Huff et al., 2023). The challenge lies not only in securing AI algorithms but also in maintaining the integrity of the research data they process.

3.3. Privacy Challenges in Specialized Research Domains

Specific domains such as genomic analysis, finance, education, and healthcare are uniquely vulnerable to AI-driven privacy violations. AI-powered genomic analysis platforms, although promising for personalized medicine, pose enormous risks to medical data privacy (Daraf & Badi, 2023). Similarly, AI-powered finance introduces ethical dilemmas regarding the profiling and behavioral prediction of users (Brightwood & Jame, 2024; Gupta, Amarnani, Soanki, & Kishore, 2025).

In education, the integration of AI for student data analysis demands robust frameworks to prevent misuse and ensure ethical handling of personal records (Ismail & Aloshi, 2025). The health sector also grapples with safeguarding data used in AI-powered wellness applications, which can unintentionally leak personal health patterns if improperly secured (Dhinakaran et al., 2025).

3.4. Emerging Risks in the Metaverse and Smart Systems

The evolution of AI into extended reality platforms (e.g., metaverse) introduces novel dimensions of privacy and security risks. Infrastructures supporting AI-powered metaverse applications are susceptible to large-scale surveillance, unauthorized profiling, and breaches of private communication (Anidjar, Packin, & Panezi, 2023; Gemiharto & Masrina, 2024).

Likewise, the reliance on AI-driven smart sensors in healthcare and environmental monitoring necessitates stringent measures to avoid unauthorized tracking and data exfiltration (Bhamidipaty et al., 2025). The convergence of AI with IoT ecosystems demands privacy-preserving encoding mechanisms to mitigate large-scale security incidents (Farea, Alhazmi, Samet, & Guzel, 2024).

3.5. Ethical, Legal, and Regulatory Dimensions

As AI expands into sensitive research areas, ethical and legal dilemmas intensify. Ethical concerns arise when AI systems are used to automate decision-making processes without transparency or accountability (Almeida & Barr, 2025). Regulatory frameworks lag behind technological advancements, making it challenging for organizations to comply with global standards on data privacy and cybersecurity (Abolaji & Akinwande, 2024; Abbas & Qazi, 2024).

Researchers and policymakers advocate for AI systems that are not only technically robust but also legally and ethically aligned, promoting patient empowerment, digital fairness, and trust (Almeida & Barr, 2025; Gholami & Omar, 2024).

3.6. Synthetic Data and AI Model Security

Recent research highlights the use of synthetic data to enhance AI model efficiency and privacy-preserving capabilities. However, synthetic datasets themselves can introduce biases and vulnerabilities if not carefully validated (Gholami & Omar, 2023). Moreover, when student models trained on synthetic data are deployed at scale, their security performance must match the robustness of their teacher models to prevent exploitable weaknesses (Gholami & Omar, 2024).

3.7. Strategic Countermeasures and Future Outlook

The threat landscape necessitates an integrated approach combining AI-driven threat detection with strategic cybersecurity practices (Mbah & Evelyn, 2024; Akhtar & Rawol, 2024). Techniques such as adaptive data protection (Kumar, Lokeshwari, & Shanmugam, 2024) and AI-enhanced attack detection (Jones & Omar, 2024;

Jones, Omar, Mohammed, Nobles, & Dawson, 2023) are at the forefront of current defensive strategies.

Additionally, the publication of new frameworks and research—such as *Digital Forensics in the Age of AI* (Omar & Zangana, 2025)—emphasizes the critical need to innovate forensic practices in light of AI-driven threats.

Ongoing collaboration between AI researchers, cybersecurity professionals, and policymakers will be crucial in safeguarding the future of AI-powered research. Organizations must invest not only in technological solutions but also in ethical training, legal compliance, and governance structures to stay resilient against emerging threats.

4. PRIVACY-PRESERVING TECHNIQUES IN AI

As artificial intelligence (AI) systems become increasingly integrated into everyday life—from healthcare and finance to education and social media—the protection of sensitive data has emerged as a critical concern. The rise of AI-powered solutions necessitates the development of robust privacy-preserving techniques to ensure user trust, regulatory compliance, and ethical standards. This section explores various strategies, frameworks, and innovations designed to safeguard privacy in AI systems.

4.1. Privacy Challenges in AI-Powered Systems

AI technologies inherently require large volumes of data for training and optimization, often exposing personal, financial, and healthcare information to vulnerabilities (Alhitmi, Mardiah, Al-Sulaiti, & Abbas, 2024; Brightwood & Jame, 2024). The misuse, unauthorized access, and potential biases introduced through data-centric AI processes have amplified the urgency to adopt privacy-focused mechanisms.

For instance, in AI-driven marketing and finance, the analysis of user profiles and behaviors increases the risk of unauthorized surveillance and profiling (Alhitmi et al., 2024; Brightwood & Jame, 2024). Similarly, healthcare systems utilizing AI for diagnosis and treatment recommendations must prioritize patient confidentiality against breaches (Gawankar, Nair, Pawar, Vhatkar, & Chavan, 2024).

4.2. AI-Driven Privacy-Preserving Techniques

a.   Differential Privacy

Differential privacy introduces statistical noise to datasets, allowing AI models to learn patterns without exposing individual data entries. This method is particularly effective in sectors like education (Ismail & Aloshi, 2025) and healthcare (Arefin, 2024), where the protection of personal information is paramount.

b. Federated Learning

Federated learning enables AI models to train across decentralized devices without transferring raw data to a central server. This model has proven essential in protecting IoT environments (Farea, Alhazmi, Samet, & Guzel, 2024) and healthcare data (Bhamidipaty et al., 2025).

c. Homomorphic Encryption

Homomorphic encryption allows computations to be performed on encrypted data without needing decryption, ensuring that data remains confidential during AI processing (Farea et al., 2024).

d. Synthetic Data Generation

Another emerging approach is the creation of synthetic datasets to replace real user information. Synthetic data retains statistical properties without compromising individual privacy (Gholami & Omar, 2023). It is increasingly used to enhance AI training while mitigating privacy risks across sectors.

e. AI-Powered Adaptive Data Protection Frameworks

Kumar, Lokeshwari, and Shanmugam (2024) proposed an AI-powered privacy preservation framework that dynamically adapts data protection levels based on context and user sensitivity, ensuring real-time privacy adjustments.

Figure 3 compares the relative adoption rates of various privacy-preserving techniques in AI-powered research environments.

Figure 3. Adoption rates of privacy-preserving techniques in AI systems

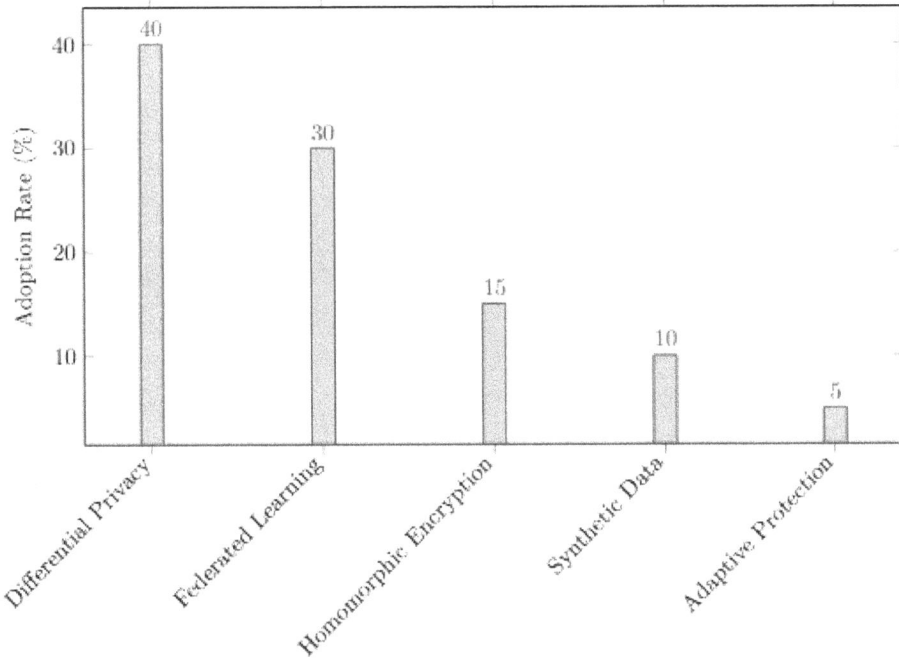

4.3. Sector-Specific Implementations

a.   Healthcare

In healthcare, AI-powered threat detection systems have become vital in protecting sensitive patient data (Arefin, 2024; Dhinakaran, Raja, Jasmine, Kumar, & Ramani, 2025). AI is leveraged to monitor anomalies and secure electronic health records while maintaining compliance with privacy regulations (Gawankar et al., 2024; Huff et al., 2023).

b.   Business and Finance

The finance sector applies AI to secure customer transactions and detect fraud while preserving client data confidentiality (Brightwood & Jame, 2024; Gupta, Amarnani, Soanki, & Kishore, 2025). Customized AI privacy configurations are also emerging on social media platforms to offer users more control over their data (Abbas & Qazi, 2024).

c. Internet of Things (IoT)

The exponential growth of IoT devices necessitates AI-driven biometric security measures (Awad, Babu, Barka, & Shuaib, 2024; Arya, Sharma, Devi, & Padmanaban, 2024). AI enhances the privacy of communication between interconnected devices through encryption and decentralized control.

d. Cloud Computing

Cloud-based AI genomic analysis applications are advancing precision medicine while emphasizing strict medical data privacy (Daraf & Badi, 2023). Similarly, AI enhances cloud security by detecting abnormal access patterns and preventing cyber threats (Gopireddy, 2021; Hamza & Omar, 2013).

e. Education

AI applications in education must navigate sensitive student data. Ismail and Aloshi (2025) emphasized solutions such as privacy-aware machine learning algorithms and secure cloud storage practices.

4.4. AI in Digital Communication and the Metaverse

The expansion of AI-powered communication systems and the metaverse has heightened concerns around privacy in virtual environments (Gemiharto & Masrina, 2024; Anidjar, Packin, & Panezi, 2023). Strategies like end-to-end encryption, identity anonymization, and digital consent frameworks are critical in safeguarding users' personal spaces.

4.5. Future Perspectives and Ethical Considerations

Emerging strategies call for a privacy-by-design approach, integrating privacy at every stage of AI development and deployment (Mbah & Evelyn, 2024). Ethical considerations, including transparency, accountability, and user empowerment, must be central to AI innovation (Almeida & Barr, 2025).

Moreover, the use of AI-enhanced intrusion detection systems like GEADD (Jones & Omar, 2024) illustrates the potential for AI not only to protect but also to revolutionize cybersecurity standards.

The role of generative AI in future healthcare (Bhamidipaty et al., 2025) and wellness management (Dhinakaran et al., 2025) emphasizes AI's dual responsibility: innovation with unwavering commitment to privacy.

Additionally, contributions from Gholami and Omar (2024) suggest that even student-level AI models, when trained responsibly, can respect and preserve data privacy while achieving high performance.

4.6. Integrated AI Privacy Solutions and Frameworks

Organizations are increasingly implementing integrated AI solutions combining encoding mechanisms, federated architectures, and real-time threat detection (Farea et al., 2024; Jones, Omar, Mohammed, Nobles, & Dawson, 2023). AI must act not only as a tool for data utilization but as an active defender of privacy (Mbah & Evelyn, 2024).

Furthermore, works such as *Digital Forensics in the Age of AI* (Omar & Zangana, 2025) stress the importance of digital forensic readiness in AI-driven systems to trace, audit, and rectify any privacy violations swiftly.

4.7. Summary

In conclusion, the evolution of AI privacy-preserving techniques is driven by an urgent need to balance technological advancement with the protection of individual rights. Strategies like differential privacy, federated learning, synthetic data generation, and AI-adaptive frameworks offer promising pathways. Sector-specific applications demonstrate the real-world relevance and necessity of privacy-preserving AI, while future frameworks call for even deeper integration of ethical, technical, and regulatory safeguards.

By embedding privacy into the very fabric of AI innovation, stakeholders can ensure a future where technology empowers rather than endangers users.

5. SECURITY STANDARDS AND FRAMEWORKS

In the rapidly evolving landscape of artificial intelligence (AI) and digital technologies, establishing robust **security standards and frameworks** is crucial to protect data integrity, user privacy, and organizational resilience. Security standards serve as a structured set of guidelines, ensuring systems are developed, deployed, and maintained with adequate defenses against modern threats (Abbas & Qazi, 2024; Abolaji & Akinwande, 2024). In particular, the integration of AI into systems has amplified both opportunities and vulnerabilities, necessitating innovative frameworks that not only address traditional cybersecurity concerns but also AI-specific risks.

5.1 Evolution of Security Standards in the Age of AI

The inclusion of **AI-powered mechanisms** in various domains — including finance, healthcare, IoT, and education — has transformed traditional security models (Akhtar & Rawol, 2024; Alhitmi et al., 2024). Frameworks are no longer static checklists but dynamic, evolving protocols that adapt in real-time to threats detected by AI-enhanced systems (Arya et al., 2023; Arya et al., 2024).

The emergence of **adaptive security frameworks** powered by machine learning allows continuous learning from threats, significantly improving proactive defense mechanisms (Gopireddy, 2021; Mbah & Evelyn, 2024). These AI-driven frameworks emphasize **predictive security** and **risk-based authentication** as core components.

5.2 Key AI-Powered Security Frameworks

Several innovative frameworks have been proposed and adopted:

- **AI-Powered Threat Detection Frameworks**: Utilizing machine learning to monitor patterns and identify anomalies, especially in healthcare data protection (Arefin, 2024; Gawankar et al., 2024).
- **Biometric Authentication Standards for IoT**: AI-driven biometrics systems, such as facial recognition and voice authentication, are emerging as industry standards for IoT security (Awad et al., 2024).
- **Cloud Security Enhancement**: AI-enabled security protocols safeguard cloud-based infrastructures against evolving threats, focusing on encryption, anomaly detection, and intelligent access management (Hamza & Omar, 2013; Mohammed et al., 2018).

Figure 4 illustrates the layered architecture of a typical AI-powered security framework, showcasing the interaction between core AI functionalities and traditional cybersecurity measures.

Figure 4. AI-powered security framework layers

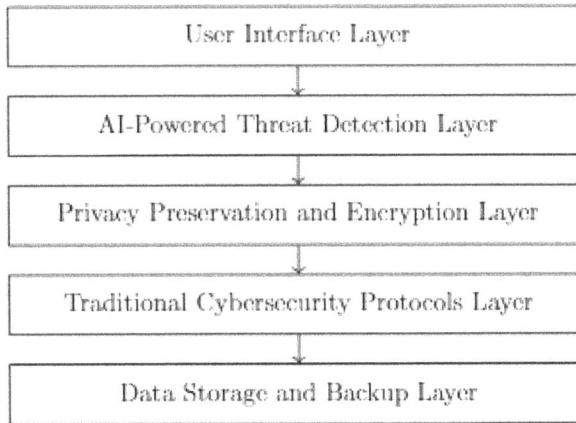

```
┌─────────────────────────────────────────┐
│           User Interface Layer            │
└─────────────────────────────────────────┘
                     │
                     ▼
┌─────────────────────────────────────────┐
│      AI-Powered Threat Detection Layer    │
└─────────────────────────────────────────┘
                     │
                     ▼
┌─────────────────────────────────────────┐
│   Privacy Preservation and Encryption Layer│
└─────────────────────────────────────────┘
                     │
                     ▼
┌─────────────────────────────────────────┐
│   Traditional Cybersecurity Protocols Layer│
└─────────────────────────────────────────┘
                     │
                     ▼
┌─────────────────────────────────────────┐
│        Data Storage and Backup Layer      │
└─────────────────────────────────────────┘
```

5.3 Privacy-by-Design and Ethical AI Security Frameworks

Given increasing public concern about data misuse, security frameworks are progressively embedding **Privacy-by-Design** principles (Brightwood & Jame, 2024; Almeida & Barr, 2025). Ethical considerations are now a formal component of security standards, especially in healthcare, finance, and education sectors (Almeida & Barr, 2025; Dhinakaran et al., 2025; Ismail & Aloshi, 2025).

Additionally, the **AI-powered Privacy Preservation Framework** proposed by Kumar et al. (2024) focuses on adaptive data protection, enabling context-aware privacy settings that can dynamically adjust based on user behavior and threat landscape.

5.4 Sector-Specific Standards and Case Studies

Healthcare: The healthcare sector, driven by sensitive patient data, has adopted frameworks emphasizing AI-powered privacy and precision medicine protections (Daraf & Badi, 2023; Bhamidipaty et al., 2025; Dhinakaran et al., 2025). Threat detection frameworks specifically target anomalies in medical data transactions.

Finance: AI in finance demands stringent standards for fraud detection, risk assessment, and client data privacy (Brightwood & Jame, 2024; Gupta et al., 2025).

IoT Ecosystem: Security frameworks for IoT are increasingly AI-enhanced, utilizing encoding mechanisms for improved privacy, security, and device performance (Farea et al., 2024).

Education: As AI transforms education systems, frameworks are being developed to address student data privacy and the responsible use of AI analytics (Ismail & Aloshi, 2025).

5.5 The Role of International and National Standards

Frameworks are increasingly aligned with international standards, including **ISO/IEC 27001**, **NIST AI Risk Management Framework**, and GDPR compliance for AI applications. These global standards are evolving to encompass AI-specific risks, including model explainability, data poisoning, and adversarial attacks (Anidjar et al., 2023; Gholami & Omar, 2024).

5.6 Challenges in Implementing AI-Driven Security Frameworks

Despite advancements, implementing AI-powered security frameworks faces several challenges:

- **Data Infrastructure Complexity**: The interconnectedness of systems like the AI-Powered Metaverse introduces unprecedented infrastructure privacy concerns (Anidjar et al., 2023).
- **Synthetic Data Usage**: The efficiency of synthetic data to train AI models while maintaining privacy without introducing bias is still under exploration (Gholami & Omar, 2023).
- **Zero-Day Threats**: New frameworks like the **GPT-2 Enhanced Attack Detection and Defense (GEADD)** have been proposed to counter zero-day attacks leveraging AI's speed and learning capabilities (Jones & Omar, 2024).

5.7 Future of AI-Powered Security Standards

The future points towards **hybrid security frameworks** combining AI, blockchain, and quantum computing for next-generation resilience (Omar & Zangana, 2025). Strategic approaches, such as those discussed by Mbah and Evelyn (2024), suggest that multi-layered defense combined with AI's predictive analytics will soon become industry standards.

Moreover, research by Jones et al. (2023) highlights that machine learning's speed and accuracy, when aligned with cybersecurity protocols, offers substantial advancements in detection and mitigation strategies. Privacy-preserving AI techniques, such as **federated learning** and **differential privacy**, will likely become embedded into frameworks across sectors (Gemiharto & Masrina, 2024).

5.8 Summary

Security standards and frameworks must continuously evolve alongside techno-logical innovations. As AI continues to redefine possibilities, so must our approaches to safeguarding data, ensuring ethical governance, and fostering global trust. The need for **adaptive**, **ethical**, and **comprehensive** AI-powered security frameworks has never been more critical, setting the foundation for safer digital ecosystems worldwide (Huff et al., 2023; Omar & Zangana, 2025).

6. DATA GOVERNANCE AND ACCESS CONTROL MODELS

In the evolving digital landscape shaped by artificial intelligence (AI), establish-ing robust **data governance** and **access control models** has become increasingly critical for ensuring data privacy, security, and ethical use. AI's capability to process vast amounts of data demands frameworks that can maintain transparency, account-ability, and resilience against breaches (Abolaji & Akinwande, 2024; Gholami & Omar, 2023).

6.1. The Foundations of Data Governance in AI-Powered Systems

Data governance involves the overall management of data availability, usability, integrity, and security. The incorporation of AI necessitates a shift from traditional governance to more dynamic, intelligent systems capable of adapting to new threats and privacy needs (Almeida & Barr, 2025; Brightwood & Jame, 2024). In sectors such as healthcare and finance, where data sensitivity is paramount, ethical and regulatory concerns are driving innovations in governance models (Anidjar, Packin, & Panezi, 2023; Alhitmi, Mardiah, Al-Sulaiti, & Abbas, 2024).

In AI-driven ecosystems, data governance also incorporates mechanisms to ensure ethical use of machine learning models, as improper data usage can result in algorithmic biases or unintended security vulnerabilities (Ismail & Aloshi, 2025; Gawankar et al., 2024).

6.2. AI-Powered Data Access Control Models

Access control models define who can access data and under what conditions. Traditional models such as **Role-Based Access Control (RBAC)** are increasingly being enhanced by AI to predict and adapt permissions dynamically based on user behavior and contextual information (Akhtar & Rawol, 2024; Jones & Omar, 2024).

- **Attribute-Based Access Control (ABAC)** is being improved through AI, enabling systems to consider multiple dynamic attributes for granting access (Kumar, Lokeshwari, & Shanmugam, 2024).
- **AI-Powered Adaptive Access Control (AI-AAC)** models use machine learning to predict risks and adapt policies in real time, offering more resilience against evolving cybersecurity threats (Mbah & Evelyn, 2024; Arya, Sharma, Devi, & Padmanaban, 2024).

Emerging frameworks, such as the **GEADD model**, leverage large language models for zero-day threat detection and dynamic access adaptation (Jones & Omar, 2024).

6.3. Securing Data Across Critical Sectors

The healthcare sector presents a prime example where AI and data governance intersect deeply. Solutions integrating AI with privacy-enhancing technologies ensure patient data confidentiality while allowing AI to drive innovations in treatment (Bhamidipaty et al., 2025; Bhamidipaty et al., 2025; Gopireddy, 2021).

Advanced AI-powered threat detection systems are used to secure healthcare records, demonstrating the need for strong access controls aligned with governance principles (Arefin, 2024; Arya et al., 2023).

Similarly, in education, AI's integration necessitates robust governance to balance personalized learning with data privacy (Ismail & Aloshi, 2025). AI-based solutions must comply with policies regulating student data access and usage (Gemiharto & Masrina, 2024).

6.4. Challenges in AI-Enhanced Governance

Despite advancements, implementing AI in governance and access control presents significant challenges:

- **Data bias and discrimination:** AI models trained on biased datasets can reinforce inequalities, requiring vigilant governance (Anidjar et al., 2023; Daraf & Badi, 2023).
- **Security threats in decentralized systems:** In cloud and IoT ecosystems, AI-powered biometrics and encryption-enhanced governance are critical for preventing breaches (Awad, Babu, Barka, & Shuaib, 2024; Farea et al., 2024).

Moreover, the increasing sophistication of cyber-attacks, such as wormhole attacks in wireless sensor networks, underscores the importance of resilient AI-driven security measures (Mohammed, Omar, & Nguyen, 2018).

6.5. Privacy-Enhancing Innovations and Regulatory Compliance

Innovative models focus on integrating encoding mechanisms with AI to boost privacy, particularly in the Internet of Things (IoT) and smart environments (Farea et al., 2024; Awad et al., 2024). Privacy-preserving techniques such as **federated learning** and **synthetic data generation** further support secure AI deployment (Gholami & Omar, 2024).

AI-based governance frameworks must align with international regulations such as GDPR, HIPAA, and upcoming AI-specific laws, ensuring both technological innovation and legal compliance (Abbas & Qazi, 2024; Gupta, Amarnani, Soanki, & Kishore, 2025).

Regulatory frameworks increasingly advocate for explainable AI (XAI) to promote transparency and fairness, which directly influences access control systems by enabling auditability (Omar & Zangana, 2025; Huff et al., 2023).

6.6. The Role of Synthetic Data and Privacy-Preserving Techniques

Synthetic data generation is playing a critical role in supporting AI systems without compromising individual privacy (Gholami & Omar, 2023). This technique allows AI models to train effectively while minimizing risks associated with real data exposure, particularly useful in healthcare and finance (Bhamidipaty et al., 2025; Brightwood & Jame, 2024).

Moreover, **AI-based encoding and encryption** mechanisms add layers of protection in critical applications such as genomic analysis and precision medicine (Daraf & Badi, 2023; Dhinakaran, Raja, Jasmine, Kumar, & Ramani, 2025).

6.7. Future Directions for Data Governance and Access Control in AI

Looking ahead, organizations must embrace more **context-aware**, **risk-adaptive**, and **autonomous governance models**. These models should not only respond to present cybersecurity threats but also predict future vulnerabilities (Mbah & Evelyn, 2024).

Research indicates the potential for **AI-driven continuous compliance** systems that automate adherence to regulatory requirements across multiple jurisdictions (Farea et al., 2024; Mbah & Evelyn, 2024).

New strategic approaches must also consider the ethical implications of AI in digital forensics and cybersecurity investigations (Omar & Zangana, 2025; Mohammed, Omar, & Nguyen, 2018).

7. CHALLENGES AND GAPS IN CURRENT PRACTICES

The integration of artificial intelligence (AI) across sectors such as healthcare, finance, education, IoT, and digital communications has created unprecedented opportunities for innovation and efficiency. However, the current practices surrounding AI-powered privacy and security mechanisms are fraught with several challenges and gaps that need urgent attention.

7.1. Inadequate Customization and Personalization of AI Privacy Configurations

One major gap is the lack of user-centric, customized privacy settings in AI-powered platforms. Many current solutions adopt a "one-size-fits-all" approach, ignoring user-specific privacy expectations. Abbas and Qazi (2024) emphasize that social media platforms, for instance, often fail to offer adaptable AI-powered privacy configurations, leaving users vulnerable to breaches.

7.2. Fragmented Research and Lack of Unified Frameworks

While numerous studies explore AI's role in enhancing privacy (Abolaji & Akinwande, 2024; Akhtar & Rawol, 2024), there remains a significant fragmentation in efforts to create unified security frameworks. Abolaji and Akinwande (2024) note that existing AI-powered privacy mechanisms are often tailored to specific industries, making cross-domain interoperability a challenge.

7.3. Data Security Concerns in AI-Driven Marketing and Business Sectors

The aggressive use of AI in marketing raises substantial data security and ethical concerns. Alhitmi, Mardiah, Al-Sulaiti, and Abbas (2024) explore how AI-driven marketing strategies often prioritize profit over customer data protection, exposing serious gaps in consent management and transparency.

7.4. Ethical and Legal Dilemmas in Health Data Protection

Health data, among the most sensitive types of personal information, faces critical risks. Almeida and Barr (2025) argue that AI-powered health diagnosis systems must strike a delicate balance between innovation and stringent ethical compliance, yet many current practices fall short of ensuring robust legal safeguards.

7.5. Emerging Privacy Challenges in the AI-Driven Metaverse

The rapid development of the AI-powered metaverse introduces a complex matrix of privacy risks. Anidjar, Packin, and Panezi (2023) highlight that decentralized and immersive environments intensify data governance challenges, often outpacing current privacy laws.

7.6. Insufficient Threat Detection in Healthcare Systems

Healthcare systems increasingly rely on AI for threat detection; however, as Arefin (2024) points out, existing implementations are often reactive rather than proactive, lacking advanced predictive capabilities essential for safeguarding patient data.

7.7. IoT Vulnerabilities and the Need for AI-Based Threat Detection

The Internet of Things (IoT) ecosystem presents a wide attack surface. Arya, Sharma, Devi, and Padmanaban (2024, 2023) demonstrate how traditional security models are inadequate for IoT devices and stress the need for real-time AI-powered threat detection systems to plug existing vulnerabilities.

7.8. Challenges in AI-Powered Biometric Security for IoT

Although biometric solutions offer promise, Awad, Babu, Barka, and Shuaib (2024) argue that current AI-powered biometric systems for IoT still suffer from scalability issues, false positives, and inadequate resistance to spoofing attacks.

7.9. Wearable Sensors and Privacy Risks in Healthcare

AI-powered sensors in healthcare can significantly improve diagnostics (Bhamidipaty et al., 2025), yet they introduce major privacy risks by continuously collecting sensitive personal data without sufficient user control mechanisms.

7.10. Ethical Gaps in AI-Powered Financial Applications

Brightwood and Jame (2024) reveal that AI systems in finance often overlook ethical implications, such as algorithmic biases, posing threats to fairness and trust.

7.11. Genomic Data Privacy in AI-Cloud Infrastructures

AI-powered genomic analysis, as explored by Daraf and Badi (2023), enhances precision medicine but also introduces unprecedented risks related to genomic data leakage, emphasizing the need for secure cloud infrastructures.

7.12. Managing Privacy in AI-Based Wellness Systems

The adoption of AI in wellness management systems is growing. Dhinakaran et al. (2025) observe that privacy is often an afterthought, with limited standardization on how sensitive wellness data is handled.

7.13. Encoding Mechanisms for Enhanced IoT Privacy

Farea, Alhazmi, Samet, and Guzel (2024) propose advanced encoding mechanisms integrated with AI to overcome current IoT security challenges. Yet, widespread adoption remains limited due to lack of interoperability with existing IoT devices.

7.14. Inadequate Patient Privacy Safeguards in AI-Driven Healthcare

Patient data security continues to be a significant gap area. Gawankar et al. (2024) note that even with regulatory frameworks, many healthcare institutions struggle to integrate AI-driven privacy solutions effectively.

7.15. Challenges in Privacy Preservation in Digital Communication

Gemiharto and Masrina (2024) assert that AI-powered digital communication systems often inadequately address user consent and anonymity, leading to user distrust.

7.16. Efficiency and Limitations of Synthetic Data in AI Models

Although synthetic data has been proposed as a solution to privacy concerns (Gholami & Omar, 2023), its effectiveness and realism are still under debate, impacting the reliability of privacy-preserving AI models (Gholami & Omar, 2024).

7.17. Security Gaps in Cloud Environments

Cloud environments hosting AI applications often suffer from abuse and exploitation. Gopireddy (2021) and Hamza and Omar (2013) discuss how cloud services still struggle to implement effective, AI-driven security measures against insider threats and misuse.

7.18. Data Privacy Challenges in Business Applications

Gupta, Amarnani, Soanki, and Kishore (2025) reveal that businesses adopting AI often overlook crucial aspects of data governance, resulting in breaches and compliance violations.

7.19. Biotech and Healthcare Research Organizations' Security Issues

Cybersecurity threats to healthcare and biotech organizations are mounting. Huff et al. (2023) describe that many current AI practices fail to meet the sophisticated demands of these highly sensitive sectors.

7.20. Privacy Concerns in AI-Driven Educational Systems

Ismail and Aloshi (2025) discuss how AI applications in education risk compromising student data privacy without adequate security-by-design principles.

7.21. Gaps in Zero-Day Threat Detection

Although innovations like the GEADD method (Jones & Omar, 2024) are promising, mainstream systems largely lack the capabilities to detect and respond to zero-day threats quickly and autonomously.

7.22. Machine Learning Limitations in Cybersecurity Applications

Jones et al. (2023) note that while machine learning enhances cybersecurity, models still struggle with adversarial attacks and false alarms, pointing to a need for more resilient systems.

7.23. Adaptive Data Protection Frameworks for AI

Kumar, Lokeshwari, and Shanmugam (2024) propose novel frameworks for adaptive data protection, but practical, real-world deployment at scale remains a challenge.

7.24. Strategic Gaps in AI-Powered Cybersecurity Management

Mbah and Evelyn (2024) argue that many organizations lack strategic alignment between cybersecurity management practices and AI-driven tools, leading to inefficiencies.

7.25. Legacy Issues: Wireless Networks and Policy Changes

Mohammed, Omar, and Nguyen (2018) and Nguyen et al. (2018) highlight that earlier work on wireless network vulnerabilities and net neutrality policy changes still impact how AI solutions today must be designed for security.

7.26. Digital Forensics in the AI Era

Finally, Omar and Zangana (2025) underscore that digital forensics has yet to fully adapt to the AI era, resulting in significant gaps in evidence collection and analysis methods when AI-driven cybercrimes occur.

8. CASE STUDIES

8.1. AI-Powered Cybersecurity in Social Media and Communication Platforms

The rise of social media platforms has increased the urgency to develop more robust privacy and security measures. Abbas and Qazi (2024) explored the application of customized AI-powered security and privacy configurations in social media websites, highlighting how tailored solutions can effectively manage diverse user needs. Similarly, Gemiharto and Masrina (2024) emphasized the significance of user privacy preservation within AI-driven digital communication systems, addressing vulnerabilities that arise from rapid data exchanges.

The evolution of AI-powered cybersecurity frameworks has been pivotal in detecting and mitigating new forms of cyber threats. Akhtar and Rawol (2024) demonstrated that integrating AI into security mechanisms significantly enhances the defense layers against sophisticated cyberattacks. Moreover, Mbah and Evelyn (2024) discussed strategic approaches for leveraging AI to mitigate risks and safeguard user data privacy across various digital platforms.

8.2. AI and Healthcare Data Protection

Healthcare systems have increasingly adopted AI technologies, leading to new challenges related to data security and privacy. Arefin (2024) investigated the use of AI-powered threat detection to strengthen healthcare data protection, finding that predictive analytics could identify and respond to threats faster than traditional systems. Similarly, Bhamidipaty et al. (2025) discussed the impact of AI-powered sensors in revolutionizing healthcare monitoring, emphasizing how real-time patient data collection must be balanced with stringent privacy safeguards.

A growing body of work, such as that by Dhinakaran et al. (2025), has advocated for AI-powered health management systems that prioritize privacy by design. Gawankar et al. (2024) further explored patient privacy and data security challenges in AI-driven healthcare settings, proposing frameworks that ensure compliance with ethical standards.

Daraf and Badi (2023) analyzed how AI-powered genomic analysis in cloud environments enhances precision medicine but simultaneously necessitates stringent data protection protocols. Moreover, Almeida and Barr (2025) provided a comprehensive overview of health data innovation, emphasizing ethical, legal, and technological considerations in AI-driven diagnosis systems.

8.3. AI Applications in Internet of Things (IoT) Security

The Internet of Things (IoT) ecosystem is particularly vulnerable to security breaches, necessitating AI-enhanced defenses. Arya, Sharma, Devi, and Padmanaban (2024) and Arya et al. (2023) highlighted the effectiveness of AI-powered threat detection mechanisms in securing IoT environments. Their work suggests that automated, intelligent systems can mitigate vulnerabilities stemming from the interconnected nature of IoT devices.

Awad et al. (2024) reviewed the potential of AI-powered biometrics in bolstering IoT security, presenting a forward-looking vision for integrating biometric verification to prevent unauthorized access. In another important contribution, Farea et al. (2024) proposed an AI-integrated encoding mechanism aimed at enhancing the privacy, security, and performance of IoT ecosystems, demonstrating measurable improvements over traditional systems.

8.4. AI-Driven Financial and Business Data Privacy

In the finance sector, Brightwood and Jame (2024) discussed data privacy, security, and ethical concerns arising from AI adoption, underlining the need for regulatory and organizational changes to manage AI's transformative impact. Similarly, Gupta, Amarnani, Soanki, and Kishore (2025) investigated how businesses are adapting AI technologies while ensuring compliance with evolving data privacy regulations.

In the broader business landscape, Alhitmi et al. (2024) explored data security concerns of AI-driven marketing, suggesting that businesses must balance personalized customer experiences with responsible data handling practices. Additionally, Abolaji and Akinwande (2024) surveyed the current state and future directions of AI-powered privacy protection, offering critical insights for businesses navigating the complex intersection of innovation and regulation.

8.5. Ethical and Legal Challenges in AI-Driven Environments

The ethical and legal frameworks surrounding AI adoption have been explored by Anidjar, Packin, and Panezi (2023), who examined the privacy matrix within the AI-powered metaverse, highlighting the intricate web of data infrastructures that challenge traditional notions of consent and privacy.

Ismail and Aloshi (2025) extended this discourse into the field of education, addressing data privacy concerns in AI-driven learning environments and proposing solutions that prioritize student autonomy. Meanwhile, Omar and Zangana (2025) edited a comprehensive volume that delves into the implications of AI advancements for digital forensics, offering an interdisciplinary perspective on how AI reshapes evidence collection and privacy protection.

8.6. AI in Cloud Security and Computing

Cloud environments have become prime targets for cyberattacks. Gopireddy (2021) emphasized that AI-powered security solutions significantly enhance cloud data protection and threat detection, offering adaptive learning mechanisms against evolving attacks. Mohammed, Omar, and Nguyen (2018) tackled specific threats such as wormhole attacks in wireless sensor networks, advocating for AI-enhanced security protocols.

Complementing this, Nguyen et al. (2018) analyzed the effects of policy changes, such as the FCC's Net Neutrality Repeal, on cybersecurity and privacy, emphasizing the importance of adaptive AI tools to counteract emerging threats. Hamza and Omar (2013) earlier discussed abuse and nefarious uses of cloud computing, which remain relevant today in the context of AI-driven infrastructures.

8.7. Machine Learning and AI for Advanced Cybersecurity

Machine learning has been instrumental in advancing cybersecurity capabilities. Jones and Omar (2024) introduced the GPT-2 Enhanced Attack Detection and Defense (GEADD) method for countering zero-day threats, demonstrating the superiority of AI models in rapidly detecting novel attacks. Similarly, Jones et al. (2023) highlighted the benefits of integrating machine learning techniques for faster and more accurate threat detection in cybersecurity.

Huff et al. (2023) offered management practices for mitigating cybersecurity threats in biotechnology and healthcare research organizations, noting the critical role of AI-powered solutions in safeguarding sensitive information.

In the context of large language models, Gholami and Omar (2023, 2024) investigated the efficiency of synthetic data in AI training and explored whether student models could match their teachers in performance—an inquiry directly relevant to AI-driven cybersecurity training.

8.8. Future Directions and Adaptive Privacy Frameworks

Looking ahead, Kumar, Lokeshwari, and Shanmugam (2024) proposed a novel framework for adaptive data protection, driven by AI-powered privacy preservation techniques. Their approach aligns with the increasing need for flexible, context-aware privacy solutions.

Finally, the work of Mohammed et al. (2018) and the broader insights from Huff et al. (2023) underscore that, while AI offers unprecedented advancements in cybersecurity and privacy, careful governance, interdisciplinary collaboration, and continuous innovation are critical to overcoming the challenges ahead.

9. FUTURE TRENDS AND RECOMMENDATIONS

The future of AI-powered security, privacy, and data protection is dynamic and multifaceted. As the landscape continues to evolve, it demands the integration of ethical considerations, robust technological advancements, and regulatory foresight. This section explores emerging trends and offers recommendations based on recent scholarly contributions.

9.1. AI-Enhanced Personalized Privacy and Security

One emerging trend is the shift towards **customized AI-powered security configurations** tailored to individual user needs, particularly in social media and communication platforms (Abbas & Qazi, 2024; Gemiharto & Masrina, 2024). Such personalized systems can dynamically adjust privacy settings based on user behavior and threat modeling, increasing user trust and system resilience.

Recommendation: Develop AI systems capable of context-aware and user-centric privacy recommendations, particularly for social and digital communication platforms.

9.2. AI in Healthcare Privacy and Security

Healthcare is rapidly integrating AI-powered solutions, from **diagnosis systems** to **patient monitoring sensors** (Bhamidipaty et al., 2025; Almeida & Barr, 2025). However, concerns over **data breaches** and **genomic data privacy** are intensifying (Daraf & Badi, 2023; Arefin, 2024; Gawankar et al., 2024).

Recommendation: Prioritize privacy-preserving techniques such as federated learning and differential privacy in AI models for healthcare applications to ensure compliance with ethical and legal standards (Dhinakaran et al., 2025; Brightwood & Jame, 2024).

9.3. AI in IoT and Edge Security

The proliferation of IoT devices necessitates **AI-driven threat detection and biometric security systems** (Arya, Sharma, Devi, & Padmanaban, 2024; Awad, Babu, Barka, & Shuaib, 2024). Future systems must move beyond traditional encryption towards adaptive, learning-based protection frameworks.

Recommendation: Encourage research into **AI-powered integrated encoding mechanisms** to enhance IoT security, addressing both privacy and performance concerns (Farea, Alhazmi, Samet, & Guzel, 2024).

9.4. AI, Finance, and Business Privacy

In the financial and business sectors, AI's impact on data security is profound (Brightwood & Jame, 2024; Gupta, Amarnani, Soanki, & Kishore, 2025). As businesses increasingly depend on AI for decision-making, safeguarding sensitive corporate and consumer information is critical.

Recommendation: Implement AI-based adaptive data protection frameworks that incorporate real-time risk assessment and privacy-preserving computation methods (Kumar, Lokeshwari, & Shanmugam, 2024).

9.5. Ethical, Legal, and Governance Challenges

Managing the **ethical and legal challenges** of AI systems, particularly in sectors like healthcare and education, is paramount (Ismail & Aloshi, 2025; Almeida & Barr, 2025). Future regulations must address not just consent, but also algorithmic transparency and explainability (Anidjar, Packin, & Panezi, 2023).

Recommendation: Develop AI governance frameworks focusing on transparency, user rights, and cross-sector accountability, especially as AI systems increasingly impact critical societal sectors.

9.6. Advances in Synthetic Data and AI Training

There is growing interest in using **synthetic data** to improve AI training while protecting real user data (Gholami & Omar, 2023, 2024). Synthetic data offers a promising path for building efficient AI models without compromising privacy.

Recommendation: Promote research into synthetic data generation and evaluation, ensuring that synthetic datasets uphold diversity, fairness, and security standards.

9.7. AI for Cybersecurity Threat Detection

AI techniques like **GPT-enhanced attack detection** are reshaping cybersecurity strategies (Jones & Omar, 2024; Jones, Omar, Mohammed, Nobles, & Dawson, 2023). Additionally, AI is becoming central in **cloud security** environments (Gopireddy, 2021; Hamza & Omar, 2013).

Recommendation: Encourage deployment of hybrid AI models combining **signature-based** and **behavioral detection** for advanced threat detection, particularly against zero-day attacks.

9.8. Privacy in AI-Driven Education

As AI becomes embedded in education systems, data privacy for learners must become a priority (Ismail & Aloshi, 2025). Future educational platforms must ensure that sensitive learning analytics data are processed securely and ethically.

Recommendation: Apply **privacy-by-design** principles in AI educational tools, integrating consent management and minimal data collection practices.

9.9. AI for Biotechnology and Laboratory Cybersecurity

With biotech companies increasingly adopting AI, protecting research data and patient information becomes crucial (Huff et al., 2023).

Recommendation: Foster interdisciplinary collaboration between cybersecurity experts, researchers, and healthcare professionals to develop AI systems resilient against emerging cyber threats.

9.10. The Future of AI in Digital Forensics

AI's role in digital forensics is expanding, providing enhanced capabilities for evidence collection and analysis (Omar & Zangana, 2025). AI-driven tools can potentially shorten investigation times while increasing accuracy.

Recommendation: Focus on developing explainable AI models for forensic applications to meet legal admissibility standards and build trust among law enforcement agencies.

9.11. Policy Implications and Broader Societal Impact

The **repeal of net neutrality** has already raised significant concerns regarding privacy and AI-enabled discrimination (Nguyen, Mohammed, Omar, & Banisakher, 2018; Mohammed, Omar, & Nguyen, 2018). Future AI systems must be designed to mitigate these risks.

Recommendation: Advocate for policy reforms that establish stronger data protection laws and promote ethical AI usage across industries.

9.12. Strategic Approaches for Risk Mitigation

Strategic approaches for AI-powered cybersecurity will become essential to protect privacy in complex, interconnected environments (Mbah & Evelyn, 2024).

Recommendation: Encourage **AI-driven risk assessment frameworks** that adapt based on evolving threat landscapes and prioritize resilience and redundancy.

10. CONCLUSION

The rapid advancements in artificial intelligence (AI) are driving transformative changes across multiple industries, including healthcare, finance, cybersecurity, and education. As AI technologies evolve, the need for robust, AI-powered security, privacy, and ethical frameworks becomes increasingly essential. This chapter has highlighted several key trends and challenges surrounding the intersection of AI, privacy, and security, offering insights into the evolving landscape.

AI's role in enhancing security measures has demonstrated significant promise, particularly through AI-powered threat detection, predictive analytics, and the integration of advanced encryption methods. However, as organizations increasingly rely on AI-driven systems, they must be vigilant about potential privacy risks, ensuring that personal data is protected from misuse and unauthorized access. The AI-powered systems used in critical sectors such as healthcare and finance must be designed with rigorous safeguards to prevent breaches and protect sensitive information.

The recommendations for future AI developments emphasize a balanced approach, where innovation is paired with careful attention to privacy, security, and ethical considerations. Organizations must invest in the development of adaptive security measures and adopt AI technologies that not only mitigate risks but also comply with regulatory requirements and ensure transparency in decision-making processes. The integration of privacy-preserving techniques such as differential privacy and federated learning will be crucial in maintaining user trust while allowing AI systems to operate effectively.

Moreover, the increasing complexity of AI systems presents challenges in accountability and oversight. As AI continues to integrate into everyday operations, the need for a regulatory framework that addresses AI-related risks will grow. The development of global standards and regulations will be necessary to manage these risks while fostering innovation. Ethical concerns, including bias in AI algorithms and the implications of data ownership, should also be prioritized in future research and policy-making.

Ultimately, the future of AI lies in harnessing its capabilities responsibly, ensuring that security and privacy are maintained without stifling innovation. As technologies advance, the continued collaboration between researchers, policymakers, and industry professionals will be crucial in shaping a secure and ethical AI-powered future. The future directions outlined in this chapter present a roadmap for addressing the critical issues surrounding AI in security and privacy, with the goal of fostering a safer and more transparent digital landscape.

REFERENCES

Abbas, E., & Qazi, A. A. (2024). Customized Ai-powered Security and Privacy Configurations for Social MEDIA Websites. *BULLET: Jurnal Multidisiplin Ilmu*, *3*(1), 108–117.

Abolaji, E. O., & Akinwande, O. T.Elijah Oluwatoyosi AbolajiOladayo Tosin Akinwande. (2024). AI powered privacy protection: A survey of current state and future directions. *World Journal of Advanced Research and Reviews*, *23*(3), 2687–2696. DOI: 10.30574/wjarr.2024.23.3.2869

Akhtar, Z. B., & Rawol, A. T. (2024). Enhancing cybersecurity through AI-powered security mechanisms. *IT Journal Research and Development*, *9*(1), 50–67. DOI: 10.25299/itjrd.2024.16852

Alhitmi, H. K., Mardiah, A., Al-Sulaiti, K. I., & Abbas, J. (2024). Data security and privacy concerns of AI-driven marketing in the context of economics and business field: An exploration into possible solutions. *Cogent Business & Management*, *11*(1), 2393743. DOI: 10.1080/23311975.2024.2393743

Almeida, D., & Barr, N. (2025). Innovations in Health Data Protection Ethical, Legal, and Technological Perspectives in a Global Context: AI-Powered Diagnosis Systems and Health Data Innovation. In *Navigating Privacy, Innovation, and Patient Empowerment Through Ethical Healthcare Technology* (pp. 171-196). IGI Global Scientific Publishing.

Anidjar, L., Packin, N. G., & Panezi, A. (2023). The matrix of privacy: Data infrastructure in the AI-Powered Metaverse. *SSRN*, *18*, 59. DOI: 10.2139/ssrn.4363208

Arefin, S. (2024). Strengthening Healthcare Data Security with Ai-Powered Threat Detection. [IJSRM]. *International Journal of Scientific Research and Management*, *12*(10), 1477–1483. DOI: 10.18535/ijsrm/v12i10.ec02

Arya, L., Sharma, Y. K., Devi, S., & Padmanaban, H. (2024). Securing the Internet of Things: AI-Powered Threat Detection and Safety. In *Proceedings of International Conference on Recent Innovations in Computing: ICRIC 2023,* Volume 2 (Vol. 2, p. 97). Springer Nature. DOI: 10.1007/978-981-97-3442-9_7

Arya, L., Sharma, Y. K., Devi, S., Padmanaban, H., & Kumar, R. (2023, October). Securing the Internet of Things: AI-Powered Threat Detection and Safety Measures. In *The International Conference on Recent Innovations in Computing* (pp. 97-108). Singapore: Springer Nature Singapore.

Awad, A. I., Babu, A., Barka, E., & Shuaib, K. (2024). AI-powered biometrics for Internet of Things security: A review and future vision. *Journal of Information Security and Applications*, *82*, 103748. DOI: 10.1016/j.jisa.2024.103748

Bhamidipaty, V., Bhamidipaty, D. L., Guntoory, I., Bhamidipaty, K. D. P., Iyengar, K. P., Botchu, B., & Botchu, R. (2025). Revolutionizing Healthcare: The Impact of AI-Powered Sensors. *Generative Artificial Intelligence for Biomedical and Smart Health Informatics*, 355-373.

Brightwood, S., & Jame, H. (2024). *Data privacy, security, and ethical considerations in AI-powered finance. Article.* Research Gate.

Daraf, U., & Badi, S. (2023). AI-Powered Genomic Analysis in the Cloud: Enhancing Precision Medicine While Protecting Medical Data Privacy.

Dhinakaran, D., Raja, S. E., Jasmine, J. J., Kumar, P. V., & Ramani, R. (2025). The Future of Well-Being: AI-Powered Health Management with Privacy at its Core. *Wellness Management Powered by AI Technologies*, 363-402.

Farea, A. H., Alhazmi, O. H., Samet, R., & Guzel, M. S. (2024). AI-powered Integrated with Encoding Mechanism Enhancing Privacy, Security, and Performance for IoT Ecosystem. *IEEE Access : Practical Innovations, Open Solutions*, *12*, 121368–121386. DOI: 10.1109/ACCESS.2024.3449630

Gawankar, S., Nair, S., Pawar, V., Vhatkar, A., & Chavan, P. (2024, August). Patient Privacy and Data Security in the Era of AI-Driven Healthcare. In *2024 8th International Conference on Computing, Communication, Control and Automation (ICCUBEA)* (pp. 1-6). IEEE. DOI: 10.1109/ICCUBEA61740.2024.10775004

Gemiharto, I., & Masrina, D. (2024). User privacy preservation in AI-powered digital communication systems. *Jurnal Communio: Jurnal Jurusan Ilmu Komunikasi*, *13*(2), 349–359. DOI: 10.35508/jikom.v13i2.9420

Gholami, S., & Omar, M. (2023). Does Synthetic Data Make Large Language Models More Efficient? *arXiv preprint arXiv:2310.07830.*

Gholami, S., & Omar, M. (2024). Can a student large language model perform as well as its teacher? In *Innovations, Securities, and Case Studies Across Healthcare, Business, and Technology* (pp. 122-139). IGI Global. DOI: 10.4018/979-8-3693-1906-2.ch007

Gopireddy, R. R. (2021). AI-Powered Security in cloud environments: Enhancing data protection and threat detection. [IJSR]. *International Journal of Scientific Research*, *10*(11).

Gupta, A., Amarnani, M., Soanki, S., & Kishore, J. (2025, February). AI and Data Privacy in Business. In *2025 First International Conference on Advances in Computer Science, Electrical, Electronics, and Communication Technologies (CE2CT)* (pp. 109-114). IEEE.

Hamza, Y. A., & Omar, M. D. (2013). Cloud computing security: Abuse and nefarious use of cloud computing. *International Journal of Computer Engineering Research*, 3(6), 22–27.

Huff, A. J., Burrell, D. N., Nobles, C., Richardson, K., Wright, J. B., Burton, S. L., Jones, A. J., Springs, D., Omar, M., & Brown-Jackson, K. L. (2023). Management Practices for Mitigating Cybersecurity Threats to Biotechnology Companies, Laboratories, and Healthcare Research Organizations. In *Applied Research Approaches to Technology, Healthcare, and Business* (pp. 1-12). IGI Global.

Ismail, I. A., & Aloshi, J. M. R. (2025). Data Privacy in AI-Driven Education: An In-Depth Exploration Into the Data Privacy Concerns and Potential Solutions. In *AI Applications and Strategies in Teacher Education* (pp. 223-252). IGI Global.

Jones, R., & Omar, M. (2024). Revolutionizing Cybersecurity: The GPT-2 Enhanced Attack Detection and Defense (GEADD) Method for Zero-Day Threats. *International Journal of Informatics* [INJIISCOM]. *Information System and Computer Engineering*, 5(2), 178–191.

Jones, R., Omar, M., Mohammed, D., Nobles, C., & Dawson, M. (2023). Harnessing the Speed and Accuracy of Machine Learning to Advance Cybersecurity. In *2023 Congress in Computer Science, Computer Engineering, & Applied Computing (CSCE)* (pp. 418-421). IEEE. DOI: 10.1109/CSCE60160.2023.00074

Kumar, R. S., Lokeshwari, J., & Shanmugam, S. K. (2024, November). AI-Powered Privacy Preservation: A Novel Framework for Adaptive Data Protection. In *2024 2nd International Conference on Computing and Data Analytics (ICCDA)* (pp. 1-6). IEEE.

Mbah, G. O., & Evelyn, A. N. (2024). AI-powered cybersecurity: Strategic approaches to mitigate risk and safeguard data privacy.

Mohammed, D., Omar, M., & Nguyen, V. (2018). Wireless sensor network security: Approaches to detecting and avoiding wormhole attacks. *Journal of Research in Business. Economics and Management*, 10(2), 1860–1864.

Nguyen, V., Mohammed, D., Omar, M., & Banisakher, M. (2018). The Effects of the FCC Net Neutrality Repeal on Security and Privacy. [IJHIoT]. *International Journal of Hyperconnectivity and the Internet of Things*, 2(2), 21–29. DOI: 10.4018/IJHIoT.2018070102

Omar, M. (2021). New insights into database security: An effective and integrated approach for applying access control mechanisms and cryptographic concepts in Microsoft Access environments.

Omar, M. (2022). *Machine Learning for Cybersecurity: Innovative Deep Learning Solutions*. Springer Brief. https://link.springer.com/book/978303115

Omar, M. (2024). From Attack to Defense: Strengthening DNN Text Classification Against Adversarial Examples. In *Innovations, Securities, and Case Studies Across Healthcare, Business, and Technology* (pp. 174-195). IGI Global.

Omar, M. (2024). Revolutionizing Malware Detection: A Paradigm Shift Through Optimized Convolutional Neural Networks. In *Innovations, Securities, and Case Studies Across Healthcare, Business, and Technology* (pp. 196-220). IGI Global. DOI: 10.4018/979-8-3693-1906-2.ch011

Omar, M., & Zangana, H. (Eds.). (2025). *Digital Forensics in the Age of AI*. IGI Global., DOI: 10.4018/979-8-3373-0857-9

Omar, M., & Zangana, H. M. (Eds.). (2024). *Redefining Security With Cyber AI*. IGI Global., DOI: 10.4018/979-8-3693-6517-5

Omar, M., & Zangana, H. M. (Eds.). (2025). *Application of Large Language Models (LLMs) for Software Vulnerability Detection*. IGI Global., DOI: 10.4018/979-8-3693-9311-6

Omar, M., Zangana, H. M., Al-Karaki, J. N., & Mohammed, D. (2024). Harnessing LLMs for IoT malware detection: A comparative analysis of BERT and GPT-2. In *2024 8th International Symposium on Multidisciplinary Studies and Innovative Technologies (ISMSIT)* (pp. 1-6). Ankara, Turkiye. https://doi.org/DOI: 10.1109/ISMSIT63511.2024.10757249

Omar, M., Zangana, H. M., & Mohammed, D. (Eds.). (2025). *Integrating Artificial Intelligence in Cybersecurity and Forensic Practices*. IGI Global., DOI: 10.4018/979-8-3373-0588-2

Pandey, A. S., Sharma, Y., Tiwari, A., Chauhan, R., Tyagi, S., & Kumari, J. (2024, May). Ethical Implications of AI-Powered Communication Tool. In *2024 International Conference on Communication, Computer Sciences and Engineering (IC3SE)* (pp. 1857-1861). IEEE. DOI: 10.1109/IC3SE62002.2024.10593350

Patel, R., & Desai, A. (2024). Exploring the Ethical Implications of AI-Powered Security Systems. *Asian American Research Letters Journal, 1*(9), 87–95.

Rehan, H. (2023). AI-Powered Genomic Analysis in the Cloud: Enhancing Precision Medicine and Ensuring Data Security in Biomedical Research. *Journal of Deep Learning in Genomic Data Analysis, 3*(1), 37–71.

Sato, H. (2023). AI-Powered Solutions: Ensuring Data Privacy in a Transforming Digital Landscape. *Advances in Computer Sciences, 6*(1).

Thirunagalingam, A. (2024). AI-Powered Continuous Data Quality Improvement: Techniques, Benefits, and Case Studies. *Benefits, and Case Studies (August 23, 2024).*

Wright, J., Dawson, M. E.Jr, & Omar, M. (2012). Cyber security and mobile threats: The need for antivirus applications for smartphones. *Journal of Information Systems Technology and Planning, 5*(14), 40–60.

Zainab, H., Khan, A. R. A., Khan, M. I., & Arif, A. (2025). Ethical Considerations and Data Privacy Challenges in AI-Powered Healthcare Solutions for Cancer and Cardiovascular Diseases. *Global Trends in Science and Technology, 1*(1), 63–74. DOI: 10.70445/gtst.1.1.2025.63-74

Zangana, H., Al-Karaki, J., & Omar, M. (Eds.). (2025). *Revolutionizing Cybersecurity With Deep Learning and Large Language Models.* IGI Global., DOI: 10.4018/979-8-3373-3296-3

Zangana, H., & Omar, M. (Eds.). (2025). *Leveraging Large Language Models for Quantum-Aware Cybersecurity.* IGI Global., DOI: 10.4018/979-8-3373-1102-9

Zangana, H., & Omar, M. (Eds.). (2025). *Leveraging Large Language Models for Quantum-Aware Cybersecurity.* IGI Global., DOI: 10.4018/979-8-3373-1102-9

Zangana, H. M. (2024). Exploring Blockchain-Based Timestamping Tools: A Comprehensive Review. *Redefining Security With Cyber AI*, 92-110.

Zangana, H. M. (2024). Exploring the Landscape of Website Vulnerability Scanners: A Comprehensive Review and Comparative Analysis. *Redefining Security With Cyber AI*, 111-129.

Zangana, H. M., & Li, S. (2025). Future Trends in AI and Digital Forensics. In Omar, M., & Zangana, H. (Eds.), *Digital Forensics in the Age of AI* (pp. 347–380). IGI Global Scientific Publishing., DOI: 10.4018/979-8-3373-0857-9.ch013

Zangana, H. M., Luckyardi, S., Mustafa, F. M., & Li, S. (2025). Enhancing Agricultural Cybersecurity: Leveraging Deep Learning and Large Language Models for Smart Farming Protection. In Zangana, H., Al-Karaki, J., & Omar, M. (Eds.), *Revolutionizing Cybersecurity With Deep Learning and Large Language Models* (pp. 307–338). IGI Global Scientific Publishing., DOI: 10.4018/979-8-3373-3296-3.ch010

Zangana, H. M., Mohammed, A. K., Sallow, A. B., & Sallow, Z. B. (2024). Cybernetic Deception: Unraveling the Layers of Email Phishing Threats. [INJURATECH]. *International Journal of Research and Applied Technology*, 4(1), 35–47.

Zangana, H. M., & Mohammed, D. (2025). Foundations of Large Language Models in Software Vulnerability Detection. In Omar, M., & Zangana, H. (Eds.), *Application of Large Language Models (LLMs) for Software Vulnerability Detection* (pp. 41–74). IGI Global., DOI: 10.4018/979-8-3693-9311-6.ch002

Zangana, H. M., & Mustafa, F. M. (2024). Hybrid Image Denoising Using Wavelet Transform and Deep Learning. *EAI Endorsed Transactions on AI and Robotics, 3.* Advance online publication. DOI: 10.4108/airo.7486

Zangana, H. M., & Mustafa, F. M. (2025). Image Denoising Techniques for Cybersecurity and Forensic Applications: AI-Driven Approaches. In Omar, M., Zangana, H., & Mohammed, D. (Eds.), *Integrating Artificial Intelligence in Cybersecurity and Forensic Practices* (pp. 117–142). IGI Global Scientific Publishing., DOI: 10.4018/979-8-3373-0588-2.ch005

Zangana, H. M., Mustafa, F. M., & Li, S. (2025). Large Language Models in Cybersecurity: From Automation to Intelligence. In Zangana, H., & Omar, M. (Eds.), *Leveraging Large Language Models for Quantum-Aware Cybersecurity* (pp. 277–300). IGI Global Scientific Publishing., DOI: 10.4018/979-8-3373-1102-9.ch009

Zangana, H. M., Mustafa, F. M., Li, S., & Al-Karaki, J. N. (2025). Natural Language Processing for Cyber Threat Intelligence in a Quantum World. In Zangana, H., & Omar, M. (Eds.), *Leveraging Large Language Models for Quantum-Aware Cybersecurity* (pp. 345–388). IGI Global Scientific Publishing., DOI: 10.4018/979-8-3373-1102-9.ch011

Zangana, H. M., Mustafa, F. M., Mohammed, A. K., & Omar, M. (2025). The Role of Change Control Boards in Ensuring Cybersecurity Compliance for IT Infrastructure. *JITCE (Journal of Information Technology and Computer Engineering), 9*(1). Retrieved from https://jitce.fti.unand.ac.id/index.php/JITCE/article/view/303

Zangana, H. M., Mustafa, F. M., & Omar, M. (2024). A Hybrid Approach for Robust Object Detection: Integrating Template Matching and Faster R-CNN. *EAI Endorsed Transactions on AI and Robotics*, *3*. Advance online publication. DOI: 10.4108/airo.6858

Zangana, H. M., & Omar, M. (2020). Threats, Attacks, and Mitigations of Smartphone Security. *Academic Journal of Nawroz University*, *9*(4), 324–332. DOI: 10.25007/ajnu.v9n4a989

Zangana, H. M., & Omar, M. (2025). Harnessing the Power of Large Language Models for Cybersecurity: Applications, Challenges, and Future Directions. In Omar, M., & Zangana, H. (Eds.), *Application of Large Language Models (LLMs) for Software Vulnerability Detection* (pp. 1–40). IGI Global., DOI: 10.4018/979-8-3693-9311-6.ch001

Zangana, H. M., & Omar, M. (2025). Introduction to Digital Forensics and Artificial Intelligence. In Omar, M., & Zangana, H. (Eds.), *Digital Forensics in the Age of AI* (pp. 1–30). IGI Global Scientific Publishing., DOI: 10.4018/979-8-3373-0857-9.ch001

Zangana, H. M., & Omar, M. (2025). Introduction to Quantum-Aware Cybersecurity: The Need for LLMs. In Zangana, H., & Omar, M. (Eds.), *Leveraging Large Language Models for Quantum-Aware Cybersecurity* (pp. 1–28). IGI Global Scientific Publishing., DOI: 10.4018/979-8-3373-1102-9.ch001

Zangana, H. M., & Omar, M. (2025). The Role of Leadership in Advancing Inclusive Health Technologies. In Burrell, D., & Nguyen, C. (Eds.), *New Horizons in Leadership: Inclusive Explorations in Health, Technology, and Education* (pp. 203–220). IGI Global Scientific Publishing., DOI: 10.4018/979-8-3693-6437-6.ch009

Zangana, H. M., Omar, M., & Al-Karaki, J. N. (2025). Foundations of Deep Learning and Large Language Models in Cybersecurity. In Zangana, H., Al-Karaki, J., & Omar, M. (Eds.), *Revolutionizing Cybersecurity With Deep Learning and Large Language Models* (pp. 1–36). IGI Global Scientific Publishing., DOI: 10.4018/979-8-3373-3296-3.ch001

Zangana, H. M., Omar, M., Al-Karaki, J. N., & Mohammed, D. (2024). Comprehensive Review and Analysis of Network Firewall Rule Analyzers: Enhancing Security Posture and Efficiency. *Redefining Security With Cyber AI*, 15-36.

Zangana, H. M., Sallow, Z. B., & Omar, M. (2025). The Human Factor in Cybersecurity: Addressing the Risks of Insider Threats. *Jurnal Ilmiah Computer Science*, *3*(2), 76–85. DOI: 10.58602/jics.v3i2.37

Żywiołek, J. (2024). Empirical Examination Of Ai-Powered Decision Support Systems: Ensuring Trust And Transparency In Information And Knowledge Security. *Scientific Papers of Silesian University of Technology. Organization & Management/ Zeszyty Naukowe Politechniki Slaskiej. Seria Organizacji i Zarzadzanie*, (197).

Chapter 2
Ethical and Secure AI Applications in Agricultural Research:
Challenges and Opportunities

Hewa Majeed Zangana
https://orcid.org/0000-0001-7909-254X
Duhok Polytechnic University, Iraq

Senny Luckyardi
https://orcid.org/0000-0001-9954-7433
Universitas Komputer Indonesia, Indonesia

Firas Mahmood Mustafa
Duhok Polytechnic University, Iraq

Shuai Li
University of Oulu, Finland

ABSTRACT

The integration of Artificial Intelligence (AI) into agricultural research has introduced transformative innovations that promise increased productivity, sustainability, and precision in farming practices. However, the rapid advancement and deployment of AI systems in this domain also bring significant ethical and security concerns. Issues such as data privacy, algorithmic bias, equitable access, environmental impacts, and transparency present complex challenges that must be carefully navigated. This chapter explores the critical ethical considerations and security risks associated with AI applications in agricultural research, while also highlighting the immense opportunities for advancing scientific understanding, improving food security, and

DOI: 10.4018/979-8-3373-4252-8.ch002

promoting sustainable practices. By examining current trends, case studies, and regulatory frameworks, the chapter provides a comprehensive overview of how agricultural AI can be ethically and securely developed and deployed, ensuring that innovation serves both humanity and the environment responsibly.

1. INTRODUCTION

The integration of Artificial Intelligence (AI) into agricultural research marks a transformative era, offering revolutionary approaches to traditional farming challenges and contributing significantly to food security, environmental conservation, and rural development. From predictive analytics in crop management to autonomous machinery and precision agriculture, AI's potential in agriculture is expansive and profound. As Abd-Elsalam and Abdel-Momen (2023) suggest, AI technologies are rapidly reshaping scientific research methodologies, including those in the agricultural sciences, enabling more efficient data processing and predictive modeling.

However, alongside these opportunities emerge complex ethical, social, and security concerns. Ensuring that AI applications are responsibly developed and implemented in agricultural contexts demands rigorous attention to data privacy, algorithmic fairness, transparency, and equitable access. As highlighted by Ahmad et al. (2024), while AI can empower agriculture for global food security, the challenges, particularly in developing nations, remain significant and multifaceted.

Ethical concerns are increasingly pressing in agricultural AI. These range from the misuse of farmers' personal data to biased decision-making algorithms that may disproportionately disadvantage smallholder farmers. As Ali et al. (2024) and Kapoor (2025) note, the lack of robust ethical frameworks tailored for agricultural contexts can exacerbate existing inequalities and undermine trust in technological innovations. Moreover, in the context of global agrifood systems, issues such as environmental sustainability, responsible resource use, and labor rights must be central to the design and deployment of AI systems (Chen et al., 2024).

Security challenges are equally prominent. Smart farming systems, driven by AI and Internet of Things (IoT) technologies, are vulnerable to cyberattacks, data breaches, and manipulation (Ali, Mijwil, Buruga, Abotaleb, & Adamopoulos, 2024; Bhangar & Shahriyar, 2023). Protecting agricultural data — whether related to proprietary seed genetics, soil health, or market-sensitive pricing information — is critical for maintaining system integrity and farmer trust. Dembani et al. (2025) emphasize the need for secure data governance models, including federated learning approaches, to mitigate privacy risks while maximizing data utility.

In addition, the global disparities in AI access and implementation cannot be ignored. While developed economies are rapidly adopting sophisticated AI tools, many developing nations face barriers including infrastructure deficits, limited digital literacy, and regulatory gaps (Aderibigbe et al., 2023). These inequities threaten to widen the global agricultural productivity gap and highlight the urgent need for inclusive AI policies and capacity-building initiatives (Adanma & Ogunbiyi, 2024).

At the intersection of these issues — innovation, ethics, and security — lies both a challenge and an opportunity. As emphasized by Alexander, Yarborough, and Smith (2024), cultivating responsible AI practices in agriculture is not merely a technical endeavor; it requires multi-stakeholder collaboration, transparency, and a commitment to social justice. A proactive approach, informed by interdisciplinary research and grounded in ethical principles, is essential for ensuring that AI truly benefits all stakeholders in the agricultural sector — from large-scale agribusinesses to smallholder farmers and indigenous communities.

This chapter aims to critically explore the ethical and security challenges associated with AI applications in agricultural research. It also seeks to highlight emerging opportunities for building resilient, inclusive, and sustainable AI-driven agricultural systems. Drawing upon contemporary studies, case examples, and expert recommendations, it provides a comprehensive roadmap for stakeholders committed to advancing ethical and secure AI innovations in agriculture.

The following flowchart summarizes the major ethical and security challenges arising from the adoption of AI in agricultural research, as discussed in this introduction.

Figure 1. Ethical and security challenges in agricultural AI

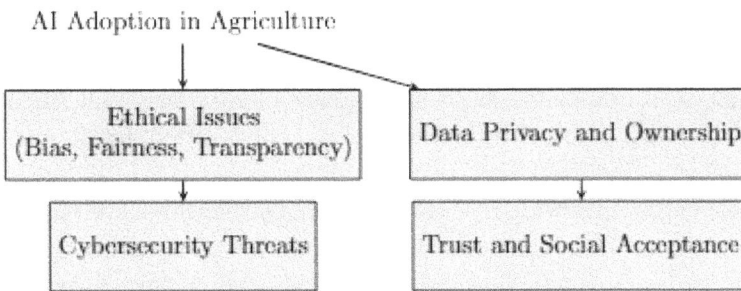

2. AI-DRIVEN INNOVATIONS IN AGRICULTURAL RESEARCH

The agricultural sector, historically dependent on manual labor and traditional practices, is undergoing a revolutionary transformation fueled by artificial intelligence (AI). Across the globe, AI-driven innovations are reshaping agricultural research, offering solutions for increasing food security, promoting sustainable practices, and addressing pressing environmental challenges.

2.1 Precision Agriculture and Smart Farming

Precision agriculture, enhanced by AI and big data analytics, is transforming farming into a highly optimized science. As Bhat and Huang (2021) outlined, AI technologies such as machine learning algorithms, satellite imaging, and IoT sensors are enabling precise monitoring of crop health, soil conditions, and pest outbreaks. This revolution is not only improving yields but also reducing resource wastage, essential for sustainable farming.

Similarly, Araújo et al. (2021) described "Agriculture 4.0" as a landscape where AI-driven systems monitor, predict, and automate agricultural processes, helping farmers make data-driven decisions. Bhangar and Shahriyar (2023) further emphasized the synergy between AI and IoT, showcasing their potential in next-generation farming systems.

2.2 Enhancing Crop Production and Food Security

Crop productivity has become a central concern amid global population growth. According to Ahmed et al. (2024), AI technologies like deep learning and computer vision are enhancing yield predictions, disease detection, and automated irrigation systems, directly contributing to agricultural revolutions. Ahmad et al. (2024) and Ali et al. (2024) further emphasized AI's role in empowering agriculture to achieve global food security, especially in developing nations where traditional methods are insufficient.

The potential of AI to bridge gaps in agricultural practices in developing countries was critically examined by Aderibigbe et al. (2023), who discussed the challenges and opportunities in AI adoption and the need for strategic investments.

2.3 Ethical, Trust, and Privacy Considerations

As AI permeates agriculture, ethical concerns arise. Kapoor (2025) discussed the necessity of forging responsible paths in AI and machine learning for the agri-food sector, stressing ethics and accountability. Dara et al. (2022) provided recommendations for the ethical and responsible use of AI in digital agriculture, highlighting the need to balance innovation with social responsibility.

Building trust in AI technologies is essential. Alexander, Yarborough, and Smith (2024) pointed out the need for responsible AI frameworks to gain farmer and consumer trust. Similarly, Gardezi et al. (2024) explored the challenges farmers face in trusting AI-driven farming solutions, especially when transparency is lacking.

Data privacy remains a major challenge. Dembani et al. (2025) and Gavai et al. (2025) emphasized that agricultural data privacy is paramount for farmer autonomy, proposing federated learning and emerging platforms as viable solutions. Devare et al. (2023) further underlined governance issues surrounding agricultural data and recommended policies for more responsible data management.

2.4 AI in Environmental Conservation and Agritourism

AI applications extend beyond productivity to sustainability and conservation. Adanma and Ogunbiyi (2024) assessed how AI contributes to environmental conservation by identifying cyber risks and promoting sustainable practices.

Furthermore, AI's impact on agritourism was investigated by Denhere and Shao (2024), who analyzed opportunities and ethical considerations, particularly in African contexts. Their research points to AI's dual potential to bolster rural economies and raise ethical dilemmas regarding technology access and equity.

2.5 Cybersecurity Challenges in Smart Agriculture

As agriculture becomes more connected, cybersecurity concerns grow. Ali et al. (2024) presented a comprehensive survey identifying AI's role in cybersecurity for smart agriculture, while Jerhamre, Carlberg, and van Zoest (2022) examined the vulnerabilities introduced by smart farming technologies. Groumpos (2023) echoed similar warnings, advocating for critical reflections on AI's threats alongside its promises.

Holzinger et al. (2021) stressed that the digital transformation needed for achieving the Sustainable Development Goals (SDGs) must prioritize security, safety, and privacy, especially in agriculture where vulnerabilities can directly affect food systems.

2.6 Emerging Applications and Future Prospects

The application of AI in agricultural research is diversifying rapidly. Gupta et al. (2024) showcased AI-powered advancements in plant sciences, with innovations ranging from genome editing to pest prediction models. Chen et al. (2024) provided a comprehensive survey of AI's progress in the agrifood sector, noting its opportunities in supply chain optimization, livestock monitoring, and precision harvesting.

In agricultural supply chains, Kanyepe et al. (2025) detailed how AI can optimize logistics, minimize waste, and ensure timely delivery from farms to markets, revolutionizing the entire food ecosystem.

To visualize the focus areas of AI applications in agricultural research, the following pie chart illustrates the relative emphasis across key domains such as precision farming, environmental conservation, supply chain optimization, and others.

Figure 2. Key focus areas of AI innovations in agriculture

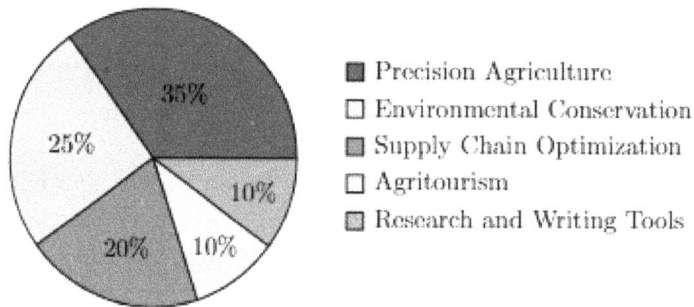

- ■ Precision Agriculture
- □ Environmental Conservation
- ▨ Supply Chain Optimization
- □ Agritourism
- ▨ Research and Writing Tools

2.7 Challenges to Implementation

Despite its promise, AI faces barriers to full-scale agricultural integration. Aggarwal, Bansal, and Goel (2024) described AI in agriculture as both a looming challenge and a gleaming opportunity, pointing to high costs, technical complexity, and lack of farmer training as persistent obstacles. Chen et al. (2024) and He et al. (2021) similarly stressed the need for human-centered AI to ensure systems are accessible and trustworthy for farmers.

The susceptibility of smart farming to technical and social vulnerabilities was further explored by Jerhamre et al. (2022), who advocated for robust resilience strategies.

2.8 AI's Role in Academic Research and Writing

Beyond field applications, AI is shaping agricultural research dissemination. Abd-Elsalam and Abdel-Momen (2023) highlighted the development and challenges of using AI in scientific writing, noting its potential to streamline publication but cautioning against ethical pitfalls. Similarly, Aithal and Aithal (2023) envisioned AI tools like ChatGPT transforming higher education and research landscapes.

Bahrini et al. (2023) also explored ChatGPT's applications in academia, raising awareness about opportunities and threats stemming from its rapid integration into scientific workflows.

2.9 Summary

AI-driven innovations are undeniably reshaping agricultural research and practice. They hold tremendous potential to improve food security, promote sustainability, and enhance farmer livelihoods globally. However, this transformation must be approached with ethical foresight, robust privacy safeguards, and strategies to build trust among stakeholders. Future research must continue to bridge technical, ethical, and governance gaps to ensure AI's benefits are equitably shared across diverse agricultural landscapes.

3. ETHICAL CONCERNS IN AI-BASED AGRICULTURAL STUDIES

The increasing adoption of Artificial Intelligence (AI) in agriculture promises revolutionary benefits, including enhanced productivity, sustainability, and global food security. However, it simultaneously raises profound ethical challenges that must be addressed for responsible deployment and public trust.

3.1. Data Privacy and Ownership

In AI-driven agriculture, large-scale data collection from sensors, drones, and satellites is common. However, issues related to data privacy, ownership, and control have emerged as major concerns (Dembani et al., 2025; Gavai et al., 2025; Devare et al., 2023). Farmers often lack clarity on who owns their operational data and how it is used or shared, leading to potential exploitation. Jerhamre, Carlberg, and van Zoest (2022) emphasized the vulnerability of smart farming systems to unauthorized data access, which could endanger both individual farmers and entire agricultural ecosystems.

3.2. Algorithmic Bias and Fairness

AI models, if trained on unrepresentative or biased datasets, risk perpetuating inequalities in agricultural access and outcomes (Kapoor, 2025; Gupta et al., 2024). For instance, marginalized communities in developing countries might not benefit equally from AI innovations due to regional or economic biases embedded in algorithms (Aderibigbe et al., 2023; Ahmad et al., 2024). Dara, Hazrati Fard, and Kaur (2022) recommended rigorous audits and transparency mechanisms to prevent bias and promote fairness in digital agriculture systems.

3.3. Trust and Transparency Challenges

Building trust in AI-driven agriculture requires transparent models and explainable AI systems (Alexander et al., 2024; Gardezi et al., 2024). Without transparency, farmers may distrust AI recommendations, potentially reducing technology adoption rates. Gardezi et al. (2024) highlighted that mistrust arises when AI decision-making processes are opaque or fail to account for farmers' experiential knowledge.

Holzinger et al. (2021) and He et al. (2021) also stressed that human-centered AI design, prioritizing user understanding and interaction, is critical to overcoming the "black box" problem and establishing trust.

3.4. Cybersecurity Threats

The integration of AI and Internet of Things (IoT) technologies in agriculture increases susceptibility to cyber threats (Ali et al., 2024; Ali, Anderson et al., 2024; Bhangar & Shahriyar, 2023). Cyberattacks on agricultural systems could disrupt supply chains, manipulate data-driven decisions, or cause physical harm to infrastructure (Adanma & Ogunbiyi, 2024; Ali et al., 2024).

Bhat and Huang (2021) stressed that cybersecurity measures must evolve alongside AI innovations to ensure resilience, particularly in the context of precision farming.

3.5. Environmental and Sustainability Concerns

Although AI offers tools for promoting environmental conservation, there are ethical debates about its real-world impacts (Adanma & Ogunbiyi, 2024; Aggarwal, Bansal, & Goel, 2024). Misapplication of AI could intensify resource exploitation, contribute to biodiversity loss, or inadvertently reinforce unsustainable agricultural practices if not carefully managed.

Chen et al. (2024) and Ahmad et al. (2024) advocated for ethical frameworks ensuring AI systems align with broader environmental sustainability goals.

3.6. Socio-Economic Disparities

AI has the potential to widen existing socio-economic gaps in rural communities (Aggarwal et al., 2024; Ahmad et al., 2024). Wealthier farmers may benefit disproportionately, while smallholders struggle with access to expensive AI-based tools and infrastructures (Ahmad et al., 2024; Alexander et al., 2024).

Furthermore, Denhere and Shao (2024) analyzed how the African agritourism sector, increasingly reliant on AI, faces a delicate balance between technological advancement and equitable opportunity distribution.

3.7. Responsibility and Accountability

Determining who is responsible for AI-driven decisions in agriculture is complex. When AI systems fail—whether by recommending incorrect pesticide usage or mismanaging irrigation—attribution of blame becomes difficult (Alexander et al., 2024; Kapoor, 2025). Aithal and Aithal (2023) emphasized that defining legal and ethical responsibility in AI outputs is critical for ensuring justice and public trust.

Bahrini et al. (2023) warned that reliance on AI without clear accountability mechanisms could exacerbate the consequences of errors, making it imperative to establish regulatory safeguards.

3.8. Ethical Governance of Agricultural AI

A coordinated global effort is necessary to develop ethical standards for AI deployment in agriculture (Kapoor, 2025; Groumpos, 2023). Gardezi et al. (2024) recommended participatory governance frameworks where farmers, policymakers, researchers, and technologists co-develop ethical guidelines.

Similarly, Araújo et al. (2021) and Dara et al. (2022) called for multistakeholder involvement to address ethical complexities, including labor displacement, farmer autonomy, and societal impacts.

3.9. Misinformation and Scientific Integrity

With the rise of AI-generated scientific content, concerns about misinformation and authenticity have also emerged (Abd-Elsalam & Abdel-Momen, 2023; Bahrini et al., 2023). In agricultural research, AI tools like ChatGPT are transforming knowledge production, but they also risk spreading inaccuracies if not properly regulated (Aithal & Aithal, 2023).

Ahmed et al. (2024) highlighted the dual role of AI as both a powerful aid and a potential threat to scientific credibility in agricultural studies.

3.10. Future Ethical Pathways

Looking forward, responsible AI in agriculture will require dynamic and context-specific ethical frameworks that prioritize human rights, environmental stewardship, and social equity (Holzinger et al., 2021; Gavai et al., 2025).

Chen et al. (2024) proposed a multidimensional ethical model encompassing technical robustness, societal benefit, and sustainability imperatives.

Moreover, Kanyepe et al. (2025) stressed the critical role of AI-powered supply chains in promoting ethical sourcing and transparency across the agri-food sector.

The following bar chart highlights the most critical ethical risks identified in AI-based agricultural studies, emphasizing data privacy, algorithmic bias, and cybersecurity threats.

Figure 3. Top ethical risks in smart agriculture

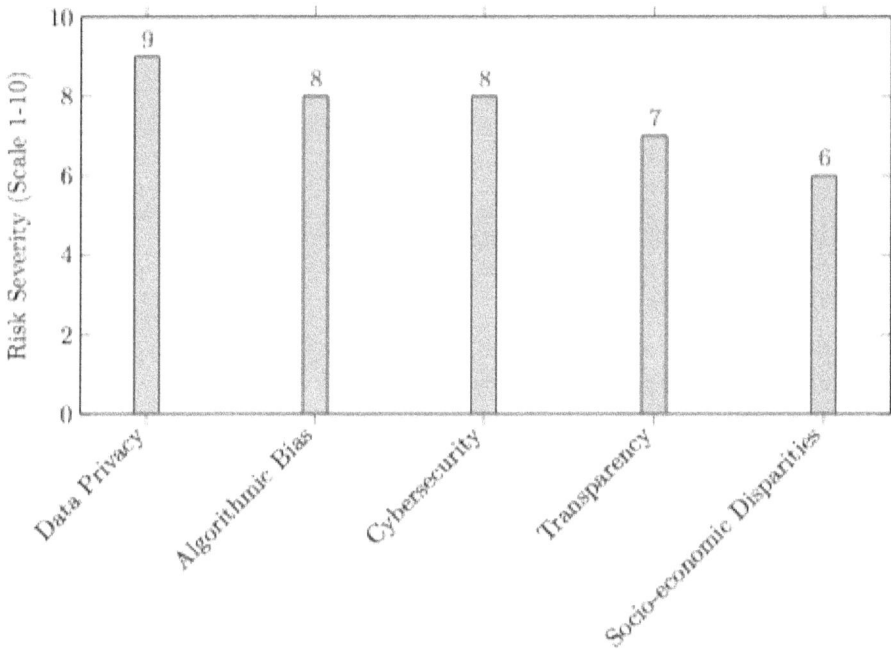

4. DATA PRIVACY AND OWNERSHIP IN AGRI-TECH RESEARCH

The rise of artificial intelligence (AI) and smart technologies in agriculture, commonly referred to as Agri-Tech, has led to the generation of vast amounts of agricultural data. This explosion of data offers immense opportunities but also raises serious concerns about data privacy, ownership, cybersecurity, and ethical use, particularly in research and development contexts.

4.1. The Critical Role of Data in Agri-Tech

Agricultural data fuels modern innovations, from precision farming to crop monitoring and supply chain optimization (Bhat & Huang, 2021; Bhangar & Shahriyar, 2023). AI-driven technologies enhance food production, resource management, and sustainability (Chen et al., 2024; Gupta et al., 2024), but they also demand extensive data collection — often involving sensitive information about farmers' land, resources, and operations.

However, the governance of agricultural data remains fragmented and lacks standardized policies (Devare et al., 2023). Studies emphasize that without clear frameworks for data management and ownership, technological adoption may deepen existing inequalities between technology providers and end-users (Ahmad et al., 2024; Ali et al., 2024).

4.2. Challenges of Data Privacy in Agri-Tech Research

The digital transformation of agriculture introduces vulnerabilities that expose data to cyber threats and misuse (Ali et al., 2024; Adanma & Ogunbiyi, 2024). Smart farming systems interconnected through IoT devices are particularly susceptible to attacks (Jerhamre et al., 2022). Furthermore, AI systems processing agricultural data often face challenges in ensuring transparency and user trust (Gardezi et al., 2024).

AI models trained on agricultural datasets can inadvertently expose sensitive operational practices, crop yields, and even personal data of smallholder farmers (Alexander et al., 2024; Holzinger et al., 2021). The lack of robust cybersecurity measures in developing nations further aggravates the situation (Aderibigbe et al., 2023; Ahmad et al., 2024).

4.3. Ownership Issues and Ethical Concerns

Who owns agricultural data remains a contentious issue. Farmers, corporations, researchers, and governments all stake claims to data collected from agricultural activities (Dembani et al., 2025; Gavai et al., 2025). As AI becomes integral to agricultural decision-making, ownership disputes could hinder innovation and fair benefit-sharing (Dara et al., 2022; Kapoor, 2025).

Kapoor (2025) points out that ethical frameworks must be established to ensure farmers' rights are protected. Similarly, Araújo et al. (2021) suggest that participatory models of data governance, where farmers co-own the data they generate, could create more equitable Agri-Tech ecosystems.

4.4. Technological Innovations and Federated Learning

To address privacy concerns, federated learning has emerged as a promising solution. Rather than centralizing sensitive agricultural data, federated learning enables model training at the edge, preserving data locality and reducing exposure risks (Dembani et al., 2025).

Ahmed et al. (2024) highlight that AI-driven agricultural systems incorporating federated learning can enhance crop yields while safeguarding farmer data. However, this requires significant infrastructural investments, which remain challenging in low-resource settings (Aderibigbe et al., 2023; Kanyepe et al., 2025).

4.5. Building Trust in AI-Driven Agriculture

Trust in AI-based agricultural technologies hinges on transparent practices, clear communication about data usage, and shared benefits (Gardezi et al., 2024; Alexander et al., 2024). Research shows that when farmers perceive AI tools as opaque or exploitative, their adoption rates decline sharply (Devare et al., 2023; Kapoor, 2025).

Educational initiatives and transparent AI policies are necessary to foster trust among stakeholders (Aithal & Aithal, 2023; Groumpos, 2023). Moreover, AI developers must design explainable systems that empower users to understand how their data are processed (He et al., 2021).

4.6. Cybersecurity Risks in Smart Agriculture

Cybersecurity is a growing concern in smart agriculture systems. Attacks targeting sensors, drones, and automated irrigation systems could compromise food supply chains and farmer livelihoods (Ali et al., 2024; Holzinger et al., 2021).

Ali et al. (2024) and Denhere and Shao (2024) emphasize the need for a comprehensive cybersecurity strategy tailored to the agricultural sector's unique requirements. This includes securing data transmission channels, implementing strong authentication mechanisms, and regularly updating system firmware.

4.7. Regulatory Efforts and Global Perspectives

While some regions have made strides toward regulating data use in Agri-Tech, global standards are still evolving (Gavai et al., 2025; Kapoor, 2025). Ethical guidelines, such as those proposed by Dara et al. (2022), recommend prioritizing farmers' data rights, mandating informed consent, and ensuring the equitable distribution of benefits arising from agricultural data exploitation.

International collaborations could play a vital role in harmonizing data governance standards, as argued by Ahmad et al. (2024) and Ali et al. (2024). Efforts must ensure that both developing and developed nations benefit fairly from AI advancements in agriculture.

4.8. Future Opportunities and Research Directions

The future of data privacy and ownership in Agri-Tech research lies in integrating technological solutions with ethical principles (Chen et al., 2024; Gupta et al., 2024). Blockchain technology, for instance, could enable tamper-proof records of data ownership and sharing agreements (Jerhamre et al., 2022; Kanyepe et al., 2025).

At the same time, there is a need for continuous interdisciplinary research involving AI experts, ethicists, farmers, and policymakers to co-create responsible AI ecosystems (Dara et al., 2022; Kapoor, 2025; Ahmad et al., 2024). Education and training on data rights and cybersecurity practices will be essential for future farmers and agricultural researchers alike (Bahrini et al., 2023; Abd-Elsalam & Abdel-Momen, 2023).

5. SECURITY RISKS IN SMART AGRICULTURE

Smart agriculture, driven by artificial intelligence (AI), the Internet of Things (IoT), big data, and robotics, is revolutionizing food production systems. However, alongside its transformative benefits, it introduces profound cybersecurity risks that can jeopardize data integrity, operational continuity, and even food security.

5.1. Cyber Threats Targeting Smart Agriculture

The integration of AI and IoT technologies in agriculture has exposed farms to new cyber threats such as ransomware attacks, data breaches, and sensor manipulation (Ali et al., 2024). Cybercriminals exploit vulnerabilities in poorly secured IoT devices and centralized agricultural databases (Ali, Mijwil, Buruga, Abotaleb, & Adamopoulos, 2024). In fact, studies have shown that smart farming systems, such as precision irrigation or autonomous tractors, are often susceptible to hacking due to outdated software and limited cybersecurity protocols (Jerhamre, Carlberg, & van Zoest, 2022).

Big data platforms in agriculture aggregate sensitive information about soil health, crop yields, and proprietary farming methods, making them attractive targets (Bhat & Huang, 2021). Consequently, a cyberattack can not only disrupt operations but also cause long-term damage to agricultural businesses and food supply chains (He et al., 2021).

The diagram below presents the main cybersecurity threat pathways targeting smart agriculture systems, from vulnerable IoT devices to compromised AI decision-making.

Figure 4. Cybersecurity threat pathways in smart agriculture

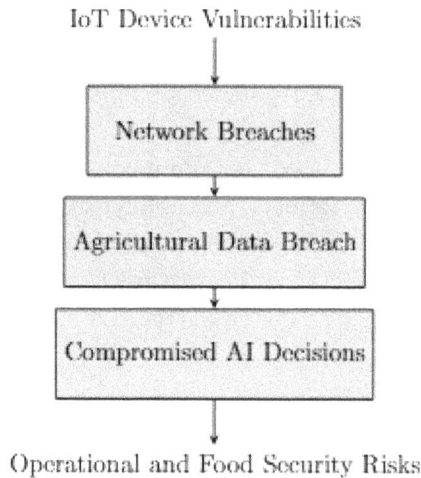

IoT Device Vulnerabilities

Network Breaches

Agricultural Data Breach

Compromised AI Decisions

Operational and Food Security Risks

5.2. Data Privacy and Governance Challenges

Smart farms increasingly depend on vast networks of sensors and AI models, leading to complex data privacy issues (Dembani et al., 2025; Gavai et al., 2025). Farmers often have limited control over the ownership and use of their data, raising concerns about exploitation and surveillance. Agricultural data is frequently stored in centralized cloud servers, creating single points of failure vulnerable to cyber intrusions (Devare, Arnaud, Antezana, & King, 2023).

To mitigate these risks, researchers recommend adopting federated learning and decentralized data governance models to ensure that farmers retain autonomy over their data while enhancing system resilience (Dembani et al., 2025; Gavai et al., 2025).

5.3. Ethical and Trust Issues

Building trust in AI-powered farming technologies is a critical security concern. If farmers do not trust the security or fairness of smart agriculture solutions, adoption rates will decline, undermining digital agriculture initiatives (Alexander, Yarborough, & Smith, 2024; Gardezi et al., 2024). As highlighted by Kapoor (2025) and Dara, Hazrati Fard, and Kaur (2022), there is a pressing need for ethical guidelines that emphasize transparency, data sovereignty, and responsible AI development in the agri-food sector.

The ethical use of AI extends beyond privacy. Issues such as algorithmic biases, unauthorized surveillance, and the opaque nature of AI decision-making processes must also be addressed to foster a secure agricultural environment (Kapoor, 2025; Adanma & Ogunbiyi, 2024).

5.4. AI Model Vulnerabilities

AI algorithms used in smart agriculture—such as those predicting crop yields or pest infestations—are themselves vulnerable to adversarial attacks (Chen et al., 2024; Groumpos, 2023). By manipulating sensor data or introducing noise, attackers can corrupt AI model outputs, leading to incorrect farming decisions and crop failures.

Holzinger, Weippl, Tjoa, and Kieseberg (2021) emphasize that cybersecurity strategies must be built into AI systems from the design stage ("security-by-design") to counter potential adversarial risks.

Moreover, generative AI tools, including large language models like ChatGPT, though offering promising benefits, also introduce potential threats such as misinformation, phishing risks, and unintended data exposure (Bahrini et al., 2023; Aithal & Aithal, 2023).

5.5. Emerging Threats with Agriculture 4.0 and IoT

The fourth agricultural revolution, often termed Agriculture 4.0, combines AI, robotics, and biotechnology. This convergence, while promising, compounds cybersecurity threats as various interconnected devices form intricate and often insecure networks (Araújo, Peres, Barata, Lidon, & Ramalho, 2021; Bhangar & Shahriyar, 2023).

A compromised drone or autonomous tractor, for example, could cause massive economic damage and even physical harm. Ahmed et al. (2024) warn that unless robust authentication and encryption mechanisms are adopted, Agriculture 4.0 systems will remain vulnerable.

Furthermore, climate-based smart farming solutions, while critical for sustainability, are not immune to these risks (Ahmad, Liew, Venturini, Kalogeras, Candiani, Di Benedetto, & Martos, 2024). Attackers can falsify environmental data to manipulate irrigation, fertilizer distribution, or pest control systems.

5.6. Global Challenges and Regional Disparities

In developing nations, the cybersecurity infrastructure needed to protect smart farms is often lacking (Ahmad et al., 2024; Ali et al., 2024; Aderibigbe et al., 2023). This digital divide not only increases vulnerability but also exacerbates inequality in access to secure agricultural technology. According to Ahmad et al. (2024), AI holds immense potential to address global food security challenges, but cybersecurity must be a foundational element of these efforts.

Denhere and Shao (2024) further highlight that African agritourism ventures adopting smart agriculture face unique ethical and security challenges due to infrastructural and regulatory gaps.

5.7. Future Outlook and Recommendations

As smart agriculture continues to evolve, proactive cybersecurity measures must be integral to system design, deployment, and maintenance. Researchers advocate for a multi-layered security approach encompassing encryption, AI model hardening, ethical guidelines, decentralized data systems, and robust regulatory frameworks (Chen et al., 2024; Gupta, Pagani, Zamboni, & Singh, 2024).

Moreover, continuous farmer education about cyber risks and trust-building initiatives are essential to enhance resilience across the agricultural sector (Gardezi et al., 2024).

To create a truly secure and sustainable future for global agriculture, a collaborative effort involving technology developers, policymakers, researchers, and farmers is indispensable (Kanyepe, Chibaro, Morima, & Moeti-Lysson, 2025).

6. REGULATORY AND GOVERNANCE FRAMEWORKS

The rapid integration of Artificial Intelligence (AI) into agriculture and food systems necessitates the development of robust regulatory and governance frameworks to address emerging challenges around ethics, data privacy, cybersecurity, trust, and equitable access. As AI reshapes agricultural practices worldwide, particularly in developing nations (Aderibigbe et al., 2023; Ahmad et al., 2024), governing bodies must ensure that AI deployments are responsible, transparent, and aligned with societal goals.

6.1. Regulatory Challenges in AI-Driven Agriculture

Despite its transformative potential, AI introduces significant governance challenges in agriculture. Existing regulations often lag behind technological advances, creating gaps that can be exploited (Alexander et al., 2024; Gardezi et al., 2024). For instance, issues such as algorithmic transparency, biases in decision-making, and ethical considerations surrounding data usage are critical but under-regulated (Dara et al., 2022; Kapoor, 2025). In agricultural contexts, the urgency is heightened as AI systems increasingly impact food security and rural livelihoods (Ahmad et al., 2024; Gupta et al., 2024).

AI regulation in agriculture must therefore consider domain-specific issues, including precision farming vulnerabilities (Jerhamre et al., 2022), agricultural data governance (Devare et al., 2023), and the safeguarding of farmer autonomy against opaque machine decisions (Chen et al., 2024). Furthermore, cyber risks linked to the adoption of AI and IoT technologies in smart farming amplify the need for cybersecurity regulations (Ali et al., 2024; Bhangar & Shahriyar, 2023).

6.2. Data Governance and Privacy Frameworks

One of the most pressing regulatory needs is the establishment of strong data governance frameworks. Agricultural operations increasingly depend on the collection and processing of vast datasets, raising concerns over ownership, access, and misuse (Gavai et al., 2025; Dembani et al., 2025). Federated learning models, which allow decentralized data usage while preserving privacy, are emerging as a promising regulatory solution (Dembani et al., 2025).

Data privacy regulations must also address the susceptibility of smart farming systems to breaches (Jerhamre et al., 2022; Groumpos, 2023). Existing frameworks in other industries provide valuable templates, but agriculture's unique dynamics — such as smallholder farm participation and seasonal variability — demand tailored approaches (Araújo et al., 2021; Gardezi et al., 2024).

6.3. Building Trust Through Governance

Trust remains foundational for AI's success in agriculture. Governance frameworks must foster public confidence by emphasizing transparency, fairness, and accountability (Gardezi et al., 2024; Alexander et al., 2024). Human-centered AI design principles, which prioritize explainability and user agency, are essential (He et al., 2021).

Agricultural AI governance should also address ethical questions surrounding data ownership, particularly when dealing with indigenous knowledge and local farming practices (Kanyepe et al., 2025). Incorporating participatory governance models — where farmers and local communities are involved in decision-making — can mitigate distrust and foster broader acceptance (Kapoor, 2025).

6.4. Cybersecurity and Ethical AI Guidelines

As agriculture becomes digitized, cybersecurity must be embedded into AI regulatory standards. AI-enabled farming systems face risks from cyberattacks that could manipulate yields, disrupt supply chains, or jeopardize food security (Ali et al., 2024; Holzinger et al., 2021). Ethical frameworks must also guide AI development to avoid reinforcing inequalities, ensuring that marginalized groups benefit equitably (Denhere & Shao, 2024; Adanma & Ogunbiyi, 2024).

Bahrini et al. (2023) highlight the dual nature of AI: offering vast opportunities while simultaneously posing significant threats if left unchecked. In this vein, ethical codes and cybersecurity protocols tailored to agriculture are critical to safeguarding systems and data integrity (Ali et al., 2024; Holzinger et al., 2021).

6.5. International Coordination and Standardization

Given the global nature of agricultural trade and supply chains, international cooperation is vital in harmonizing AI governance standards. Disparate national regulations could create fragmented markets and technological silos (Chen et al., 2024; Bhat & Huang, 2021).

Initiatives aimed at international standardization — such as the integration of Sustainable Development Goals (SDGs) into digital transformation efforts (Holzinger et al., 2021) — can provide a unified framework for ethical, secure, and sustainable AI use. Moreover, emerging global efforts around responsible AI, such as principles for transparency, fairness, and accountability in digital agriculture, must be adopted widely (Dara et al., 2022; Ahmad et al., 2024).

6.6. Recommendations for a Robust Framework

To navigate the complexities discussed, several recommendations emerge:

- **Ethical AI Principles:** Embedding fairness, accountability, and transparency as non-negotiable foundations in agricultural AI governance (Kapoor, 2025; Dara et al., 2022).
- **Farmer-Centric Data Rights:** Legislating data ownership models that prioritize farmers' rights over their data (Devare et al., 2023; Gavai et al., 2025).
- **Resilient Cybersecurity Architectures:** Mandating robust cybersecurity measures in AI agricultural systems (Ali et al., 2024; Holzinger et al., 2021).
- **Capacity Building:** Enhancing the regulatory and technical capabilities of developing countries to bridge the implementation gap (Aderibigbe et al., 2023; Ahmad et al., 2024).
- **Participatory Governance Models:** Encouraging stakeholder participation in AI development and deployment processes (Kanyepe et al., 2025; Kapoor, 2025).

By embracing a multidimensional governance approach that blends regulatory oversight, ethical considerations, cybersecurity vigilance, and participatory practices, the agriculture sector can fully realize AI's transformative potential while minimizing its associated risks.

7. CASE STUDIES AND BEST PRACTICES

7.1. Introduction

The integration of Artificial Intelligence (AI) into agriculture and related fields has revolutionized traditional practices, offering pathways toward sustainability, efficiency, and resilience. However, along with opportunities, significant challenges regarding ethical considerations, data privacy, cybersecurity, and trust have emerged (Chen et al., 2024; Groumpos, 2023). This section explores detailed case studies and best practices addressing these challenges while showcasing successful implementations.

7.2. Case Studies

7.2.1 AI in Precision Agriculture: Enhancing Productivity

A landmark study by Ahmad et al. (2024) highlighted how AI-powered tools significantly increased crop yields in developing nations by optimizing irrigation schedules and pest control methods. Similarly, Ahmed et al. (2024) demonstrated that integrating AI with remote sensing technologies enhanced real-time crop monitoring, leading to a 15–20% increase in overall agricultural productivity.

Bhat and Huang (2021) provided evidence of how big data and AI-driven models predict weather patterns and soil health with remarkable accuracy, empowering farmers to make informed decisions. Araújo et al. (2021) characterized Agriculture 4.0's evolution, emphasizing the integration of AI and IoT technologies for real-time agricultural management.

7.2.2 Ethical Challenges and Data Governance

As digital transformation accelerates, ethical governance of agricultural data has gained attention. Dara, Hazrati Fard, and Kaur (2022) recommended guidelines for the responsible use of AI in agriculture, stressing transparency, accountability, and inclusiveness.

Gardezi et al. (2024) discussed the importance of building trust among farmers, suggesting that co-designing AI tools with end-users promotes higher adoption rates. Devare, Arnaud, Antezana, and King (2023) further emphasized governance mechanisms to ensure responsible plant data management.

Moreover, Dembani et al. (2025) and Gavai et al. (2025) reviewed federated learning and emerging privacy platforms as solutions to agricultural data privacy concerns, a growing necessity given rising cyber threats.

7.2.3 Cybersecurity and AI: Addressing Risks

Ali, Mijwil, Buruga, Abotaleb, and Adamopoulos (2024) provided a comprehensive survey on cybersecurity threats facing smart agriculture, underscoring the need for robust AI-based defense mechanisms.

Adanma and Ogunbiyi (2024) explored AI's dual role in promoting sustainability and exposing environmental systems to new cyber vulnerabilities. Similarly, Jerhamre, Carlberg, and van Zoest (2022) identified vulnerabilities in smart farming, recommending layered cybersecurity frameworks.

Holzinger, Weippl, Tjoa, and Kieseberg (2021) argued that cybersecurity, safety, and privacy must form the pillars of AI integration aligned with Sustainable Development Goals (SDGs).

7.2.4 Building Responsible AI Systems

Alexander, Yarborough, and Smith (2024) outlined a responsible AI model for food systems, emphasizing interdisciplinary collaboration and ethical foresight to foster trust and acceptance.

Kapoor (2025) proposed ethical frameworks for the agrifood industry, stressing the importance of fairness, transparency, and minimizing bias in machine learning models.

He et al. (2021) introduced the concept of "human-centered AI," proposing the development of AI systems that prioritize user needs and safety in agriculture and autonomous farming equipment.

7.3. Best Practices

7.3.1 Human-Centered and Ethical AI Design

The need for human-centered AI designs is critical, as emphasized by He et al. (2021) and Kapoor (2025). Best practices involve involving farmers and local stakeholders during AI tool development, ensuring systems are adaptable to diverse socio-economic environments (Alexander et al., 2024).

Dara et al. (2022) stressed the importance of ethical audits and continuous monitoring to maintain AI systems' trustworthiness in agriculture.

7.3.2 Data Governance and Privacy Protection

Implementing federated learning models (Dembani et al., 2025) and privacy-preserving platforms (Gavai et al., 2025) is essential to safeguarding sensitive agricultural data.

Devare et al. (2023) recommended creating national-level data governance policies that align with international standards to ensure responsible agricultural data linkage.

7.3.3 Cybersecurity Resilience

Integrating AI with cybersecurity solutions tailored for smart agriculture is crucial. Ali et al. (2024) proposed using anomaly detection models and blockchain technology to safeguard agricultural networks.

Jerhamre et al. (2022) advocated for multi-layered defense strategies, while Holzinger et al. (2021) suggested embedding cybersecurity from the design stage of AI systems.

7.3.4 Leveraging AI for Environmental Sustainability

AI's role in sustainable farming practices has been extensively discussed by Adanma and Ogunbiyi (2024) and Bhangar and Shahriyar (2023), who suggested integrating AI models that minimize resource waste and optimize input use.

Aggarwal, Bansal, and Goel (2024) highlighted AI's potential in predicting climate impacts on agriculture, thus helping farmers adapt practices accordingly.

7.3.5 Education and Capacity Building

Education initiatives that enhance AI literacy among farmers and agricultural practitioners are vital. Aithal and Aithal (2023) proposed leveraging tools like ChatGPT for personalized education and advisory services in agriculture.

Abd-Elsalam and Abdel-Momen (2023) noted the importance of upskilling researchers and scientists to responsibly use AI in academic and agricultural writing.

7.4. Emerging Trends and Future Directions

Chen et al. (2024) and Ahmad et al. (2024) forecasted increasing integration of AI with biotechnology and sensor networks to address food security challenges.

Ali et al. (2024) and Gupta, Pagani, Zamboni, and Singh (2024) suggested that AI-driven advancements in plant sciences could revolutionize crop breeding and pest management for future agricultural systems.

Bahrini et al. (2023) discussed the rapid growth of generative AI models like ChatGPT in agriculture, presenting both innovative applications and potential threats.

Kanyepe, Chibaro, Morima, and Moeti-Lysson (2025) emphasized AI's transformative impact on agricultural supply chains, advocating for resilient, transparent, and decentralized supply networks.

Denhere and Shao (2024) pointed out opportunities and ethical considerations in AI-driven agritourism models across Africa, suggesting responsible innovations to promote community-based agriculture.

7.5. Summary

These case studies and best practices reveal that while AI holds immense promise for agriculture and environmental conservation, realizing its full potential requires addressing critical challenges around ethics, trust, cybersecurity, and governance. Multidisciplinary collaboration, proactive regulation, and farmer-centric designs are key to ensuring that AI technologies deliver sustainable and equitable benefits to the global agricultural community.

8. TOWARDS RESPONSIBLE AI IN AGRICULTURE

The agricultural sector is undergoing a profound transformation driven by Artificial Intelligence (AI) technologies. From precision farming to supply chain optimization, AI presents opportunities to enhance productivity, sustainability, and food security (Ahmad et al., 2024; Ali et al., 2024). However, this transformation brings a responsibility to ensure ethical deployment, address cybersecurity risks, manage agricultural data privacy, and foster trust among stakeholders (Alexander, Yarborough, & Smith, 2024; Gardezi et al., 2024). Towards building a responsible AI ecosystem in agriculture, several critical dimensions must be addressed.

8.1. Ethical Deployment of AI in Agriculture

Ethical concerns surrounding AI applications in agriculture are gaining attention. As Kapoor (2025) notes, there is a pressing need to "harvest ethics" by forging responsible paths for AI and machine learning (ML) within the agri-food industry. AI models must be transparent, fair, and non-discriminatory, avoiding biases that could marginalize smallholder farmers or exacerbate inequalities (Dara, Hazrati Fard, & Kaur, 2022).

Building human-centered AI systems is essential to gain trust from farmers and other agricultural stakeholders. He et al. (2021) emphasize the importance of designing trustworthy, human-centered AI and autonomous systems, while Holzinger et al. (2021) link the goals of digital transformation to broader Sustainable Development Goals (SDGs), stressing security, safety, and privacy.

8.2. Agricultural Data Privacy and Governance

The increasing use of sensors, drones, and IoT devices in agriculture has led to a surge in agricultural data generation. Protecting this data is vital for responsible AI development. Studies by Gavai et al. (2025) and Dembani et al. (2025) underscore that agricultural data privacy is an emerging challenge, with federated learning and privacy-preserving technologies offering promising solutions. Likewise, Devare et al. (2023) emphasize the complexities of governing agricultural data and recommend frameworks for responsible plant data linkage.

Without robust governance structures, agricultural data misuse can become a barrier to innovation and trust. Gardezi et al. (2024) argue that privacy concerns, if not properly addressed, will impede farmers' willingness to share critical data needed for AI model development.

8.3. Addressing Cybersecurity Threats in Smart Agriculture

Cybersecurity is another dimension of responsible AI deployment. As agriculture becomes increasingly digitized, it becomes more vulnerable to cyber threats. Ali et al. (2024) provide an in-depth survey of cybersecurity challenges in smart agriculture, calling for stronger AI-based security measures to detect and counteract cyber risks.

Moreover, Adanma and Ogunbiyi (2024) highlight how environmental conservation efforts relying on AI must simultaneously account for cyber vulnerabilities. Jerhamre, Carlberg, and van Zoest (2022) further explore the susceptibility of smart farming to cyber-attacks, urging pre-emptive measures to enhance system resilience.

8.4. Building Trust and Public Acceptance

Public skepticism towards AI systems in agriculture stems from concerns about data misuse, job displacement, and transparency (Gardezi et al., 2024; Alexander, Yarborough, & Smith, 2024). Trust can be fostered by implementing transparent AI decision-making processes, ensuring explainability, and actively involving farmers in technology design (Chen et al., 2024).

Furthermore, historical insights by Groumpos (2023) stress the necessity of learning from past AI failures and societal responses to ensure a more inclusive and responsible future. In line with this, Aithal and Aithal (2023) argue that advanced AI tools like ChatGPT, when applied responsibly, can revolutionize agricultural education and research.

Building trust in AI applications for agriculture requires a multi-step process, as outlined in the following flowchart, emphasizing transparency, inclusivity, and regulatory safeguards.

Figure 5. Steps to building trust in AI-driven agriculture

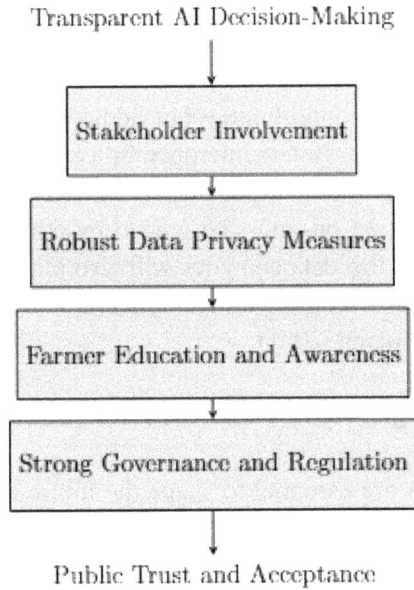

Transparent AI Decision-Making

↓

Stakeholder Involvement

↓

Robust Data Privacy Measures

↓

Farmer Education and Awareness

↓

Strong Governance and Regulation

↓

Public Trust and Acceptance

8.5. Opportunities and Challenges in Developing Nations

Developing countries face both unique opportunities and hurdles in AI adoption in agriculture. Ahmad et al. (2024) and Aderibigbe et al. (2023) stress the potential of AI to revolutionize food security if infrastructural and educational gaps are addressed. However, a critical gap remains between potential and actual implementation, mainly due to lack of access, affordability issues, and insufficient policy support.

Aggarwal, Bansal, and Goel (2024) further elaborate on how AI in agriculture represents both a looming challenge and a gleaming opportunity, especially in resource-constrained environments. Addressing these gaps will require international collaboration, investments in digital literacy, and local capacity building.

8.6. Integrating AI and IoT for Sustainable Agriculture

Emerging paradigms like Agriculture 4.0 emphasize the fusion of AI with IoT technologies to enhance sustainability. Araújo et al. (2021) characterize Agriculture 4.0 as a landscape rich with opportunities but riddled with challenges such as data management complexity and system interoperability.

Bhangar and Shahriyar (2023) argue that AI and IoT integration will be critical for next-generation farming but must be guided by ethical standards and robust cybersecurity protocols. Big data analytics will also play a significant role in advancing precision agriculture, but with inherent risks if data security and privacy are neglected (Bhat & Huang, 2021).

8.7. Policy and Regulatory Frameworks for Responsible AI

Policy interventions are essential to shape the future of responsible AI in agriculture. Dara, Hazrati Fard, and Kaur (2022) recommend ethical guidelines and regulatory mechanisms to prevent misuse and foster trust. Similarly, Kapoor (2025) advocates for integrating ethical considerations into every stage of AI development and deployment in the food system.

Kapoor's notion of "Harvesting Ethics" echoes across other studies that call for international standards on agricultural data protection, cybersecurity, and transparency in AI models (Holzinger et al., 2021; Gardezi et al., 2024).

8.8. Future Prospects and Recommendations

To ensure responsible AI adoption in agriculture, stakeholders must collaborate across sectors — from policymakers and researchers to farmers and technology developers. Emphasis must be placed on:

- **Transparency**: Clear explainability of AI-driven decisions.
- **Privacy Protection**: Adopting federated learning and encrypted data-sharing models.
- **Cybersecurity**: Building resilient smart farming systems.
- **Ethical Standards**: Embedding fairness, accountability, and inclusivity in AI models.
- **Capacity Building**: Bridging digital divides through education and infrastructure investments.

Studies by Ahmad et al. (2024), Gupta et al. (2024), and Kanyepe et al. (2025) all converge on the idea that AI can empower the agricultural sector sustainably if a responsible innovation framework is embraced early and consistently.

In closing, while AI promises transformative impacts on agriculture, this promise can only be fulfilled through an unwavering commitment to ethics, data privacy, cybersecurity, and inclusive innovation. As AI continues to evolve, so must the structures that govern its development and application to safeguard agriculture's future.

9. CONCLUSION

The integration of Artificial Intelligence (AI) into agriculture presents a transformative opportunity to address critical global challenges such as food security, climate resilience, and sustainable farming. As this chapter has demonstrated, AI technologies—from precision farming and predictive analytics to intelligent supply chains and smart environmental monitoring—hold immense potential to enhance productivity, reduce waste, and optimize resource usage across the agricultural value chain (Aggarwal, Bansal, & Goel, 2024; Ahmad et al., 2024; Gupta et al., 2024).

Yet, the path toward responsible AI in agriculture is fraught with complex challenges. Ethical concerns, including biases in data, transparency, and the accountability of automated decisions, remain pressing (Alexander, Yarborough, & Smith, 2024; Dara, Hazrati Fard, & Kaur, 2022). Issues of cybersecurity, privacy, and the vulnerability of interconnected smart farming systems to cyber threats further complicate the landscape (Ali et al., 2024; Dembani et al., 2025; Gavai et al., 2025). The digital divide between developed and developing regions also exacerbates disparities in AI adoption, threatening to widen global inequalities rather than bridge them (Aderibigbe et al., 2023; Ahmad et al., 2024).

Trust emerges as a pivotal cornerstone in ensuring the sustainable deployment of AI in agriculture (Gardezi et al., 2024; He et al., 2021). Building trust requires transparent governance frameworks, robust standards for data sharing and privacy, participatory design approaches that engage farmers and rural communities, and the cultivation of AI literacy across stakeholders (Alexander et al., 2024; Devare et al., 2023). Furthermore, frameworks for the ethical use of AI must prioritize human-centered values, environmental stewardship, and social justice (Kapoor, 2025; Holzinger et al., 2021).

Notably, emerging technologies such as federated learning, explainable AI, and blockchain offer promising pathways to address some of these challenges, enhancing security, accountability, and data sovereignty in agricultural systems (Araújo et al., 2021; Jerhamre, Carlberg, & van Zoest, 2022; Gavai et al., 2025). At the same time, interdisciplinary collaboration between technologists, farmers, policymakers, ethicists, and environmental scientists is crucial to ensure that AI innovations are aligned with the broader goals of sustainability and food justice (Chen et al., 2024; Bhat & Huang, 2021).

In conclusion, while the road toward responsible AI in agriculture is complex and multifaceted, it is not insurmountable. By embedding ethical considerations, inclusive governance, and a human-centered approach at the core of AI development and deployment, stakeholders can unlock AI's transformative potential while safeguarding against unintended consequences. Achieving this balance will be essential not only for advancing agricultural innovation but also for promoting a more equitable, sustainable, and resilient food system for generations to come.

REFERENCES

Abd-Elsalam, K. A., & Abdel-Momen, S. M. (2023). Artificial intelligence's development and challenges in scientific writing. *Egyptian Journal of Agricultural Research*, *101*(3), 714–717. DOI: 10.21608/ejar.2023.220363.1414

Adanma, U. M., & Ogunbiyi, E. O.Uwaga Monica AdanmaEmmanuel Olurotimi Ogunbiyi. (2024). Artificial intelligence in environmental conservation: Evaluating cyber risks and opportunities for sustainable practices. *Computer Science & IT Research Journal*, *5*(5), 1178–1209. DOI: 10.51594/csitrj.v5i5.1156

Aderibigbe, A. O., Ohenhen, P. E., Nwaobia, N. K., Gidiagba, J. O., & Ani, E. C.Adebayo Olusegun AderibigbePeter Efosa OhenhenNwabueze Kelvin Nwaobia-Joachim Osheyor GidiagbaEmmanuel Chigozie Ani. (2023). Artificial intelligence in developing countries: Bridging the gap between potential and implementation. *Computer Science & IT Research Journal*, *4*(3), 185–199. DOI: 10.51594/csitrj.v4i3.629

Aggarwal, S., Bansal, S., & Goel, R. (2024). AI In Agriculture: A Looming Challenge, A Gleaming Opportunity. *International Journal of Engineering Science and Humanities, 14*(Special Issue 1), 43-52.

Ahmad, A., Liew, A. X., Venturini, F., Kalogeras, A., Candiani, A., Di Benedetto, G., Ajibola, S., Cartujo, P., Romero, P., Lykoudi, A., De Grandis, M. M., Xouris, C., Lo Bianco, R., Doddy, I., Elegbede, I., D'Urso Labate, G. F., García del Moral, L. F., & Martos, V. (2024). AI can empower agriculture for global food security: Challenges and prospects in developing nations. *Frontiers in Artificial Intelligence*, *7*, 1328530. DOI: 10.3389/frai.2024.1328530 PMID: 38726306

Ahmad, A., Liew, A. X., Venturini, F., Kalogeras, A., Candiani, A., Di Benedetto, G., ... & Martos, V. (2024). AI can empower agriculture for.

Ahmed, M. N., Singh, A. P., Hussain, M. R., Rasool, M. A., Khan, I. M., & Dildar, M. S. (2024, July). Enhancing Crop Production using Artificial Intelligence in Agricultural Revolution. In *2024 IEEE 7th International Conference on Advanced Technologies* [ATSIP]. *Signal and Image Processing : an International Journal*, *1*, 432–437.

Aithal, P. S., & Aithal, S. (2023). Application of ChatGPT in higher education and research–A futuristic analysis. [IJAEML]. *International Journal of Applied Engineering and Management Letters*, *7*(3), 168–194. DOI: 10.47992/IJAEML.2581.7000.0193

Alexander, C. S., Yarborough, M., & Smith, A. (2024). Who is responsible for 'responsible AI'?: Navigating challenges to build trust in AI agriculture and food system technology. *Precision Agriculture*, *25*(1), 146–185. DOI: 10.1007/s11119-023-10063-3

Ali, A., Anderson, L., Venturini, F., Athanasios, K., Candiani, A., Di Benedetto, G., ... & Martos, V. (2024). AI can empower agriculture for global food security: challenges and prospects in developing nations.

Ali, G., Mijwil, M. M., Buruga, B. A., Abotaleb, M., & Adamopoulos, I. (2024). A survey on artificial intelligence in cybersecurity for smart agriculture: State-of-the-art, cyber threats, artificial intelligence applications, and ethical concerns. *Mesopotamian Journal of Computer Science*, *2024*, 53–103. DOI: 10.58496/MJCSC/2024/007

Araújo, S. O., Peres, R. S., Barata, J., Lidon, F., & Ramalho, J. C. (2021). Characterising the agriculture 4.0 landscape—Emerging trends, challenges and opportunities. *Agronomy (Basel)*, *11*(4), 667. DOI: 10.3390/agronomy11040667

Bahrini, A., Khamoshifar, M., Abbasimehr, H., Riggs, R. J., Esmaeili, M., Majdabadkohne, R. M., & Pasehvar, M. (2023, April). ChatGPT: Applications, opportunities, and threats. In *2023 Systems and Information Engineering Design Symposium (SIEDS)* (pp. 274-279). IEEE. DOI: 10.1109/SIEDS58326.2023.10137850

Bhangar, N. A., & Shahriyar, A. K. (2023). Iot and ai for next-generation farming: Opportunities, challenges, and outlook. *International Journal of Sustainable Infrastructure for Cities and Societies*, *8*(2), 14–26.

Bhat, S. A., & Huang, N. F. (2021). Big data and ai revolution in precision agriculture: Survey and challenges. *IEEE Access : Practical Innovations, Open Solutions*, *9*, 110209–110222. DOI: 10.1109/ACCESS.2021.3102227

Chen, T., Lv, L., Wang, D., Zhang, J., Yang, Y., Zhao, Z., & Tao, D. (2024). Empowering agrifood system with artificial intelligence: A survey of the progress, challenges and opportunities. *ACM Computing Surveys*, *57*(2), 1–37.

Dara, R., Hazrati Fard, S. M., & Kaur, J. (2022). Recommendations for ethical and responsible use of artificial intelligence in digital agriculture. *Frontiers in Artificial Intelligence*, *5*, 884192. DOI: 10.3389/frai.2022.884192 PMID: 35968036

Dembani, R., Karvelas, I., Akbar, N. A., Rizou, S., Tegolo, D., & Fountas, S. (2025). Agricultural data privacy and federated learning: A review of challenges and opportunities. *Computers and Electronics in Agriculture*, *232*, 110048. DOI: 10.1016/j.compag.2025.110048

Denhere, V., & Shao, D. (2024). Artificial Intelligence in Agritourism: Utilitarian Analysis of Opportunities, Challenges, and Ethical Considerations in the African Context. *Agritourism in Africa*, 17-36.

Devare, M., Arnaud, E., Antezana, E., & King, B. (2023). Governing agricultural data: Challenges and recommendations. *Towards Responsible Plant Data Linkage: Data Challenges for Agricultural Research and Development, 201*.

Gardezi, M., Joshi, B., Rizzo, D. M., Ryan, M., Prutzer, E., Brugler, S., & Dadkhah, A. (2024). Artificial intelligence in farming: Challenges and opportunities for building trust. *Agronomy Journal*, *116*(3), 1217–1228. DOI: 10.1002/agj2.21353

Gavai, A. K., Bouzembrak, Y., Xhani, D., Sedrakyan, G., Meuwissen, M. P., Souza, R. G. S., ... & van Hillegersberg, J. (2025). Agricultural data Privacy: Emerging platforms & strategies. *Food and Humanity*, 100542.

Groumpos, P. P. (2023, July). A critical historic overview of artificial intelligence: Issues, challenges, opportunities, and threats. *Artificial Intelligence and Applications (Commerce, Calif.)*, *1*(4), 181–197. DOI: 10.47852/bonviewAIA3202689

Gupta, D. K., Pagani, A., Zamboni, P., & Singh, A. K. (2024). AI-powered revolution in plant sciences: Advancements, applications, and challenges for sustainable agriculture and food security. *Exploration of Foods and Foodomics*, *2*(5), 443–459. DOI: 10.37349/eff.2024.00045

He, H., Gray, J., Cangelosi, A., Meng, Q., McGinnity, T. M., & Mehnen, J. (2021). The challenges and opportunities of human-centered AI for trustworthy robots and autonomous systems. *IEEE Transactions on Cognitive and Developmental Systems*, *14*(4), 1398–1412. DOI: 10.1109/TCDS.2021.3132282

Holzinger, A., Weippl, E., Tjoa, A. M., & Kieseberg, P. (2021, August). Digital transformation for sustainable development goals (sdgs)-a security, safety and privacy perspective on ai. In *International cross-domain conference for machine learning and knowledge extraction* (pp. 1-20). Cham: Springer International Publishing.

Jerhamre, E., Carlberg, C. J. C., & van Zoest, V. (2022). Exploring the susceptibility of smart farming: Identified opportunities and challenges. *Smart Agricultural Technology*, *2*, 2. DOI: 10.1016/j.atech.2021.100026

Kanyepe, J., Chibaro, M., Morima, M., & Moeti-Lysson, J. (2025). AI-Powered Agricultural Supply Chains: Applications, Challenges, and Opportunities. *Integrating Agriculture, Green Marketing Strategies, and Artificial Intelligence*, 33-64.

Kapoor, P. (2025). Harvesting Ethics: Forging Responsible Paths in AI and ML for the Agri-Food Industry. In *Food and Industry 5.0: Transforming the Food System for a Sustainable Future* (pp. 305–315). Springer Nature Switzerland. DOI: 10.1007/978-3-031-76758-6_19

Kaushik, K., Khan, A., Kumari, A., Sharma, I., & Dubey, R. (2024). Ethical considerations in AI-based cybersecurity. In *Next-generation cybersecurity: AI, ML, and Blockchain* (pp. 437–470). Springer Nature Singapore. DOI: 10.1007/978-981-97-1249-6_19

Kochupillai, M., Kahl, M., Schmitt, M., Taubenböck, H., & Zhu, X. X. (2022). Earth observation and artificial intelligence: Understanding emerging ethical issues and opportunities. *IEEE Geoscience and Remote Sensing Magazine, 10*(4), 90–124. DOI: 10.1109/MGRS.2022.3208357

Kochupillai, M., & Köninger, J. (2023). Creating a digital marketplace for agrobiodiversity and plant genetic sequence data: Legal and ethical considerations of an ai and blockchain based solution. *Towards Responsible Plant Data Linkage: Data Challenges for Agricultural Research and Development*, 223.

Leong, Y. M., Lim, E. H., Subri, N. F. B., & Jalil, N. B. A. (2023, September). Transforming agriculture: Navigating the challenges and embracing the opportunities of artificial intelligence of things. In *2023 IEEE International Conference on Agrosystem Engineering, Technology & Applications (AGRETA)* (pp. 142-147). IEEE. DOI: 10.1109/AGRETA57740.2023.10262747

Malhotra, K., & Firdaus, M. (2022). Application of artificial intelligence in IoT security for crop yield prediction. *ResearchBerg Review of Science and Technology, 2*(1), 136–157.

Mark, R. (2019). Ethics of using AI and big data in agriculture: The case of a large agriculture multinational. *The ORBIT Journal, 2*(2), 1–27. DOI: 10.29297/orbit.v2i2.109

Neupane, S., Mitra, S., Fernandez, I. A., Saha, S., Mittal, S., Chen, J., Pillai, N., & Rahimi, S. (2024). Security considerations in ai-robotics: A survey of current methods, challenges, and opportunities. *IEEE Access : Practical Innovations, Open Solutions, 12*, 22072–22097. DOI: 10.1109/ACCESS.2024.3363657

Otieno, M. (2023). An extensive survey of smart agriculture technologies: Current security posture. *World J. Adv. Res. Rev, 18*(3), 1207–1231. DOI: 10.30574/wjarr.2023.18.3.1241

Pandey, D. K., & Mishra, R. (2024). *Towards sustainable agriculture: Harnessing AI for global food security*. Artificial Intelligence in Agriculture.

Pedro, F., Subosa, M., Rivas, A., & Valverde, P. (2019). Artificial intelligence in education: Challenges and opportunities for sustainable development.

Senoo, E. E. K., Anggraini, L., Kumi, J. A., Luna, B. K., Akansah, E., Sulyman, H. A., & Aritsugi, M. (2024). IoT solutions with artificial intelligence technologies for precision agriculture: Definitions, applications, challenges, and opportunities. *Electronics (Basel), 13*(10), 1894. DOI: 10.3390/electronics13101894

Shafik, W., Tufail, A., De Silva, C. L., Haji, R. A. A., & Apong, M. (2025). The Role, Application, and Impact of Artificial Intelligence in the Agriculture Industry. In *Future Tech Startups and Innovation in the Age of AI* (pp. 36–60). CRC Press.

Sood, A., Sharma, R. K., & Bhardwaj, A. K. (2022). Artificial intelligence research in agriculture: A review. *Online Information Review, 46*(6), 1054–1075. DOI: 10.1108/OIR-10-2020-0448

Sparrow, R., & Howard, M. (2021). Robots in agriculture: prospects, impacts, ethics, and policy. *precision agriculture, 22*, 818-833.

Sparrow, R., Howard, M., & Degeling, C. (2021). Managing the risks of artificial intelligence in agriculture. *NJAS: Impact in Agricultural and Life Sciences, 93*(1), 172–196. DOI: 10.1080/27685241.2021.2008777

Tzachor, A., Devare, M., King, B., Avin, S., & Ó hÉigeartaigh, S. (2022). Responsible artificial intelligence in agriculture requires systemic understanding of risks and externalities. *Nature Machine Intelligence, 4*(2), 104–109. DOI: 10.1038/s42256-022-00440-4

Uplaonkar, S. S., Veershetty, R., Bahar, Z., & Sangeeta, G. (2024). The role of productive AI: A supporter or challenger in the future of agricultural librarianship. *IJAR, 10*(4), 46–53.

Wei, H., Xu, W., Kang, B., Eisner, R., Muleke, A., Rodriguez, D., deVoil, P., Sadras, V., Monjardino, M., & Harrison, M. T. (2024). Irrigation with artificial intelligence: Problems, premises, promises. *Human-Centric Intelligent Systems, 4*(2), 187–205. DOI: 10.1007/s44230-024-00072-4

Williamson, H. F., Brettschneider, J., Caccamo, M., Davey, R. P., Goble, C., Kersey, P. J., May, S., Morris, R. J., Ostler, R., Pridmore, T., Rawlings, C., Studholme, D., Tsaftaris, S. A., & Leonelli, S. (2023). Data management challenges for artificial intelligence in plant and agricultural research. *F1000 Research, 10*, 324. DOI: 10.12688/f1000research.52204.2 PMID: 36873457

Zangana, H. M., Luckyardi, S., Mustafa, F. M., & Li, S. (2025). Enhancing Agricultural Cybersecurity: Leveraging Deep Learning and Large Language Models for Smart Farming Protection. In Zangana, H., Al-Karaki, J., & Omar, M. (Eds.), *Revolutionizing Cybersecurity With Deep Learning and Large Language Models* (pp. 307–338). IGI Global Scientific Publishing., DOI: 10.4018/979-8-3373-3296-3.ch010

Chapter 3
Contract Expiry in AI–Driven STM Research:
Legal Challenges, Risks, and Regulatory Considerations

Dilshad Ahmad Mhia-alddin
https://orcid.org/0009-0005-4046-1607
University of Mosul, Iraq

Akram Mahmoud Hussein
University of Mosul, Iraq

ABSTRACT

The integration of Artificial Intelligence (AI) into Scientific, Technical, and Medical (STM) research has transformed traditional workflows, enabling accelerated discovery and innovation. However, the expiry of contracts governing AI-driven STM projects introduces complex legal challenges, potential risks, and evolving regulatory demands. Issues such as intellectual property ownership, data governance, confidentiality, and the continuity of AI systems post-contract termination require careful consideration. This chapter explores these multifaceted legal concerns, examines real-world case studies, identifies risks emerging from insufficient contract management, and proposes regulatory frameworks to mitigate negative outcomes. Through a critical analysis, it underscores the necessity for adaptive legal strategies that uphold the integrity, security, and ethical standards of STM research in the AI era.

DOI: 10.4018/979-8-3373-4252-8.ch003

1. INTRODUCTION

The advancement of Artificial Intelligence (AI) technologies has significantly transformed Scientific, Technical, and Medical (STM) research, leading to unprecedented innovation and productivity. AI-driven systems now play a central role in automating data analysis, enhancing predictive modeling, and accelerating the discovery process across STM fields (Leontidis, 2024). However, these benefits are accompanied by new legal complexities, particularly in the context of contract expiry. As AI tools become integral to research, the expiration of agreements that govern their use can expose institutions and researchers to significant legal, ethical, and operational risks (Mhia-Alddin & Hussein, 2025).

A major challenge emerges from the evolving nature of AI-infused contracts, where traditional terms often fail to address the dynamic, learning, and adaptive characteristics of AI systems (Act et al., 2024). Unlike static tools, AI algorithms continue to evolve, raising questions about ownership of improvements, maintenance responsibilities, data governance, and continuity of services after contract termination (Mbah, n.d.). In the context of STM research, these challenges are even more critical given the high sensitivity of scientific data and the necessity for reproducibility, accuracy, and ethical compliance (Awofala et al., 2025).

Moreover, the risks associated with contract expiry are not confined to operational disruptions. Legal disputes over intellectual property rights (Mahoney, 2017), unauthorized data usage (Chung et al., 2022), regulatory non-compliance (Butt, 2024), and breaches of confidentiality (Harper, Ellis, & Tucker, 2022) can severely damage the integrity of STM research projects. Researchers often rely on AI solutions provided by third parties or developed under collaborative agreements; thus, clear contractual terms are essential to ensure that rights and obligations remain enforceable post-expiry.

The regulatory environment is rapidly evolving to address these emerging challenges. The European Union's Artificial Intelligence Act (Butt, 2024) and the General Data Protection Regulation (Act et al., 2024) are examples of frameworks aiming to balance innovation with accountability and transparency. Additionally, specific sectors such as healthcare (Horgan, Romao, Morré, & Kalra, 2020; Nordlinger, Villani, & Rus, 2020) and emergency care (Hosseini et al., 2023) have initiated their own AI governance protocols to mitigate risks associated with contract expiration and system failures.

Yet, regulatory responses often lag behind technological innovation. As AI systems deployed in STM fields become increasingly complex—incorporating natural language processing (Haney, 2020), unstructured data mining (de Haan et al., 2024), and explainable AI approaches (Kosasih, Papadakis, Baryannis, & Brintrup, 2024)—the need for proactive legal strategies becomes even more pressing.

Scholars highlight the vulnerabilities arising from poorly managed AI lifecycle contracts, emphasizing the importance of lifecycle-based contracting models that extend beyond initial deployment and encompass maintenance, updating, and system retirement phases (Mhia-Alddin & Hussein, 2025).

Ethical challenges also accompany legal risks. AI's role in STM research raises critical concerns around data biases, transparency, accountability, and surveillance (Baskara, 2024; Harper et al., 2022). Without clear contractual clauses addressing ethical obligations, AI applications in research may perpetuate systemic inequalities or produce invalid results, undermining the core principles of scientific inquiry (Biesbroek, Wright, Eguren, Bonotto, & Athanasiadis, 2022; Nunkoo, Sharma, Rana, Dwivedi, & Sunnassee, 2023).

Furthermore, international considerations complicate matters. Differences in legal interpretations regarding AI ownership, liability, and privacy across jurisdictions (Cho & Crompvoets, 2019; Leal & Musgrave, 2023) necessitate that STM institutions adopt a global perspective when drafting and managing AI-related contracts. Issues such as cross-border data flows, the treatment of zero-day vulnerabilities (Leal & Musgrave, 2023), and export controls on AI technologies all impact contract terms and expiry management.

Finally, the emergence of smart contracts powered by blockchain and AI introduces a new dimension to contract law, transforming traditional licensing and service agreements into automated, self-enforcing instruments (Mbah, n.d.). While these smart contracts offer greater efficiency, they also present new legal uncertainties regarding termination, modification, and dispute resolution, particularly in complex STM collaborations.

In light of these complexities, this chapter aims to offer a comprehensive analysis of contract expiry in AI-driven STM research. It will explore legal challenges, identify associated risks, examine current regulatory responses, and propose frameworks for effective contract management and risk mitigation. Special attention will be given to case studies from healthcare, education, and defense sectors, where AI adoption is reshaping traditional research practices (Chavan, Paul, & Kolekar, 2024; Caso, 2024; Caggiano, Gatt, Mollo, Izzo, & Troisi, 2024).

The legal challenges arising from AI-driven contract expirations are multifaceted, involving intellectual property disputes, regulatory compliance risks, and data governance issues. Figure 1 illustrates the major categories of legal concerns associated with AI contract expiry in STM research.

Figure 1. Major legal challenges emerging from AI contract expiry

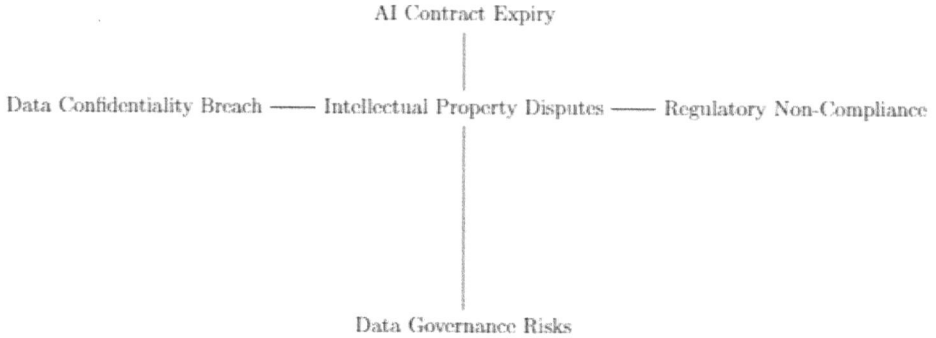

AI Contract Expiry

Data Confidentiality Breach —— Intellectual Property Disputes —— Regulatory Non-Compliance

Data Governance Risks

By addressing these critical issues, this chapter contributes to ensuring that STM research in the AI era remains legally sound, ethically responsible, and resilient against the challenges of contractual discontinuities.

2. UNDERSTANDING CONTRACT EXPIRY IN RESEARCH CONTEXTS

Contract expiry represents a critical phase in the research and innovation eco-system, impacting legal, operational, ethical, and strategic dimensions of research projects. Especially in an era where artificial intelligence (AI) and digitalization are increasingly embedded into contractual frameworks (Act et al., 2023; Butt, 2024), understanding the dynamics of contract expiry is vital for managing compliance, mitigating risks, and ensuring the continuity of knowledge and resources.

2.1 The Legal and Regulatory Foundation

Contract expiry in research is intricately connected with broader legal principles. The emergence of AI-specific regulations, such as the EU Artificial Intelligence Act (Butt, 2024), reinforces the need for a systematic approach to managing contract lifecycles. Legal aspects tied to AI applications in research contracts, especially regarding data protection and disclosure (Act et al., 2023), are influencing new practices around termination and renewal conditions.

Cho and Crompvoets (2019) highlight the challenges associated with the legal ownership and control of data produced during research projects, particularly when contracts expire. Meanwhile, Mahoney (2017) discusses how contract structure influences organizational performance, reinforcing the idea that poorly managed expiries can have long-term effects.

2.2 Ethical, Security, and Compliance Considerations

As digital tools become entrenched in research management (Cox & Thelwall, 2025; Haney, 2020), ethical considerations in contract expiration have grown. Harper, Ellis, and Tucker (2022) emphasize the covert risks of surveillance that persist even after a research contract concludes, highlighting the need for secure data disposition policies.

Furthermore, Mhia-Alddin and Hussein (2025) introduce AI-powered models for managing contract security, expiry, and compliance, demonstrating how deep learning and large language models (LLMs) can predict expiry risks and propose mitigation strategies.

Security in contracts related to defense and autonomous technologies (Gunawan et al., 2022) or sensitive sectors like healthcare (Nordlinger, Villani, & Rus, 2020) also presents new challenges. In military and healthcare research, confidentiality obligations often survive beyond contract termination, requiring careful post-expiry management (Caso, 2024; Horgan et al., 2020).

2.3 AI and the Dynamics of Contract Expiry

AI is transforming the way contracts are monitored and managed. AI-infused contracting systems (Act et al., 2023) now automatically flag approaching expirations, suggest renewals, or initiate new compliance assessments. Mbah (n.d.) discusses how smart contracts and AI revolutionize licensing agreements, affecting how expiry clauses are drafted and enforced.

Explainable AI systems are also being integrated into contract management (Kosasih et al., 2024), offering transparency in the decision-making processes associated with contract renewals or terminations. According to Hosseini et al. (2023), AI in emergency care illustrates the urgency of understanding contract expiry, especially when critical services depend on uninterrupted partnerships.

Additionally, recent work by de Haan et al. (2024) stresses the importance of structuring research around unstructured data governance, as such data must be managed appropriately once contracts lapse.

Effective contract lifecycle management is vital for mitigating legal, operational, and ethical risks associated with AI applications in STM research. Figure 2 presents a structured view of the key stages in AI-driven contract lifecycle management.

Figure 2. Contract lifecycle management in AI-driven STM research

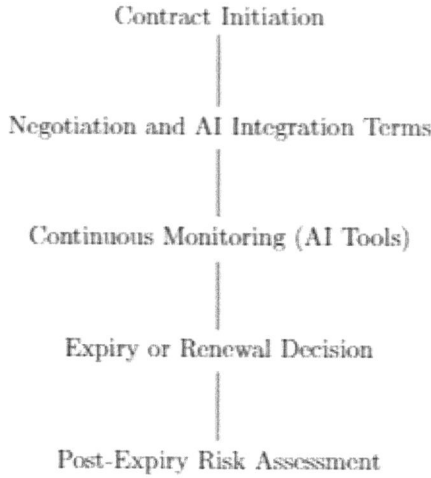

Contract Initiation

Negotiation and AI Integration Terms

Continuous Monitoring (AI Tools)

Expiry or Renewal Decision

Post-Expiry Risk Assessment

2.4 Impact on Scientific and Educational Research

In academic settings, the expiry of research contracts directly impacts ongoing scientific discovery. Researchers like Leontidis (2024) argue that AI is fundamentally altering scientific methodology, meaning contractual terms must now account for the management of AI-generated outputs post-expiry.

Baskara (2024) and Kendall and Teixeira da Silva (2024) both point out risks tied to academic integrity when AI tools are not properly governed after a research agreement concludes. Issues of authorship, data ownership, and publication rights often resurface during expiry phases.

Moreover, Awofala et al. (2025) show that AI adoption among educators requires sustainable contractual frameworks, emphasizing the importance of expiry provisions that do not disrupt tool access or pedagogical continuity.

2.5 Practical Approaches and Case Studies

Real-world examples underscore the necessity of robust expiry management. Harwich and Laycock (2018) explore how the UK's NHS incorporates AI while planning for contract succession and expiry to prevent service gaps. Similarly, Chavan, Paul, and Kolekar (2024) in food safety, and Caggiano et al. (2024) in aerospace health, underline how expiry mismanagement can endanger critical supply chains and missions.

In the domain of cybersecurity, Leal and Musgrave (2023) emphasize that failure to manage zero-day vulnerabilities tied to expiring contracts can expose organizations to severe risks.

Tourism research by Guttentag (2015) and Nunkoo et al. (2023) further illustrates the economic impact of poorly handled contract expirations, especially in sectors relying heavily on agile, short-term agreements.

2.6 Future Directions and Technology Integration

Looking ahead, innovative AI techniques like deep learning for expiry prediction (Mhia-Alddin & Hussein, 2025) and neurosymbolic explainable models (Kosasih et al., 2024) will increasingly dominate contract management.

Nicholls and Culpepper (2021) suggest that computational approaches for media framing can be repurposed to understand and predict contractual narrative shifts over time, adding a proactive layer to expiry management.

Furthermore, international collaborations in sectors like nuclear policy (Nriezedi-Anejionu, 2024) and sustainability initiatives (Biesbroek et al., 2022) call for harmonized expiry policies to ensure continuous cooperation post-contract.

2.7 Challenges and Limitations

Despite these advancements, challenges remain. As highlighted by Moussa and Teixeira da Silva (2023), distinguishing between legitimate and predatory practices in contract renewal offers a cautionary tale for research institutions.

Moreover, the vulnerability of AI systems to misuse (Kendall & Teixeira da Silva, 2024) complicates automated expiry monitoring. Harper et al. (2022) also warn of ethical blind spots that arise when expiry protocols are handled solely by algorithmic systems without human oversight.

Finally, as articulated by Horgan et al. (2020) and Nordlinger et al. (2020), an overreliance on technology without transparent, human-centered policies risks undermining the very objectives of sustainable and ethical research practices.

3. RISKS ASSOCIATED WITH CONTRACT EXPIRY IN AI-ENRICHED ENVIRONMENTS

The integration of Artificial Intelligence (AI) into contracting processes introduces complex new challenges related to contract expiry. Traditional risks, such as service disruption, data loss, and compliance breaches, are exacerbated in AI-enriched environments, where automated processes, predictive analytics, and intelligent systems interact with human-driven operations.

3.1. Automated Dependencies and Service Disruptions

AI systems embedded in contracts often automate critical operational functions, creating strong dependencies that can be abruptly severed upon contract expiry (Hamza et al., 2024). For example, when contracts involving AI-based accounting or logistics systems lapse without renewal or a clear transition plan, organizations may face immediate operational paralysis. This risk is magnified in sectors like healthcare, where AI-driven decision support systems are vital (Nordlinger, Villani, & Rus, 2020; Horgan et al., 2020).

Moreover, as emphasized by Mhia-Alddin and Hussein (2025), without proactive AI-powered contract security measures, the end of a contract term can trigger compliance failures, service gaps, and potential legal violations. Deep learning and Large Language Models (LLMs) can mitigate these risks by predicting expiry threats, but not all systems are currently robust enough.

3.2. Data Ownership, Integrity, and Portability

The expiry of AI-enriched contracts raises urgent questions about data custody, transfer, and integrity (de Haan et al., 2024; Cho & Crompvoets, 2019). AI systems often generate and process vast volumes of proprietary or sensitive information. As contracts terminate, disputes may arise over data ownership, exacerbated by unclear clauses regarding cloud-stored or algorithmically processed data (Caggiano et al., 2024).

Additionally, handling "unstructured" data (Chung et al., 2022) becomes a challenge if contract terms do not adequately anticipate data migration needs or privacy protections, risking breaches of regulations such as the GDPR and emerging AI governance laws (Butt, 2024; Act et al., 2024).

3.3. Ethical, Legal, and Surveillance Concerns

AI-infused environments heighten ethical and legal vulnerabilities upon contract expiry, especially where covert data surveillance or algorithmic decision-making was integral to service delivery (Harper, Ellis, & Tucker, 2022; Christensen, 2021). Terminated contracts may expose both parties to litigation if covert AI operations persist or if personal data governance deteriorates post-expiry.

Gunawan et al. (2022) point out parallels in autonomous systems' accountability in international law, emphasizing that lapsing contracts leave a dangerous accountability gap. Additionally, the misuse of zero-day vulnerabilities during transitional periods poses cybersecurity threats (Leal & Musgrave, 2023).

3.4. Emergent Risks in Specific Sectors

Contract expiry risks vary significantly across sectors:

- **Education**: As AI becomes embedded in learning management systems, abrupt contract endings can jeopardize educational continuity and data privacy for learners (Awofala et al., 2025).
- **Food Safety**: AI managing food production and supply chains must ensure that post-expiry transitions do not compromise safety standards (Chavan, Paul, & Kolekar, 2024).
- **Healthcare**: Early termination of AI-driven diagnostics or patient monitoring systems can endanger lives (Hosseini et al., 2023).

Caso (2024) highlights that in defense and space sectors, AI contract terminations could expose sensitive technologies, while Harwich and Laycock (2018) emphasize the need for public sector resilience in AI transitions, particularly in healthcare.

The severity of risks linked to AI contract expirations varies across sectors such as healthcare, education, defense, and food safety. Figure 3 illustrates the relative risk levels in these domains based on existing research findings.

Figure 3. Sectoral risk levels associated with AI contract expiry

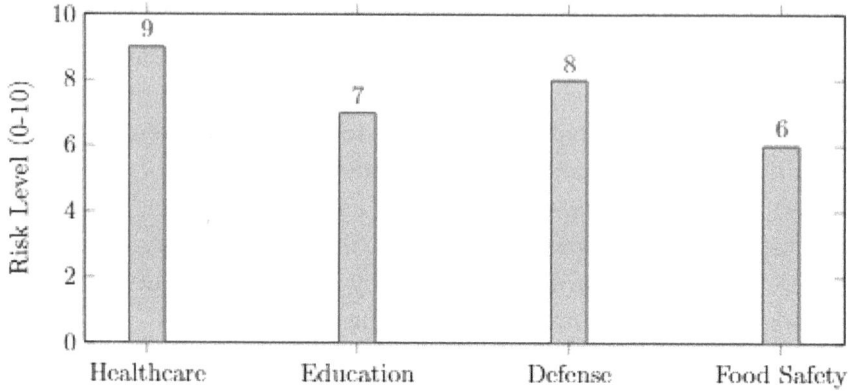

3.5. Intellectual Property and Licensing Risks

Mbah (n.d.) notes that AI-driven smart contracts and licensing agreements often involve intricate intellectual property (IP) stipulations. At expiry, questions arise over the continued use of AI models trained during the contractual period. Disputes over algorithmic ownership, rights to derivative works, and continuing royalties can lead to costly litigation if not properly addressed.

Similarly, Leontidis (2024) warns that the evolving nature of AI research, where models continually adapt, complicates the traditional understanding of contract finality and IP closure.

3.6. Deception, Fraud, and Predatory Practices

The expiry of AI contracts also opens avenues for predatory behavior. Entities might manipulate LLM-generated outputs to obfuscate obligations or exploit ambiguities for fraudulent gains (Kendall & Teixeira da Silva, 2024; Moussa & Teixeira da Silva, 2023). Furthermore, Haney (2020) points to natural language processing tools being misused in legal practices to extend influence beyond contract terms.

In tourism and hospitality, platforms like Airbnb have shown how disruptive innovations create informal sectors, often leading to unregulated contract expirations and consumer risk (Guttentag, 2015; Nunkoo et al., 2023).

3.7. Transparency and Explainability Challenges

Kosasih et al. (2024) explain that the lack of transparency ("black-box" AI) further complicates managing expiry risks. When systems are opaque, identifying the full scope of obligations and liabilities that persist post-expiry becomes daunting.

Structured approaches to unstructured data and explainable AI are crucial to ensure orderly transition and regulatory compliance (de Haan et al., 2024; Cox & Thelwall, 2025).

3.8. Regulatory Pressures and Evolving Legal Landscapes

New regulations, such as the EU AI Act (Butt, 2024) and sector-specific guidelines (Chavan, Paul, & Kolekar, 2024; Biesbroek et al., 2022), require firms to reassess how contract expiry is handled. Failure to comply could result in severe penalties, reputational damage, and operational risks.

Mhia-Alddin and Hussein (2025) stress that AI-enhanced contract management must now include continuous compliance monitoring, even after contract expiration, to prevent downstream regulatory breaches.

3.9. Broader Policy and Societal Implications

At the macro level, expired contracts in AI-enriched systems may influence broader policy goals, from cybersecurity stability (Leal & Musgrave, 2023) to sustainable energy transitions (Nriezedi-Anejionu, 2024). Nicholls and Culpepper (2021) argue that computational framing of media narratives can influence public perceptions of AI contract disputes, affecting political outcomes.

Finally, as Baskara (2024) warns, academic and organizational integrity risks linked to AI misuse during transitions must be managed carefully to protect trust and societal resilience.

4. CASE STUDIES AND REAL-WORLD INCIDENTS

The rapid integration of artificial intelligence (AI) across industries has introduced profound changes, benefits, and challenges. A range of real-world incidents and case studies highlights the opportunities and complexities in deploying AI responsibly, ethically, and effectively. This section provides an in-depth exploration of several key cases, citing contemporary research and professional analyses.

4.1. AI in Education: Adoption and Integrity Challenges

The education sector has seen significant AI adoption, but also faces issues surrounding academic integrity. In Nigeria, the integration of educational AI tools among science, technology, and mathematics teachers revealed varying degrees of adoption influenced by structural factors (Awofala et al., 2025). Similarly, concerns regarding AI's impact on academic honesty were exposed by Baskara (2024), who explored the narrative of "Deus Ex Machina" and the erosion of traditional academic values.

Moreover, text and data mining rights under directives like the Digital Single Market Directive are influencing how AI can be used for developing creative educational tools (Christensen, 2021). These developments call for clearer guidelines to balance innovation and integrity.

4.2. AI in Healthcare: Opportunities and Risks

The healthcare industry showcases both the promise and peril of AI deployment. Studies such as those by Horgan et al. (2020) demonstrate that AI can significantly enhance public health genomics and healthcare delivery. Conversely, challenges exist in emergency care settings, where Hosseini et al. (2023) mapped various risks tied to AI integration.

Recent explorations, including Caggiano et al. (2024) on health assessments for space exploration, and Nordlinger et al. (2020) on AI's broader healthcare impacts, demonstrate the sector's complexity. In Saudi Arabia, AI's impact on accounting systems of health companies was investigated by Hamza et al. (2024), underscoring the need for robust regulatory mechanisms.

4.3. Legal and Ethical Challenges of AI Systems

The legislative landscape for AI is evolving. Butt (2024) analyzed the world's first Artificial Intelligence (AI) Act by the European Union, setting a critical precedent. Gunawan et al. (2022) addressed the responsibility of autonomous weapons under international humanitarian law, reflecting broader legal debates.

Additionally, AI's implications in contracting are explored through Act et al. (AI-infused contracts/contracting, 2024) and the role of smart contracts in intellectual property management (Mbah, 2025). Managing AI-driven contract security has been discussed by Mhia-Alddin and Hussein (2025), highlighting how deep learning ensures compliance and mitigates risks.

4.4. Surveillance, Security, and Data Privacy

AI-driven surveillance brings significant ethical dilemmas. Harper et al. (2022) discussed covert aspects of surveillance research, while Haney (2020) explored natural language processing applications in law practices that may cross ethical lines.

Data privacy, especially in geospatial information (Cho & Crompvoets, 2019) and emergency pandemic responses (Chung et al., 2022), underscores the tension between technological capability and ethical responsibility.

The assessment by Biesbroek et al. (2022) further shows how governments are addressing climate vulnerabilities, often relying on AI for predictive analytics, raising new ethical debates.

4.5. AI in Business, Tourism, and Supply Chains

Disruptive innovation due to AI is profoundly altering business models. Guttentag (2015) illustrated this with Airbnb's rise, transforming traditional hospitality industries. In supply chains, Kosasih et al. (2024) reviewed the use of explainable AI (XAI), calling for transparent AI models in logistics and production sectors.

Leal and Musgrave (2023) analyzed public evaluations concerning cybersecurity vulnerabilities, stressing the importance of public trust, while de Haan et al. (2024) urged more structured approaches in using unstructured business data.

Meanwhile, the role of AI in sustainable tourism research (Nunkoo et al., 2023) exemplifies interdisciplinary efforts to advance Sustainable Development Goals (SDGs).

4.6. Risks of AI and Predatory Practices

Concerns around AI misuse are rising. Kendall and Teixeira da Silva (2024) highlighted risks of large language models (LLMs) being exploited for predatory publishing practices, while Moussa and Teixeira da Silva (2023) tested the robustness of COPE's criteria on predatory publishers.

Nicholls and Culpepper (2021) detailed how AI can computationally frame media narratives, posing risks of manipulation and bias.

4.7. AI's Role in Governance and Policy

AI is increasingly embedded in policymaking and governance. Cox and Thelwall (2025) emphasized how AI reshapes knowledge management, while Leontidis (2024) addressed the transformation of scientific methodologies through AI.

Climate governance studies (Biesbroek et al., 2022) and calls for new treaties around emerging nuclear technologies (Nriezedi-Anejionu, 2024) also involve AI's predictive capabilities.

Moreover, Harwich and Laycock (2018) proposed that AI could be a game-changer in healthcare governance, particularly within the NHS.

4.8. Military and Space: AI in Emerging Frontiers

Caso (2024) discussed how emerging technologies, including AI, are increasingly vital in military space operations for training and education. Similarly, international humanitarian legal frameworks are being tested by AI-driven military tools (Gunawan et al., 2022).

Space law now must consider ethical assessments of AI-driven health monitoring systems in extraterrestrial environments (Caggiano et al., 2024).

4.9. Smart Contracts and the Future of Licensing

Smart contracts powered by AI are transforming licensing agreements in the tech industry (Mbah, 2025). The fusion of blockchain, AI, and intellectual property law is expected to enhance security, compliance, and transparency.

In addition, Mhia-Alddin and Hussein (2025) proposed AI-based lifecycle management tools for contracts to automate risk mitigation and ensure regulatory adherence.

5. INTELLECTUAL PROPERTY AND OWNERSHIP POST-EXPIRY

The evolution of artificial intelligence (AI) technologies has significantly transformed the landscape of intellectual property (IP) ownership, especially regarding rights post-expiry. Traditionally, intellectual property protections—such as patents, copyrights, and trademarks—were designed to incentivize innovation by granting temporary monopolies. Once these rights expire, the protected work enters the public domain. However, the dynamic, self-improving nature of AI, combined with new forms of data-driven creation, has complicated the post-expiry IP regime.

5.1 Shifting Nature of Ownership and Expired Rights

Post-expiry, traditional IP rules assume that works can be freely reused. However, AI-generated outputs blur this boundary. For instance, systems trained on expired works may create new derivative outputs, challenging notions of originality and ownership (Christensen, 2021; Cox & Thelwall, 2025). Furthermore, policies like the Digital Single Market Directive emphasize open access for text and data mining (TDM) while still balancing rights for AI developers (Christensen, 2021).

Emerging laws, like the European Union's Artificial Intelligence Act (Butt, 2024), introduce layered obligations that indirectly impact IP reuse post-expiry by setting standards for AI transparency, risk assessment, and data governance. Meanwhile, studies highlight that unstructured data usage, including that derived from public domain works, must be approached with structured, ethical considerations (de Haan et al., 2024).

5.2 Challenges with AI-Generated Content Post-Expiry

AI's capacity to continually generate new outputs from expired IP raises important questions around derivative rights. Harwich and Laycock (2018) discuss how healthcare AI applications must balance innovation with ownership transparency, a principle applicable across domains. Similarly, in aerospace health projects, integrating legal, ethical, and sustainability frameworks is critical when dealing with IP sourced materials (Caggiano et al., 2024).

Academic contexts reflect similar tensions. Baskara (2024) notes that AI's influence on academic integrity challenges traditional authorship models, particularly when expired materials inform AI training. Moreover, the potential misuse of large language models (LLMs) for uncredited derivative works, as flagged by Kendall and Teixeira da Silva (2024), emphasizes the need for clearer frameworks in handling post-expiry IP scenarios.

5.3 Regulatory and Ethical Dimensions

A global regulatory overview suggests fragmented approaches to post-expiry IP governance in the AI age. Biesbroek et al. (2022) note variability in national policies regarding adaptation and climate impacts, drawing parallels to how IP regulations evolve unevenly across jurisdictions. Gunawan et al. (2022) argue that autonomous technologies must respect international humanitarian law even when operating with freely available data—a principle extendable to expired IP utilization.

Additionally, covert surveillance technologies raise ethical questions regarding the silent, continued use of expired or public-domain IP (Harper et al., 2022). Ethical deployment of AI in domains like cybersecurity (Leal & Musgrave, 2023) and emergency healthcare (Hosseini et al., 2023) necessitates renewed focus on respecting even "free" knowledge assets.

Recent scholarship also stresses the dangers of predatory practices in post-expiry environments, especially in scientific publishing (Moussa & Teixeira da Silva, 2023). Without rigorous IP governance, expired works might be exploited unethically to inflate AI training datasets, resulting in lower trust in AI outputs.

5.4 AI-Infused Contracts and Post-Expiry Management

The emerging practice of AI-powered smart contracts is reshaping how post-expiry IP is managed. As Mhia-Alddin and Hussein (2025) argue, deep learning models and LLMs enable dynamic contract frameworks that automatically address expiry and compliance, minimizing human error and ensuring transparent handling of formerly protected works.

Similarly, Mbah (n.d.) emphasizes that smart contracts infused with AI principles can revolutionize licensing agreements, making it easier to enforce obligations around expired works, even when such materials are integrated into complex AI-driven systems.

The Act et al. (n.d.) outlines how AI-based contracts increasingly rely on disclosure principles and standardized assessment, which must also evolve to accommodate the nuances of post-expiry intellectual property usage.

5.5 Opportunities and Future Trends

Leveraging expired IP offers vast opportunities for innovation if managed ethically. Leontidis (2024) suggests that AI is fundamentally transforming scientific inquiry, often building upon foundations laid by expired research outputs. However, scholars like Caso (2024) warn that without clear ethical frameworks, military and training applications using historical, expired data could create accountability challenges.

Chavan, Paul, and Kolekar (2024) discuss how food safety policies integrate AI, indicating that expired data sets—properly managed—can enhance safety protocols. Similarly, Nunkoo et al. (2023) highlight that interdisciplinarity, supported by open-access and expired research, can accelerate sustainable development goals.

Nonetheless, concerns persist regarding the covert use of expired IP in the supply chain domain (Kosasih et al., 2024) and informal tourism sectors (Guttentag, 2015), calling for continued vigilance.

The integration of AI in law (Haney, 2020) and knowledge systems (Cox & Thelwall, 2025) demands robust, forward-looking IP frameworks that not only honor the letter of expiry laws but also their spirit—ensuring fair access, ethical reuse, and sustainable innovation.

6. REGULATORY AND COMPLIANCE FRAMEWORKS

The growing integration of Artificial Intelligence (AI) across multiple sectors demands robust regulatory and compliance frameworks. As AI systems increasingly influence decision-making, ethical standards, accountability structures, and legal protections must evolve concurrently. This section explores key regulations, global frameworks, and critical research developments surrounding AI compliance.

A complex web of regulatory frameworks governs AI applications in STM research. Figure 4 visualizes key global, sectoral, and ethical regulatory layers impacting contract expiry management.

Figure 4. Multi-layered regulatory frameworks for AI contract management

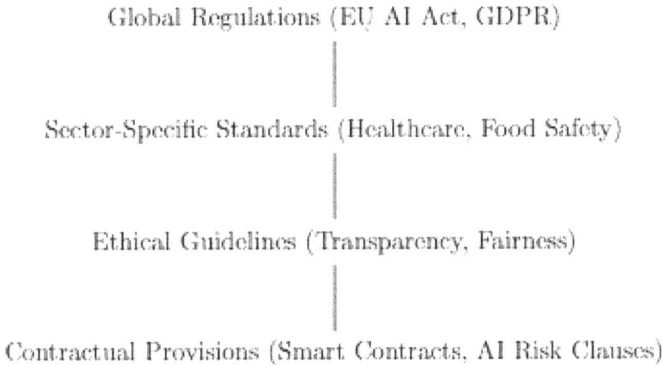

Global Regulations (EU AI Act, GDPR)

|

Sector-Specific Standards (Healthcare, Food Safety)

|

Ethical Guidelines (Transparency, Fairness)

|

Contractual Provisions (Smart Contracts, AI Risk Clauses)

6.1. Global AI Regulatory Initiatives

A significant development in AI regulation is the European Union's Artificial Intelligence Act, regarded as the world's first comprehensive AI law (Butt, 2024). The Act classifies AI systems based on risk categories—unacceptable, high-risk, limited risk, and minimal risk—and imposes varying obligations accordingly. For instance, high-risk AI systems must undergo rigorous conformity assessments, demonstrating transparency, accountability, and safety.

Additionally, the General Data Protection Regulation (GDPR) remains a foundational pillar in protecting individuals' data rights when interacting with AI (Act, Agent, Assessment, Act, Directive, & Regulation, 4(1), 652). Specific GDPR provisions, such as the "right to explanation," directly impact AI systems using personal data.

The Digital Single Market Directive also emphasizes fair data access, particularly regarding text and data mining rights for AI-driven creativity and innovation (Christensen, 2021).

6.2. Sector-Specific Compliance Considerations

6.2.1 Healthcare

In healthcare, AI must align with stringent ethical, legal, and safety requirements. According to Nordlinger, Villani, and Rus (2020), trustworthiness and transparency are essential for AI applications in diagnosis and treatment. Horgan, Romao, Morré, and Kalra (2020) further highlight that while AI can revolutionize healthcare delivery, it must respect patient autonomy and privacy.

The integration of AI in emergency care also underscores the need for specialized compliance measures to ensure patient safety and data confidentiality (Hosseini, Hosseini, Qayumi, Ahmady, & Koohestani, 2023). Similarly, Caggiano, Gatt, Mollo, Izzo, and Troisi (2024) argue for embedding ethical and sustainability assessments within space health initiatives.

6.2.2 Food Safety

Chavan, Paul, and Kolekar (2024) explore how AI is reshaping food safety policies, calling for updated frameworks that integrate AI-driven quality monitoring systems to enhance public health outcomes.

6.2.3 Accounting and Business

The impact of AI on financial reporting and accounting systems necessitates compliance with traditional and emerging regulations. Hamza et al. (2024) observed that Saudi companies face both opportunities and risks when incorporating AI into their accounting systems, demanding robust internal control mechanisms.

Similarly, in business applications, de Haan, Padigar, El Kihal, Kübler, and Wieringa (2024) emphasized structured approaches for managing unstructured data to comply with information governance standards.

6.3. Intellectual Property and AI-Driven Contracting

Smart contracts powered by AI introduce new challenges in intellectual property rights and compliance (Mbah, n.d.). Moreover, AI-infused contracting must adapt to emerging regulatory standards to address issues like disclosure, contract expiry, and compliance risk (Mhia-Alddin & Hussein, 2025). Managing the lifecycle of smart contracts through deep learning and LLMs is increasingly critical.

Mahoney (2017) also cautioned about market structures influencing contracting practices in private industries, while Moussa and Teixeira da Silva (2023) examined concerns around predatory practices in academic publishing—a caution relevant to AI's influence on legal documentation and scholarly communication.

6.4. Ethical Concerns and Surveillance

AI-driven surveillance systems pose ethical dilemmas, especially when covert operations are involved (Harper, Ellis, & Tucker, 2022). Data privacy regulations must balance the need for security with the protection of individual freedoms.

Gunawan, Aulawi, Anggriawan, and Putro (2022) examined the role of autonomous weapons and the command responsibility under international humanitarian law, raising broader questions about compliance with ethical warfare principles.

Nicholls and Culpepper (2021) demonstrated the significance of computational approaches in identifying media biases, further complicating regulatory discussions about AI's role in shaping public opinion.

6.5. Educational, Scientific, and Technological Contexts

In education, regulatory frameworks are adapting to AI's influence on pedagogy and academic integrity. Awofala et al. (2025) studied the adoption of AI educational tools by Nigerian science and mathematics teachers, suggesting that compliance with educational standards must be dynamic and culturally sensitive.

Baskara (2024) also critically addressed academic integrity in the AI age, warning against over-reliance on machine-generated content.

Leontidis (2024) emphasized that AI is reshaping the very nature of scientific inquiry, necessitating updated ethical guidelines. Similarly, Kosasih, Papadakis, Baryannis, and Brintrup (2024) reviewed explainable AI frameworks in supply chain management, underscoring the need for transparency regulations.

6.6. Climate and Sustainability Regulations

AI's role in environmental policy is increasingly recognized. Biesbroek et al. (2022) assessed national responses to climate impacts and suggested that AI could facilitate compliance with climate adaptation frameworks.

Nriezedi-Anejionu (2024) called for international treaties to manage the deployment of small modular reactors (SMRs), illustrating how AI can support but also complicate nuclear energy governance.

Nunkoo et al. (2023) emphasized the interdisciplinary role of AI in advancing Sustainable Development Goals (SDGs), stressing the need for integrated regulatory approaches across fields like tourism and environmental science.

6.7. Future Directions in AI Regulation

Haney (2020) and Cox and Thelwall (2025) both argued that the future of AI regulation must include stronger natural language processing compliance frameworks, particularly in legal and knowledge management contexts.

Chung, Rodriguez, Lanier, and Gibbs (2022) illustrated how structural topic modeling could be utilized to enhance compliance monitoring by analyzing unstructured survey data.

Harwich and Laycock (2018) discussed NHS initiatives integrating AI thoughtfully within regulatory bounds, while Caso (2024) explored compliance aspects for AI applications in military space operations.

Leal and Musgrave (2023) discussed the controversial use of zero-day vulnerabilities in cybersecurity, revealing the urgent need for better ethical and legal regulations in digital warfare.

Lastly, Kendall and Teixeira da Silva (2024) warned about the abuse of large language models like ChatGPT in scientific publishing, reinforcing the call for stronger compliance frameworks against misuse in knowledge production.

7. BEST PRACTICES FOR MANAGING CONTRACT EXPIRY IN AI RESEARCH

Effective management of contract expiry in AI research is critical for ensuring compliance, safeguarding intellectual property, maintaining ethical standards, and supporting the continuous advancement of innovation. This section outlines best practices informed by recent literature and global regulatory developments.

Establishing best practices for managing contract expiry in AI research environments is critical for resilience and compliance. Figure 5 provides a visual summary of recommended strategies.

Figure 5. Best practices for managing AI contract expiry

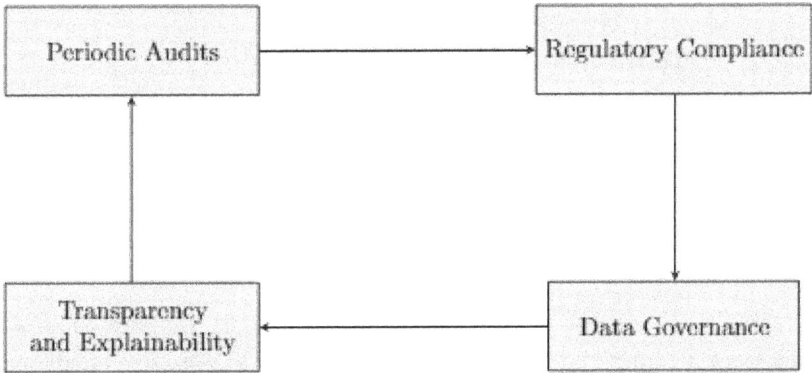

7.1. Proactive Contract Lifecycle Management

A proactive approach to managing contracts in AI research involves early monitoring of expiration dates and timely renegotiations to prevent disruptions. AI-powered contract management systems, utilizing large language models (LLMs) and deep learning, can automate monitoring and risk detection (Mhia-Alddin & Hussein, 2025). These technologies can predict expiry risks and compliance gaps well in advance.

As highlighted by Mahoney (2017), understanding the structure of contract markets and adjusting management strategies accordingly is crucial, especially where private and semi-public AI research partnerships are involved. Automation tools can enhance contract lifecycle visibility (Haney, 2020; Cox & Thelwall, 2025).

7.2. Regulatory Compliance and Ethical Considerations

Managing AI contracts must align with evolving regulatory frameworks such as the EU Artificial Intelligence Act (Butt, 2024) and the GDPR (Act et al., 4(1)). These regulations emphasize data protection, ethical use, and transparency, requiring that contracts explicitly address data rights, usage limitations, and compliance protocols.

Health, military, and education sectors show that integrating legal and ethical assessments at the contract stage is necessary (Caggiano et al., 2024; Caso, 2024; Harwich & Laycock, 2018). Especially in sensitive domains like healthcare, AI contract clauses must account for patient safety and ethical data usage (Nordlinger, Villani, & Rus, 2020; Horgan et al., 2020).

7.3. Data Governance and Intellectual Property Protection

AI research contracts often involve sensitive data and intellectual property (IP) concerns. Clear stipulations about data ownership, usage rights, and IP clauses are essential for post-expiry protections (Mbah, n.d.). Contracts should anticipate disputes over AI-generated outputs and embed dispute resolution frameworks accordingly (Cho & Crompvoets, 2019; Christensen, 2021).

The risks of misuse and ownership ambiguities are magnified with AI systems handling large unstructured datasets (de Haan et al., 2024), reinforcing the need for well-defined IP clauses.

7.4. Explainability and Transparency Clauses

Recent reviews emphasize the growing importance of explainable artificial intelligence (XAI) in AI-based decision-making (Kosasih et al., 2024). Contracts should require researchers to maintain transparency in model outputs and provide documentation that remains accessible even after contract termination. Transparency is also linked to public trust and acceptance, particularly in critical domains like food safety and emergency care (Chavan et al., 2024; Hosseini et al., 2023).

7.5. Ethical AI Use and Sustainability Commitments

Contracts should explicitly bind parties to ethical AI use, avoiding covert or un-intended surveillance applications (Harper, Ellis, & Tucker, 2022). They should also encourage sustainability, especially as AI research consumes significant resources (Biesbroek et al., 2022; Nunkoo et al., 2023).

Environmental considerations are increasingly relevant in sectors ranging from climate change adaptation (Biesbroek et al., 2022) to energy policy, suggesting the inclusion of sustainability metrics in AI research contracts (Nriezedi-Anejionu, 2024).

7.6. Periodic Risk Assessments and Audits

The dynamic nature of AI means risks evolve over time. Contracts should require periodic AI system audits and compliance reviews, ensuring continuous alignment with current laws and ethical standards (Gunawan et al., 2022; Leal & Musgrave, 2023).

Such provisions are crucial to addressing issues like the unauthorized use of zero-day vulnerabilities and potential abuses of autonomous systems (Gunawan et al., 2022; Leal & Musgrave, 2023).

7.7. Leveraging AI for Contract Analysis

AI and text-mining techniques can support efficient contract management by analyzing legal language, obligations, and risk patterns (Chung et al., 2022; Nicholls & Culpepper, 2021). Natural language processing (NLP) methods can detect ambiguities and ensure that contract terms remain understandable and enforceable (Haney, 2020; Cox & Thelwall, 2025).

Emerging research underscores that computational tools must be carefully managed to avoid reinforcing biases or undermining legal precision (Leontidis, 2024).

7.8. Managing Academic and Research Integrity

Contract clauses must reinforce research integrity, particularly in the context of AI-driven publications and collaborations. Issues such as predatory publishing and authorship fraud (Kendall & Teixeira da Silva, 2024; Moussa & Teixeira da Silva, 2023) necessitate clear definitions of publication responsibilities and ethical standards.

In education sectors, proper management ensures responsible adoption of AI tools by teachers and researchers (Awofala et al., 2025; Baskara, 2024).

7.9. Preparing for AI-Driven Disruptions

Given AI's disruptive nature (Guttentag, 2015), contracts should be flexible to accommodate technological evolution. Parties should define provisions for contract modification and renewal based on technological shifts (Karim, Galar, & Kumar, 2023).

An interdisciplinary approach to contract management can help address emerging challenges in cybersecurity, healthcare, and other critical infrastructures (Harwich & Laycock, 2018; Horgan et al., 2020).

7.10 Summary

Managing contract expiry in AI research is not merely an administrative task; it is a strategic necessity. Best practices include proactive monitoring, regulatory alignment, data governance, transparency, ethical safeguards, periodic audits, computational analysis, academic integrity reinforcement, and disruption planning. These strategies are essential for mitigating risks, enhancing innovation, and ensuring that AI research continues to benefit society responsibly and sustainably.

8. FUTURE DIRECTIONS AND LEGAL INNOVATIONS

As artificial intelligence (AI) technologies rapidly evolve, the intersection of law, ethics, and technological innovation becomes increasingly complex and critical. Future directions in this space demand a blend of legislative innovation, interdisciplinary research, and proactive governance to manage emerging challenges while fostering innovation.

8.1. Legal Innovations: New Regulatory Frameworks for AI

The world's first comprehensive legislation on AI, the **EU Artificial Intelligence (AI) Act**, offers a foundational step towards harmonized regulation. As Butt (2024) emphasizes, the Act aims to balance innovation with fundamental rights protection, categorizing AI systems based on risk levels and imposing corresponding obligations. The **General Data Protection Regulation (GDPR)** and associated directives have also evolved, as noted by Act et al. (n.d.), to better accommodate AI-infused contracting and data processing in the digital economy.

Similarly, Caggiano et al. (2024) discuss how legal frameworks are being expanded to include sustainability and ethical assessments, particularly in emerging fields like health and space operations. In the context of military applications, Caso

(2024) highlights the necessity for specific legal standards to regulate AI use in military training and space operations, ensuring ethical compliance.

International humanitarian law is also adapting, particularly concerning autonomous weapons. Gunawan et al. (2022) stress the importance of reinterpreting **command responsibility** principles to accommodate AI systems' unique challenges in conflict scenarios.

Moreover, Cho and Crompvoets (2019) raise critical legal concerns regarding the "prod-users" of geospatial information, pointing out the need for better legal clarity around user-generated spatial data in AI applications.

8.2. Ethical and Compliance Challenges: AI's Impact on Governance

Harper, Ellis, and Tucker (2022) point out the covert aspects of AI-driven surveillance, emphasizing the need for stricter ethical regulations around data collection and privacy. Furthermore, Kendall and Teixeira da Silva (2024) warn about the misuse of AI in scientific publishing, including predatory practices and "paper mills," which require immediate policy responses.

Moussa and Teixeira da Silva (2023) critically assess how AI and machine learning technologies could exacerbate weaknesses in scholarly publishing, underscoring the necessity for more robust and transparent governance frameworks.

Leal and Musgrave (2023) highlight the cybersecurity implications, particularly around the public's perception of government use of zero-day vulnerabilities. Future regulatory directions must address public trust, technological sovereignty, and ethical use.

In healthcare, Nordlinger, Villani, and Rus (2020) stress that integrating AI technologies must be accompanied by rigorous legal and ethical oversight to ensure equitable healthcare delivery, a sentiment echoed by Horgan et al. (2020).

8.3. AI, Intellectual Property, and Smart Contracting

The evolving landscape of **AI-infused contracts** (Act et al., n.d.) and **smart contracts** (Mbah, n.d.) presents unique legal challenges. AI-powered contracts must address issues of compliance, expiry management, and risk mitigation, as discussed by Mhia-Alddin and Hussein (2025).

As Mahoney (2017) notes in the context of private contracting in the military industry, understanding market structure dynamics is essential for drafting AI-related agreements that protect company performance and national security interests.

Moussa and Teixeira da Silva (2023) further highlight intellectual property risks associated with AI-generated content, urging the development of adaptive licensing and copyright models.

8.4. Technological Developments and Their Legal Implications

Haney (2020) provides practical insights into how **Natural Language Processing (NLP)** technologies are transforming legal practice, offering tools for predictive analytics, contract review, and case outcome forecasting. Meanwhile, Christensen (2021) notes that exceptions granted under the **Digital Single Market Directive** for **Text and Data Mining (TDM)** are crucial for AI development, suggesting that legislative bodies must adapt to promote responsible innovation.

De Haan et al. (2024) stress the importance of structured methodologies when using unstructured business data, a key issue as AI tools grow more powerful in interpreting complex datasets.

In the context of sustainable development, Nunkoo et al. (2023) call for interdisciplinarity in research, which must also reflect in future legal frameworks regulating AI's environmental and societal impacts.

8.5. Future Research and Interdisciplinary Needs

Biesbroek et al. (2022) illustrate that despite the increasing global attention to climate change, adaptation strategies often lag in policy documents. AI could aid in adaptation planning, but it will require new governance structures that integrate climate science and machine learning.

Leontidis (2024) underscores that AI is fundamentally reshaping the method of scientific inquiry, calling for a reexamination of academic and research ethics policies.

Baskara (2024) warns that AI's role in education and research raises concerns about academic integrity, highlighting the need for regulatory bodies to establish AI-specific academic guidelines.

Awofala et al. (2025) find that teacher adoption of AI tools depends heavily on perceived ease of use and policy support, suggesting that future education policies must be AI-literate.

Chavan, Paul, and Kolekar (2024) discuss AI's role in food safety, calling for international cooperation on setting AI-driven standards for food production and monitoring.

Nriezedi-Anejionu (2024) proposes that international treaties are essential for managing carbon reduction technologies like **small modular reactors (SMRs)**, setting a precedent for AI governance in other critical infrastructure domains.

8.6. AI, Media, and Public Discourse

Nicholls and Culpepper (2021) explain how computational methods like media frame analysis can shape public opinion about AI technologies, suggesting that future regulations must account for the media's influence in the adoption and perception of AI tools.

Chung et al. (2022) demonstrate how text-mining methods provide insights into societal coping mechanisms during crises like COVID-19, offering a model for real-time public policy formulation using AI.

Guttentag (2015) shows that disruptive innovations, such as Airbnb, thrive in regulatory gaps—highlighting the need for proactive, forward-looking policies that anticipate AI's disruptions rather than react to them.

Harwich and Laycock (2018) argue that AI in healthcare must be able to "think on its own," proposing a significant overhaul of traditional NHS frameworks to embrace AI while ensuring patient safety and privacy.

Karim, Galar, and Kumar (2023) emphasize that **AI factories**—where AI systems learn and evolve—pose unique legal and ethical challenges that future policymakers must prepare for, particularly in terms of transparency and accountability.

Cox and Thelwall (2025) foresee a future where **AI for knowledge** will redefine how businesses, governments, and academia operate, necessitating dynamic legal frameworks to govern knowledge generation, ownership, and distribution.

8.7. Summary

Looking forward, the legal landscape must not only regulate but also enable innovation. As technologies like AI, blockchain, and autonomous systems converge, regulatory bodies will need to collaborate across disciplines, industries, and nations. The future demands a **proactive, ethical, and dynamic approach** to legal innovation—balancing the twin imperatives of safeguarding rights and catalyzing progress.

9. CONCLUSION

As artificial intelligence (AI) technologies continue to evolve and permeate diverse sectors, the need for comprehensive, forward-thinking legal and policy frameworks becomes increasingly critical. The preceding discussion has explored how emerging regulations—such as the EU Artificial Intelligence Act (Butt, 2024) and global initiatives related to AI governance—are beginning to shape the future legal landscape. This transformation is not just regulatory but also ethical, techno-logical, and societal, necessitating interdisciplinary collaboration and innovation.

The integration of AI into critical fields like healthcare (Horgan et al., 2020; Nordlinger et al., 2020), education (Awofala et al., 2025), supply chains (Kosasih et al., 2024), and emergency services (Hosseini et al., 2023) demonstrates the vast opportunities AI offers. However, it also raises serious concerns regarding account-ability (Gunawan et al., 2022), intellectual property (Mbah, n.d.), and scientific integrity (Kendall & Teixeira da Silva, 2024; Leontidis, 2024). Innovative solutions such as AI-powered contract security systems (Mhia-Alddin & Hussein, 2025) and explainable AI models in business (de Haan et al., 2024) are crucial responses to these challenges, showcasing proactive efforts to align technological advancement with regulatory compliance and ethical standards.

Moreover, as new AI-driven phenomena like smart contracts and autonomous decision-making become commonplace (Mahoney, 2017; Mbah, n.d.), future-oriented legal innovations must prioritize transparency, fairness, and sustainability. Initiatives in fields like space law (Caggiano et al., 2024; Caso, 2024) and cybersecurity (Leal & Musgrave, 2023) exemplify how policy needs to adapt to novel technological contexts.

Cross-sectoral dialogues will be essential to ensure that AI technologies are developed and deployed responsibly. Future directions point toward integrating AI ethics into everyday operations (Harper et al., 2022; Haney, 2020), promoting AI transparency in public services (Harwich & Laycock, 2018), ensuring responsible data usage (Cho & Crompvoets, 2019; Christensen, 2021), and fortifying safeguards against the misuse of generative technologies (Moussa & Teixeira da Silva, 2023; Baskara, 2024).

The path forward demands vigilance, agility, and creativity. Policymakers, technologists, businesses, and society at large must work together to craft adaptive legal frameworks that not only regulate but also foster innovation. Emphasizing human-centric AI principles, reinforcing ethical standards, and committing to con-tinuous assessment and revision of legal mechanisms will be key to navigating the complex AI-driven future.

In sum, while AI brings unprecedented opportunities, it also mandates profound legal and ethical innovation. By learning from ongoing developments (Biesbroek et al., 2022; Nunkoo et al., 2023), addressing emergent challenges in real-time, and fostering resilient regulatory ecosystems, we can ensure that the future of AI benefits humanity as a whole—safely, equitably, and sustainably.

REFERENCES

Act, A. A., Agent, P., Assessment, D. P. I., Act, P. E. A. I., Directive, P. P. W., & Regulation, G. D. P. AI-infused contracts/contracting (cont.). *disclosure, 4*(1), 652.

Awofala, A. O. A., Bazza, M. B., Ojo, O. T., Oladipo, A. J., Olabiyi, O. S., & Arigbabu, A. A. (2025). Structural equation modeling of Nigerian science, technology and mathematics teachers' adoption of educational artificial in-telligence tools. *Digital Education Review*, (46), 51–64. DOI: 10.1344/der.2025.46.51-64

Baskara, F. R. (2024, March). Deus Ex Machina: Unraveling Academic Integrity in the AI Narrative. In *Proceeding International Conference on Religion. Science and Education, 3*, 393–402.

Biesbroek, R., Wright, S. J., Eguren, S. K., Bonotto, A., & Athanasiadis, I. N. (2022). Policy attention to climate change impacts, adaptation and vulnerability: A global assessment of National Communications (1994–2019). *Climate Policy, 22*(1), 97–111. DOI: 10.1080/14693062.2021.2018986

Butt, J. (2024). Analytical study of the world's first EU Artificial Intelligence (AI) Act. *International Journal of Research Publication and Reviews, 5*(3), 7343–7364. DOI: 10.55248/gengpi.5.0324.0914

Caggiano, I. A., Gatt, L., Mollo, A. A., Izzo, L., & Troisi, E. (2024, June). Health in Space: Integrating Legal, Ethical and Sustainability Assessments. In *2024 11th International Workshop on Metrology for AeroSpace (MetroAeroSpace)* (pp. 296-302). IEEE.

Caso, S. (2024). Emerging technologies in military space operations: current applications and future research for educational and training purposes. *International Journal of Training Research*, 1-16.

Chavan, Y., Paul, K., & Kolekar, N. (2024). Food safety and hygiene: Current policies, quality standards, and scope of artificial intelligence. In *Food production, diversity, and safety under climate change* (pp. 319–331). Springer Nature Switzerland. DOI: 10.1007/978-3-031-51647-4_26

Cho, G., & Crompvoets, J. (2019). Prod-users of geospatial information: Some legal perspectives. *Journal of Spatial Science, 64*(2), 341–358. DOI: 10.1080/14498596.2018.1429330

Christensen, K. (2021). A European solution for Text and Data Mining in the development of creative Artificial Intelligence: With a specific focus on articles 3 and 4 of the Digital Signel Market Directive.

Chung, G., Rodriguez, M., Lanier, P., & Gibbs, D. (2022). Text-mining open-ended survey responses using structural topic modeling: A practical demonstration to understand parents' coping methods during the COVID-19 pandemic in Singapore. *Journal of Technology in Human Services, 40*(4), 296–318. DOI: 10.1080/15228835.2022.2036301

Cox, A., & Thelwall, M. (2025). *AI for Knowledge*. CRC Press. DOI: 10.1201/9781003545163

de Haan, E., Padigar, M., El Kihal, S., Kübler, R., & Wieringa, J. E. (2024). Unstructured data research in business: Toward a structured approach. *Journal of Business Research, 177*, 114655. DOI: 10.1016/j.jbusres.2024.114655

Gunawan, Y., Aulawi, M. H., Anggriawan, R., & Putro, T. A. (2022). Command responsibility of autonomous weapons under international humanitarian law. *Cogent Social Sciences, 8*(1), 2139906. DOI: 10.1080/23311886.2022.2139906

Guttentag, D. (2015). Airbnb: Disruptive innovation and the rise of an informal tourism accommodation sector. *Current Issues in Tourism, 18*(12), 1192–1217. DOI: 10.1080/13683500.2013.827159

Hamza, R. A. E. M., Ahmed, N. H., Mohamed, A. M. E., Bennaceur, M. Y., Elhefni, A. H. M., & Elshaabany, M. M. (2024). The impact of artificial intelligence (AI) on the accounting system of Saudi companies. *WSEAS Transactions on Business and Economics, 21*(January), 499–511. DOI: 10.37394/23207.2024.21.42

Haney, B. (2020). *Applied natural language processing for law practice*. Brian S. Haney, Applied Natural Language Processing for Law Practice.

Harper, D. J., Ellis, D., & Tucker, I. (2022). Covert aspects of surveillance and the ethical issues they raise. *Ethical Issues in Covert. Security and Surveillance Research Advances in Research Ethics and Integrity, 35*(2), 177–197.

Harwich, E., & Laycock, K. (2018). *Thinking on its own: AI in the NHS*. Reform Research Trust.

Horgan, D., Romao, M., Morré, S. A., & Kalra, D. (2020). Artificial intelligence: Power for civilisation–and for better healthcare. *Public Health Genomics, 22*(5-6), 145–161. DOI: 10.1159/000504785 PMID: 31838476

Hosseini, M. M., Hosseini, S. T. M., Qayumi, K., Ahmady, S., & Koohestani, H. R. (2023). The aspects of running artificial intelligence in emergency care; a scoping review. *Archives of Academic Emergency Medicine, 11*(1), e38. PMID: 37215232

Karim, R., Galar, D., & Kumar, U. (2023). *AI factory: theories, applications and case studies*. CRC Press. DOI: 10.1201/9781003208686

Kendall, G., & Teixeira da Silva, J. A. (2024). Risks of abuse of large language models, like ChatGPT, in scientific publishing: Authorship, predatory publishing, and paper mills. *Learned Publishing*, *37*(1), 55–62. DOI: 10.1002/leap.1578

Kosasih, E. E., Papadakis, E., Baryannis, G., & Brintrup, A. (2024). A review of explainable artificial intelligence in supply chain management using neurosymbolic approaches. *International Journal of Production Research*, *62*(4), 1510–1540. DOI: 10.1080/00207543.2023.2281663

Leal, M. M., & Musgrave, P. (2023). Backwards from zero: How the US public evaluates the use of zero-day vulnerabilities in cybersecurity. *Contemporary Security Policy*, *44*(3), 437–461. DOI: 10.1080/13523260.2023.2216112

Leontidis, G. (2024). Science in the age of ai: How artificial intelligence is changing the nature and method of scientific research.

Mahoney, C. W. (2017). Buyer beware: How market structure affects contracting and company performance in the private military industry. *Security Studies*, *26*(1), 30–59. DOI: 10.1080/09636412.2017.1243912

Mbah, G. O. (2024). Smart Contracts, Artificial Intelligence and Intellectual Property: Transforming Licensing Agreements in the Tech Industry.

Mhia-Alddin, D. A., & Hussein, A. M. (2025). AI-Powered Contract Security: Managing Expiry, Compliance, and Risk Mitigation Through Deep Learning and LLMs. In Zangana, H., Al-Karaki, J., & Omar, M. (Eds.), *Revolutionizing Cybersecurity With Deep Learning and Large Language Models* (pp. 37–64). IGI Global Scientific Publishing., DOI: 10.4018/979-8-3373-3296-3.ch002

Moussa, S., & Teixeira da Silva, J. A. (2023). Testing the robustness of COPE's characterization of predatory publishing on a COPE member publisher (Academic and Business Research Institute). *Publishing Research Quarterly*, *39*(4), 337–367. DOI: 10.1007/s12109-023-09967-9

Nicholls, T., & Culpepper, P. D. (2021). Computational identification of media frames: Strengths, weaknesses, and opportunities. *Political Communication*, *38*(1-2), 159–181. DOI: 10.1080/10584609.2020.1812777

Nordlinger, B., Villani, C., & Rus, D. (Eds.). (2020). *Healthcare and artificial intelligence*. Springer. DOI: 10.1007/978-3-030-32161-1

Nriezedi-Anejionu, C. (2024). Carbon reduction and nuclear energy policy U-turn: The necessity for an international treaty on small modular reactors (SMR) new nuclear technology. *Carbon Management*, *15*(1), 2396585. DOI: 10.1080/17583004.2024.2396585

Nunkoo, R., Sharma, A., Rana, N. P., Dwivedi, Y. K., & Sunnassee, V. A. (2023). Advancing sustainable development goals through interdisciplinarity in sustainable tourism research. *Journal of Sustainable Tourism*, *31*(3), 735–759. DOI: 10.1080/09669582.2021.2004416

Pakarinen, A. (2025). Consent, control and compliance: legal perspectives on processing health data in the development of AI through anonymization and pseudonymization under the GDPR.

Pellecchia, R. (2022). Leveraging AI via speech-to-text and LLM integration for improved healthcare decision-making in primary care.

Prearo, M., & Scopelliti, A. (2025). Do LGBTIQ+ issues matter for populist radical right? An analysis of Italian parties' social media narratives. *South European Society & Politics*, ●●●, 1–22. DOI: 10.1080/13608746.2025.2463915

Rashid, A. B., & Kausik, A. K. (2024). AI revolutionizing industries worldwide: A comprehensive overview of its diverse applications. *Hybrid Advances*, 100277.

Rizun, N., Revina, A., & Edelmann, N. (2025). Text analytics for co-creation in public sector organizations: A literature review-based research framework. *Artificial Intelligence Review*, *58*(4), 1–45. DOI: 10.1007/s10462-025-11112-1

Sarferaz, S. (2024). *Embedding Artificial Intelligence into ERP Software*. Springer Nature. DOI: 10.1007/978-3-031-54249-7

Simola, J. (2022). *Effects and factors of the hybrid emergency response model in public protection and disaster relief*. JYU Dissertations.

Singhal, N., Goyal, S., & Singhal, T. (2024). Decentralized Insurance Platforms: Innovation and Technology for Trust and Efficiency. In *Potential, Risks, and Ethical Implications of Decentralized Insurance* (pp. 95–163). Springer Nature Singapore. DOI: 10.1007/978-981-97-5894-4_3

Stevens, R., Taylor, V., Nichols, J., Maccabe, A. B., Yelick, K., & Brown, D. (2020). *AI for science: Report on the department of energy (doe) town halls on artificial intelligence (ai) for science* (No. ANL-20/17). Argonne National Lab.(ANL), Argonne, IL (United States).

Wright, J., Avouris, A., Frost, M., & Hoffmann, S. (2022). Supporting academic freedom as a human right: Challenges and solutions in academic publishing. *International Journal of Human Rights*, 26(10), 1741–1760. DOI: 10.1080/13642987.2022.2088520

Yao, Z., & Yu, H. (2025). A Survey on LLM-based Multi-Agent AI Hospital.

Chapter 4
Threat Detection in Multi-Cloud Environments

Noble Worlanyo Antwi
https://orcid.org/0009-0000-5123-5051
Illinois Institute of Technology, USA

ABSTRACT

This chapter investigates the critical role of threat detection in securing multi-cloud environments, a rapidly evolving area as organizations adopt platforms such as Amazon Web Services (AWS), Microsoft Azure, and Google Cloud Platform (GCP). It analyzes traditional security mechanisms, including firewalls and intrusion detection systems, highlighting their limitations in cloud-native infrastructures. The chapter explores advanced practices such as Artificial Intelligence (AI)-driven analytics, Machine Learning (ML), User and Entity Behavior Analytics (UEBA), Zero Trust Architecture (ZTA), and Extended Detection and Response (XDR). A vendor-neutral threat detection architecture is proposed for centralized monitoring and automated incident response. Ethical, legal, and compliance considerations are also discussed, aligning security practices with standards like GDPR and HIPAA. The chapter concludes with recommendations for holistic, intelligence-driven security and identifies future research opportunities in cross-cloud frameworks and explainable AI.

DOI: 10.4018/979-8-3373-4252-8.ch004

INTRODUCTION

Overview of Cloud Computing and the Rise of Multi-Cloud Strategies

Cloud computing has transformed how businesses manage their IT resources. Instead of maintaining physical servers and storage devices, companies now access computing power, storage, and applications over the internet. This shift allows for greater flexibility, cost savings, and the ability to scale operations quickly. The National Institute of Standards and Technology (NIST) outlines five key features of cloud computing: on-demand self-service, broad network access, resource pooling, rapid elasticity, and measured service. These elements set cloud computing apart from traditional IT setups and have led to its widespread use across various sectors (NIST, 2011). Typically, cloud services are divided into three main categories:

- **Infrastructure as a Service (IaaS):** This model offers virtualized computing resources online, enabling users to rent virtual machines, storage, and networks as needed.
- **Platform as a Service (PaaS):** Provides a platform for customers to develop, run, and manage applications without dealing with the underlying infrastructure, streamlining the development and deployment process.
- **Software as a Service (SaaS):** Delivers software applications over the internet on a subscription basis, removing the need for installations and maintenance.

Recently, many organizations have started using multiple cloud providers simultaneously, a strategy known as a multi-cloud approach. This method takes advantage of the unique strengths of different providers to enhance performance, reduce costs, and increase resilience. Figure 1 below shows an architectural representation of multi-cloud environments

The benefits of a multi-cloud strategy include:

MULTI-CLOUD

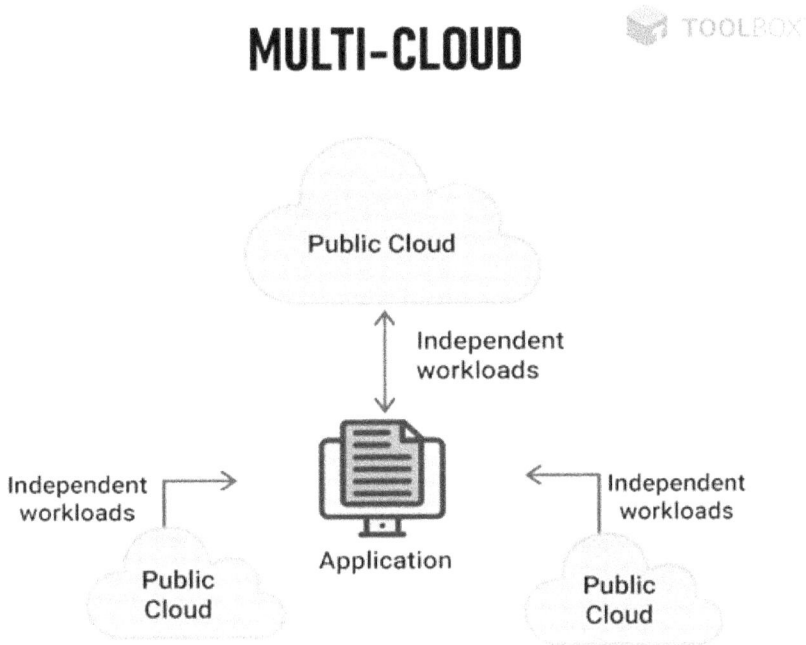

- **Avoiding Vendor Lock-In:** By working with multiple providers, companies can avoid becoming too dependent on a single vendor, giving them more flexibility and bargaining power.
- **Improved Resilience and Redundancy:** Spreading workloads across several clouds can enhance disaster recovery capabilities and minimize the risk of service disruptions.
- **Better Performance and Cost Efficiency:** Organizations can choose services from different providers based on specific needs, leading to improved performance and cost savings (Saxena, Gupta, & Singh, 2021).

However, adopting a multi-cloud strategy also brings challenges, such as increased management complexity, potential security issues, and the need for different platforms to work together seamlessly. To overcome these hurdles, companies should implement strong cloud management practices, establish clear governance policies, and use tools designed for multi-cloud coordination (Zhao, Benomar, Pfandzelter, & Georgantas, 2022).

The growing use of multi-cloud strategies highlights a broader trend toward flexibility and agility in IT operations, allowing organizations to customize their cloud infrastructures to meet specific business requirements and respond more effectively to changing market conditions (Chhabra & Singh, 2022).

The Rise of Complex Cyber Threats Targeting Multi-Cloud Infrastructures

As multi-cloud adoption grows, organizations are facing a wider variety of advanced cyber threats. The inclusion of services from a variety of cloud providers enhances flexibility and resilience but expands the attack surface, thereby complicating security management.

A significant challenge is the lack of consistency in the security measures of different cloud platforms. Each provider uses a different security protocol, which can create gaps and weaknesses when these services are combined. This fragmentation can be exploited by attackers to gain unauthorized access or disrupt services (Reece et al., 2023).

Moreover, the constantly changing dynamics of multi-cloud environments, as typified by continuous scaling and resource allocation, pose challenges for threat detection. Conventional security solutions may be challenged to detect and analyze the high volumes of diverse data flows created across different platforms, hence making it difficult to identify malicious activity in real-time (Manzoor et al., 2022).

To counter such threats, organizations are becoming more open to adopting artificial intelligence (AI) and machine learning (ML) solutions. These advancements can automate processes related to threat detection and response, scanning patterns across various cloud services to detect anomalies that may signify a possibility of cyber threats. Through AI and ML, companies can enhance their security model, thereby facilitating wider defense against the evolving nature of cyber threats (Kumari 2022).

RESEARCH PROBLEM

Organizations face unique issues when it comes to identifying threats and managing security concerns when they use setups that utilize several clouds. The process becomes more complicated when services from many cloud providers are integrated, which can result in possible security problems.

Challenges in Threat Detection

One of the primary concerns is the non-standard security configurations of different cloud platforms. Each provider has its own set of security tools and configurations, which makes it hard to implement a common security strategy. This diversity could lead to configuration mistakes, which are among the predominant causes of security breaches in cloud environments (Asthana, 2024).

In addition, the absence of a single view makes it difficult to monitor and respond to incidents. Security teams find it difficult to achieve a complete view of their systems because there are different logging and monitoring systems across different cloud providers. The segmentation makes it hard to detect threats in real-time and respond accordingly (Sasovets, 2024).

Specific Vulnerabilities in Multi-Cloud Environments

Figure 2. Top multi-cloud security threats. (Basan 2023)

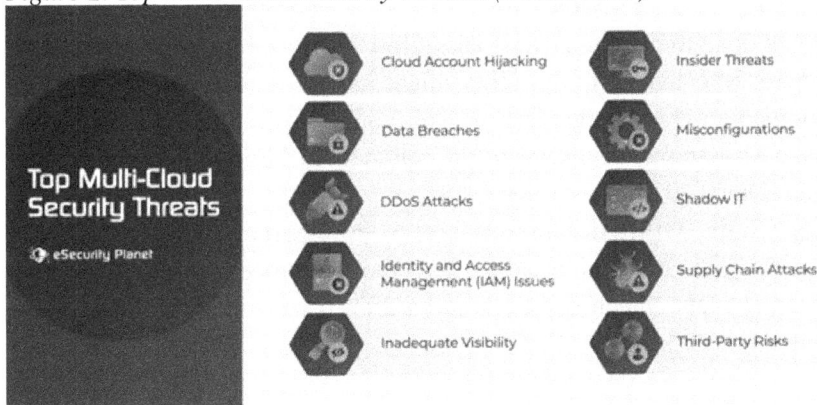

Multi-cloud setups are susceptible to several vulnerabilities some of which are listed below:

- **Insecure APIs:** APIs are utilized in integrating with cloud services, but they can be manipulated if not well secured. Insecure APIs can grant attackers unauthorized entry to cloud resources (Asthana, 2024).
- **Shadow IT:** Using apps or services not approved by an organization is capable of circumventing the predefined security policies, offering weaknesses and elevating the risks of data breaches (Asthana, 2024).
- **Poor Access Management:** Poor Access Management: Inadequate control of user access privileges can lead to unauthorized data access. Strong iden-

tity and access management policies need to be in place to thwart this threat (Asthana, 2024).

- **Data Protection and Compliance Issues:** It is challenging to maintain data security and comply with laws on different cloud platforms as there are different data protection policies and laws. Organizations need to implement strong data protection practices and be informed about compliance laws to protect sensitive data (Piskorz, n.d.)
- **Cloud Account Hijacking:** This involves attackers gaining unauthorized access to cloud accounts, either through phishing or weak passwords. Once in control, these accounts can be used to modify information, eavesdrop on transactions, and direct clients to fake websites (Morrow, 2018).
- **Identity and Access Management (IAM) Issues:** Weak or improperly configured IAM policies can allow unauthorized access. IAM policies need to be configured correctly and audited regularly to ensure security (Wiz Experts Team, 2024)
- **DDoS Attacks:** Attackers can inundate cloud services with overwhelming traffic, which is not desirable. Monitoring traffic and creating response plans should be done to combat DDoS risks (Horev, 2024). Insider Threats: Some employees who indeed have access might misuse their privileges to cause intentional harm to the organization. This could involve theft of information, sabotage of systems, or disclosure of confidential information without permission (Quist, 2023).
- **Supply Chain Attacks:** Attackers breach third-party vendors or software updates to penetrate cloud systems. This indirect process relies on trust between companies, which makes them hard to detect and avoid such attacks (SentinelOne, 2024).

To combat such vulnerabilities, we need strong security measures, regular checking, and an active mechanism to detect and respond to threats.

RESEARCH OBJECTIVES

Explore the Role of Threat Detection Systems in Multi-Cloud Environments

With more organizations moving towards multi-cloud strategies, security management becomes more complicated across different platforms. Conventional security practices are inadequate given the different protocols and settings of cloud service providers. Current research stresses the necessity for adaptive threat detection approaches suited for such intricate infrastructures. For example, Reece et al. (2023) call for in-depth security analysis to determine weaknesses emanating from the incorporation of various cloud services.

Investigate Tools and Techniques for Detecting, Preventing, and Responding to Threats Across Multiple Cloud Providers (AWS, Azure, GCP, etc.)

Securing a multi-cloud environment entails utilizing tools and methods to detect, stop, and mitigate threats on AWS, Azure, and Google Cloud Platform (GCP) and other such solutions. Every provider has security services of their own. AWS, for instance, has Amazon GuardDuty, Azure has Microsoft Defender, and GCP has Security Command Center. They assist in locating and responding to threats. Sysdig (n.d.) has a comparison that examines the functionality and what to consider with these cloud threat detection tools. It illustrates how they assist in keeping security in various cloud services.

Additionally, implementing Security Information and Event Management (SIEM) systems can enhance security monitoring in cloud environments. Alotaibi et al. (2023) introduce an innovative SIEM-based approach that offers automated visibility of cloud resources, improving threat detection and response efficiency.

Furthermore, the application of artificial intelligence (AI) in automating threat detection and response processes has gained attention. Farzaan et al. (2024) explore AI-enabled systems for efficient cyber incident detection and response in cloud environments, demonstrating the potential of AI to enhance security measures across multiple cloud platforms.

In addition to the native threat detection services available to each Public Cloud Porvider, there are number of security services each cloud provider makes available to thier customers of which Figure 3 below illustrates.

Figure 3. Comparison of security services between GCP, AWS and Azure (Cloud-withease, n.d.)

CLOUDWITHEASE A Rendezvous with Cloud Technologies	Cloud Security Comparison: AWS vs Azure vs GCP		
Security Service	**AWS**	**Azure**	**Google Cloud**
Physical Security	Numerous diversified data centers across the globe that ensure • redundancy • availability • capacity planning aws	Uses 58 meticulously chosen regions across the globe in 140 countries and/ or regions that ensure • resiliency • compliance • sovereignty • data residency Microsoft Azure	Numerous data centers spread across 22 regions and 61 zones that ensure • single failure circumvention • data residency Google Cloud Platform
Authentication & Authorization	IAM (Identity & Access Management)	Azure AD with Single Sign-On support	OAuth 2.0 protocol with SSO support
Firewall	Web App Firewall	App Gateway	App Gateway
Protection	Shield	DDoS	Google Cloud Armor
Secret Access & Storage	AWS Secret Manager	Azure Key Vault	GCP Secret Manager
Data Encryption	KMS (Key Management Service)	SSE (Storage Service Encryption)	KMS (Key Management Service)
VPN Gateway	• point to site • site to site • Limit of 10 site-to-site connections per VPN gateway	• point to site • site to site • Limit of 30 site-to-site connections per VPN gateway	Only site to site
Identity Management	Amazon Cognito	Active Directory B2C	Unified Management Console
SaaS	Amazon Inspector	Azure security centr	Trust and security centre

SIGNIFICANCE OF THE STUDY

Contribution to Improving Security Strategies and Frameworks in Multi-Cloud Environments

The objective of this study is to improve security strategies and frameworks in multi-cloud environments by identifying and examining some of the security problems that arise from using different cloud service providers. By examining these problems, this study aims at creating holistic security frameworks that provide uniform protection across various cloud platforms. Standards of this kind are essential for organizations to maintain a robust security posture, as they provide standardized procedures and best practices that are tailor-made to tackle the complexities of multi-cloud setups. Adoption of such strategies can strengthen risk management and reduce vulnerabilities associated with multi-cloud deployments.

Potential for Enhancing Cloud Security Tools, Platforms, and Processes for Organizations with Multi-Cloud Setups

The study also concentrates on assessing and enhancing cloud security platforms and tools to more effectively support organizations that work in multi-cloud ecosystems. By assessing current security tools and identifying shortcomings, this study aims to lead the development of more effective tools and methods that can seamlessly integrate with various cloud services. This process includes improving security monitoring systems, incident response protocols, and compliance management tools. It is important to optimize these elements to allow organizations to effectively handle security exposures and defend the integrity of their applications and data on various cloud platforms.

LITERATURE REVIEW

This chapter explores key concepts, challenges, and advancements related to threat detection in multi-cloud environments. The discussion includes the definition and characteristics of multi-cloud environments, cyber threats unique to such settings, existing threat detection techniques, security challenges, major security tools, and recent advances in threat detection.

Multi-Cloud Environments

Definition and Characteristics of Multi-Cloud Environments

Multi-cloud environments refer to the use of multiple cloud computing services from different providers within a single heterogeneous architecture (Saxena et al., 2021). These environments are characterized by their ability to distribute workloads across various cloud platforms, each offering unique strengths and capabilities. According to Zhao et al. (2022), multi-cloud strategies enable organizations to avoid vendor lock-in, enhance resilience, and optimize performance by leveraging the best features of different cloud services.

Key characteristics of multi-cloud environments include:

- Distributed workloads across multiple cloud service providers
- Increased flexibility and scalability
- Enhanced disaster recovery capabilities
- Potential for cost optimization through strategic resource allocation

- Redundancy and Failover Capabilities – Using multiple providers reduces the risk of downtime due to a single vendor failure.
- Avoidance of Vendor Lock-in – Organizations maintain greater control over their data and applications by diversifying cloud providers.
- Customizable Workloads – Multi-cloud enables businesses to select the most efficient services from different providers for various workloads.
- Security Challenges – The integration of multiple cloud services complicates security management, requiring robust governance frameworks (Chhabra & Singh, 2022).

Comparison Between Single-Cloud and Multi-Cloud Environments

While single-cloud environments rely on a sole provider for all cloud services, multi-cloud setups leverage multiple providers to meet diverse organizational needs. Chhabra and Singh (2022) highlight that multi-cloud architectures offer greater flexibility and resilience compared to single-cloud solutions. However, they also note that multi-cloud environments introduce additional complexity in terms of management and security.

The table 1 below summarizes key differences between single-cloud and multi-cloud environments:

Table 1. Differences between single cloud and multi-cloud environment.

Aspect	Single-Cloud	Multi-Cloud
Vendor Dependency	High (potential lock-in)	Low (distributed across providers)
Flexibility	Limited to one provider's offerings	High (best-of-breed services from multiple providers)
Complexity	Lower management overhead	Higher complexity in integration and management
Resilience	Vulnerable to provider-specific outages	Enhanced through provider diversity
Cost Optimization	Limited to single provider's pricing	Potential for cost savings through provider competition

CYBER THREATS IN MULTI-CLOUD

Unique Threats and Vulnerabilities in Multi-Cloud Environments

Multi-cloud environments inherently introduce a range of complex security challenges due to their distributed architecture and the integration of diverse cloud services. One of the most significant issues is misconfiguration, where errors in security settings across various cloud platforms can leave critical systems exposed (Reece et al., 2023). This problem is further compounded by inconsistent security policies between different providers, complicating the establishment of a uniform defense strategy.

The distributed nature of multi-cloud setups results in an increased attack surface. With multiple entry points, threat actors can exploit vulnerabilities not only through misconfigurations but also via insecure APIs that provide avenues for manipulating cloud workloads (Asthana, 2024). Moreover, the lack of centralized visibility in these environments can lead to data breaches, as it becomes challenging to monitor and secure sensitive information effectively (Manzoor et al., 2022).

Adding to these complexities, Manzoor et al. (2022) emphasize that the dynamic nature of multi-cloud environments further complicates threat detection, as traditional security tools may struggle to monitor and analyze diverse data streams across multiple platforms. This dynamic behavior necessitates advanced, adaptive security solutions capable of real-time analysis across heterogeneous data sources.

Another emerging threat is the potential for cross-cloud attacks, where adversaries leverage weaknesses in the interconnectivity of cloud platforms to move laterally across environments, thereby amplifying risk. Additionally, supply chain attacks pose significant risks; attackers can target third-party services integrated within the cloud ecosystem to gain unauthorized access or disrupt operations (SentinelOne, 2024).

Ultimately, concerns around data sovereignty and compliance exacerbate the security framework. Conducting operations in many jurisdictions necessitates compliance with numerous regulatory frameworks, frequently resulting in deficiencies in protection and enforcement.

The diverse vulnerabilities in multi-cloud environments, including misconfigurations, inconsistent policies, an enlarged attack surface, data breaches, challenges in dynamic threat detection, cross-cloud and supply chain attacks, and jurisdictional compliance issues - require a thorough, coordinated security strategy to address the evolving threats in the contemporary cloud-driven landscape.

Examples of Recent Cyberattacks Targeting Multi-Cloud Setups

The increasing adoption of multi-cloud architectures has introduced new vulnerabilities, making them attractive targets for cybercriminals. Several high-profile attacks have demonstrated the risks associated with misconfigurations, weak security policies, and supply chain compromises in multi-cloud environments. Below are some significant cyberattacks those exposed weaknesses in multi-cloud security.

SolarWinds Attack (2020)

The SolarWinds attack was one of the most devastating supply chain attacks in history, affecting thousands of enterprises, government agencies, and cloud service providers worldwide. The attack, attributed to nation-state threat actors, involved the compromise of SolarWinds Orion, an IT monitoring platform widely integrated into AWS, Azure, and GCP environments (SentinelOne, 2024). Attackers inserted a malicious backdoor, known as SUNBURST, into SolarWinds Orion software updates, allowing them to infiltrate networks of government agencies, defense contractors, and cloud service providers (Horev, 2024). This breach provided persistent access to sensitive networks, leading to data exfiltration, espionage, and severe operational disruptions (Microsoft, 2021). The attack underscored the critical importance of supply chain security, emphasizing the need for zero-trust security frameworks and real-time anomaly detection in multi-cloud architectures (CISA, 2021).

Capital One Data Breach (2019)

The Capital One data breach exposed the sensitive financial data of over 100 million customers, making it one of the largest cloud-related breaches ever recorded. The breach was caused by a misconfigured AWS S3 bucket, which allowed an ex-AWS employee to exploit a vulnerability in the bank's firewall settings (Asthana, 2024). The attacker leveraged a server-side request forgery (SSRF) vulnerability, gaining unauthorized access to Capital One's cloud-hosted databases (Reece et al., 2023). This breach resulted in identity theft, regulatory fines, and lawsuits, costing Capital One hundreds of millions in damages (Chhabra & Singh, 2022). The incident highlighted the dangers of misconfigured cloud storage, the importance of proactive security auditing, and the need for robust Identity and Access Management (IAM) policies (Saxena et al., 2021).

Microsoft Exchange Attacks (2021)

The Microsoft Exchange attacks targeted Microsoft Exchange servers, which are widely used for email and collaboration services. The attack was part of a widespread hacking campaign that impacted multi-cloud email infrastructures and was attributed to Chinese state-sponsored groups (Wiz Experts Team, 2024). Attackers exploited vulnerabilities known as ProxyLogon in on-premises Microsoft Exchange servers, allowing them to deploy web shells for persistent access (Microsoft, 2021). Over 250,000 organizations worldwide had their emails and confidential data compromised, including government agencies and multinational corporations (SentinelOne, 2024). This attack demonstrated the importance of patch management, network segmentation, and cloud-based security monitoring (CISA, 2021).

Codecov Supply Chain Attack (2021)

The Codecov supply chain attack targeted software development pipelines, affecting multi-cloud infrastructures by injecting malicious code into automated testing tools used by developers worldwide. The breach was particularly concerning because it compromised GitHub repositories, AWS environments, and CI/CD workflows (Farzaan et al., 2024). Attackers exploited a vulnerability in Codecov's Bash Uploader script, modifying it to steal credentials, API keys, and sensitive cloud configurations (Horev, 2024). This attack impacted thousands of organizations, including prominent technology firms, financial institutions, and government agencies (Morrow, 2018). The breach emphasized the importance of securing DevOps pipelines, monitoring API interactions, and enforcing stronger authentication mechanisms (Alotaibi et al., 2023).

Uber Data Breach (2022)

The Uber data breach in September 2022 was a massive cloud security incident that exposed internal Slack messages, financial records, and engineering documentation. The breach demonstrated how poor multi-cloud identity security can lead to devastating cyberattacks (Sasovets, 2024). The attacker performed social engineering on an Uber contractor, tricking them into approving multi-factor authentication (MFA) prompts. Once inside, the attacker accessed Uber's AWS, GCP, and Azure environments (Quist, 2023). The breach resulted in the attacker obtaining administrator-level access across Uber's cloud infrastructure, threatening sensitive corporate data and customer privacy (SentinelOne, 2024). This breach reinforced the need for stronger MFA policies, phishing-resistant authentication, and strict IAM role-based access controls (Wiz Experts Team, 2024).

Table 2 provides a comparative summary of key characteristics from recent high-profile cyberattacks that targeted multi-cloud infrastructures, such as the SolarWinds supply chain attack, the Capital One data breach, and the Uber data breach. These cases highlight the evolving nature of cloud security threats and underscore the need for proactive threat detection and robust security measures across multi-cloud environments.

Table 2. Comparative summary of major cyberattacks on multi-cloud environments

Attack Name	Year	Attack Vector	Target	Impact
SolarWinds Attack	2020	Supply Chain Attack (malicious update to SolarWinds Orion software)	Government agencies, tech companies, and cloud service providers	- Compromised thousands of organizations globally - Data exfiltration and espionage - Persistent access for attackers
Capital One Data Breach	2019	Misconfiguration of AWS S3 bucket	Capital One (bank)	- Exposure of 100+ million customer records - Data theft (social security numbers, credit scores) - Financial penalties
Uber Data Breach	2022	Social engineering, MFA bypass	Uber (ride-sharing company)	- Compromised internal systems, including GitHub and AWS - Sensitive customer and corporate data exposed
Microsoft Exchange Attacks	2021	Exploitation of ProxyLogon vulnerability	Microsoft Exchange servers	- Compromise of 250,000+ organizations worldwide - Unauthorized access to email and confidential data
Codecov Supply Chain Attack	2021	Malicious code in CI/CD pipeline (Codecov's Bash Uploader script)	Technology firms, financial institutions, and government agencies	- Compromise of source code repositories - Leakage of credentials, API keys, and sensitive configurations
Uber Data Breach (Revisited)	2022	Social engineering to bypass MFA	Uber	- Exposed sensitive internal data and records

Key Takeaways from Multi-Cloud Cyberattacks

The growing number of multi-cloud breaches highlights several critical security challenges. Misconfigurations, as seen in the Capital One breach and the Microsoft Exchange attack, demonstrate the importance of regular security audits and cloud-native monitoring tools (Asthana, 2024). Supply chain attacks, such as SolarWinds and Codecov, show how compromised third-party software can infiltrate cloud environments (Horev, 2024). Identity and Access Management (IAM) weaknesses, as seen in the Uber breach, underscore the need for stringent IAM policies, phishing-resistant MFA, and behavioral analytics for login attempts (Quist, 2023). Insider threats and social engineering remain prevalent risks, with multiple attacks succeeding due to weak internal security training and human error, highlighting the need for security awareness programs (Reece et al., 2023). Organizations must adopt zero-trust architectures, cloud workload protection platforms (CWPPs), and automated threat detection solutions to strengthen their multi-cloud security posture (Saxena et al., 2021).

THREAT DETECTION TECHNIQUES

Traditional threat detection methods typically rely on signature-based and rule-based approaches, including firewalls, intrusion detection systems (IDS), and log-based anomaly detection (Farzaan et al., 2024). While these techniques have been effective in more static environments, they face significant limitations in multi-cloud settings due to increased complexity and dynamic threat landscapes.

Challenges of Traditional Security in Multi-Cloud Environments

The adoption of multi-cloud environments, in which organizations utilize services from multiple cloud providers, has gained significant traction due to its advantages in redundancy, operational flexibility, and cost efficiency. However, securing such environments presents unique challenges that traditional security measures, such as firewalls and Intrusion Detection Systems (IDS), may not adequately address (Reece et al., 2023). The inherent complexity of multi-cloud architectures necessitates a reevaluation of conventional security strategies, given that they were primarily designed for centralized network structures rather than decentralized and distributed computing frameworks:

Distributed Architecture

Multi-cloud environments encompass multiple cloud infrastructures, each characterized by distinct security configurations, authentication mechanisms, and access controls. Traditional security solutions, primarily designed for single-cloud or on-premises networks, often lack the flexibility to provide uniform security oversight across disparate cloud providers (Ibrahim et al., 2016).

For instance, an enterprise might employ Amazon Web Services (AWS) for compute workloads, Microsoft Azure for AI-driven analytics, and Google Cloud Platform (GCP) for database management. Given the fundamental differences in network policies, logging formats, and identity management between these platforms, conventional security measures may struggle to ensure consistent security enforcement (Süß et al., 2024). The fragmentation of security controls across cloud providers may result in policy inconsistencies, increasing the likelihood of misconfigurations and vulnerabilities (Kumari, 2022).

Scalability Limitations

Cloud infrastructures are inherently elastic, allowing workloads to scale dynamically in response to demand fluctuations. However, traditional security solutions, particularly on-premises firewalls and legacy IDS tools, often fail to adapt to real-time changes in cloud environments (Reece et al., 2023).

For example, in serverless computing or Kubernetes-based containerized environments, workloads are transient and ephemeral, meaning they can be instantiated and terminated in seconds. Security tools that rely on manual configuration or static rule sets may be ineffective in addressing such rapid transitions. Consequently, security gaps may emerge when new cloud instances remain unmonitored or lack adequate access controls, leaving them exposed to potential cyber threats (Paul, 2024).

Limited Threat Correlation and Context Awareness

Another critical limitation of traditional security mechanisms in multi-cloud environments is their inability to correlate security events across disparate cloud platforms. Each cloud provider utilizes distinct logging standards, monitoring frameworks, and data aggregation models, making it challenging for traditional IDS and firewalls to detect coordinated cyberattacks that span multiple cloud services (Reece et al., 2023).

An adversary might conduct reconnaissance in an organization's AWS environment while concurrently exploiting vulnerabilities in an Azure-hosted application. Traditional security solutions, typically designed for monitoring within a singular context, may inadequately identify attack patterns, therefore diminishing their efficacy in identifying Advanced Persistent Threats (APTs) (Süß et al., 2024). The absence of cross-cloud visibility impairs an organization's capacity for real-time threat detection, forensic investigation, and incident response (Alotaibi et al., 2023).

Increased Complexity in Configuration and Security Management

Effective security governance in multi-cloud settings necessitates the manual deployment of security policies for each cloud provider, thereby substantially elevating administrative costs and the potential for misconfigurations. Industry surveys reveal that misconfigurations are the predominant cause of cloud security breaches, mostly attributable to the intricate management of diverse security rules across various cloud platforms (Ibrahim et al., 2016).

An organization may enforce stringent firewall regulations and network access controls in AWS but neglect to mirror the same security measures in GCP or Azure, leading to insecure access points. The lack of centralized security management intensifies the risk of compliance violations, as enterprises find it challenging to implement consistent security standards across various cloud services (Reece et al., 2023).

Regulatory and Compliance Challenges

Entities functioning within regulated areas, including healthcare, banking, and government, are required to comply with rigorous regulatory frameworks such as GDPR, HIPAA and SOC 2. Ensuring regulatory compliance across various cloud environments remains a persistent difficulty due to the discrepancies in data residency regulations, security standards, and audit requirements among cloud providers (Paul, 2024).

GDPR for instance requires enterprises to encrypt personal data and implement stringent access controls. The utilization of several cloud platforms by a corporation, each possessing distinct data governance rules, complicates the consistent execution of encryption policies and access restrictions (Ibrahim et al., 2016). Conventional security frameworks lack the necessary interoperability for centralized compliance management, heightening the risk of regulatory non-compliance and potential legal liability.

Confronting these problems necessitates the advancement of conventional security technologies and the implementation of cutting-edge security procedures tailored for seamless operation in multi-cloud systems. As cyber threats evolve in sophistication and cloud infrastructures become more intricate, dependence on traditional security measures is inadequate. Organizations must incorporate automated security frameworks, proactive threat intelligence, adaptive access controls, and policy-driven security models to provide comprehensive protection, real-time reaction, and regulatory compliance in distributed cloud environments. These contemporary methodologies, detailed in the following sections, empower organizations to successfully minimize risks while preserving agility, scalability, and operational efficiency in cloud security management.

AI, Machine Learning, and Big Data Analytics for Threat Detection

Artificial Intelligence (AI) and Machine Learning (ML) have become formidable instruments for improving threat detection in multi-cloud settings. Contemporary methodologies utilize AI-driven systems to automate threat detection and response across many cloud platforms, facilitating the analysis of extensive and diversified data streams to discern anomalies suggestive of cyber threats (Farzaan et al., 2024). Alongside these AI-driven techniques, big data analytics is essential as it utilizes the vast data produced in cloud settings, offering enterprises enhanced insights into potential security threats and irregular activities (Sasovets, 2024).

Key AI-driven capabilities include:

- **AI-Driven SIEM (Security Information and Event Management):** AI-powered SIEM systems correlate security logs across multiple clouds, significantly enhancing the accuracy of threat detection.
- **Behavioral Analytics:** Machine learning algorithms analyze user behavior to uncover suspicious activities, offering a more nuanced understanding of potential threats (Alotaibi et al., 2023).
- **Automated Incident Response:** AI tools enable the automation of security responses, thereby reducing reaction times and facilitating a proactive security posture.

These integrated approaches not only enhance security automation and threat intelligence but also enable proactive monitoring and rapid response in dynamic multi-cloud settings. As a result, they are critical in mitigating evolving cyber threats and maintaining a robust security posture across complex cloud infrastructures (Piskorz, n.d.).

To provide a clearer comparison of these methods, Table 3 below contrasts the strengths and weaknesses of each system. This visual representation helps to highlight the differences in speed, adaptability, and detection capabilities between traditional and AI-based systems, underlining the evolving nature of threat detection technologies.

Table 3. Comparison of traditional vs. AI-driven threat detection systems

Criteria	Traditional Threat Detection Systems	AI-Driven Threat Detection Systems
Detection Method	Signature-based (e.g., pattern recognition)	Anomaly detection (e.g., behavior-based analysis)
Speed of Detection	Slower (requires updates to signatures for new threats)	Faster (real-time detection using predictive analytics)
Adaptability to New Threats	Low (limited to known threats and patterns)	High (can learn from new data and adapt to evolving threats)
Scalability	Low (manual configuration, harder to scale across complex environments)	High (can automatically scale with cloud environments and adapt to increasing data)
False Positive Rate	High (due to rigid, rule-based approaches)	Lower (AI systems can filter out non-threatening anomalies)
Detection of Novel Attacks	Poor (struggles with zero-day attacks or novel threats)	Excellent (can detect new, unknown threats based on behavior)
Configuration Complexity	High (requires manual configuration and updates)	Low (self-learning models that require less manual intervention)
Resource Consumption	Moderate (typically CPU-heavy)	High (may require more computational resources for AI models)
Example Technologies	Firewalls, Intrusion Detection Systems (IDS), Antivirus	Machine Learning, Anomaly Detection Systems, SIEM with AI, Deep Learning

CHALLENGES IN MULTI-CLOUD SECURITY

Data Silos, Inconsistent Security Policies, and Lack of Visibility

A fundamental concern in multi-cloud security is the formation of data silos among various cloud providers. Asthana (2024) underscores that these divisions may result in disparate security rules and an absence of cohesive visibility over the entire cloud architecture. This fragmentation obstructs efficient threat detection and incident response.

Complexity in Managing Security Tools Across Different Cloud Providers

The management of security technologies across many cloud providers entails considerable complexity. Each supplier may present unique security features and tools, complicating the maintenance of a uniform security posture. The Wiz Experts Team (2024) emphasizes the necessity of establishing centralized security management systems to tackle this issue.

Integration of Threat Detection Tools
Across Disparate Cloud Platforms

The integration of threat detection systems across various cloud platforms continues to be a significant challenge for many enterprises. According to Horev (2024), it is essential to ensure that security technologies that operate in multiple cloud environments are able to communicate and share data in a smooth manner in order to effectively detect and respond to threats.

KEY PLAYERS AND TOOLS IN MULTI-CLOUD SECURITY

In the domain of multi-cloud security management, prominent manufacturers have created advanced solutions to address the complex difficulties of protecting data and applications across various cloud platforms. These products integrate functionalities such as AI-powered threat detection, real-time analytics, and centralized monitoring to deliver extensive security in distributed cloud environments.

Palo Alto Networks: Prisma Cloud

Prisma Cloud is a cloud-native security technology that ensures extensive protection for hosts, containers, and serverless deployments across all cloud environments. It utilizes AI-driven risk prioritization to assess vulnerabilities, allowing security teams to effectively discover and mitigate intricate dangers. Prisma Cloud offers continuous compliance monitoring, guaranteeing that cloud workloads conform to industry standards and best practices (Palo Alto Networks, n.d.).

Cisco Secure Cloud Analytics

Cisco Secure Cloud Analytics employs sophisticated anomaly detection and network security analysis to improve visibility in multi-cloud scenarios. Its advanced data correlation capabilities allow enterprises to swiftly detect anomalous behaviors and potential breaches, hence enhancing overall situational awareness and enabling rapid incident response (Cisco Systems, n.d.).

Splunk Cloud Security Analytics

Splunk's cloud security solutions consolidate and examine data from many cloud platforms, providing real-time security insights. This comprehensive strategy enables security teams to rapidly identify, analyze, and mitigate threats, guaranteeing ongoing surveillance and safeguarding across various cloud settings (Splunk Inc., n.d.).

IBM Cloud Pak for Security

IBM Cloud Pak for Security focuses on integrating security information and event management (SIEM) tailored for multi-cloud infrastructures. By centralizing security data from various sources, it facilitates automated threat analysis and streamlines incident response, thereby improving visibility and control over complex cloud ecosystems. This solution enables organizations to uncover hidden threats and make informed decisions based on comprehensive data insights (IBM, n.d.).

Wiz

Wiz is a cloud-native application protection platform (CNAPP) that provides agentless, comprehensive insight into multi-cloud systems, including AWS, Azure, Google Cloud, and Oracle Cloud. Wiz facilitates swift and thorough scanning of cloud setups, workloads, and identities through API integrations instead of agent deployment. A notable feature is the Security Graph, which correlates vulnerabilities, misconfigurations, secrets, and identity permissions to illustrate exploitable attack vectors. This contextual understanding allows security teams to prioritize remediation efforts efficiently, concentrating on the most significant issues. Wiz offers ongoing compliance evaluations in accordance with standards such as CIS, PCI DSS, NIST, and HIPAA, rendering it exceptionally effective in regulated sectors (Wiz, 2024).

Trend Micro Cloud One

Trend Micro Cloud One provides an extensive array of services aimed at securing cloud-native applications and infrastructure across AWS, Azure, and Google Cloud. It consolidates Workload Security, File Storage Security, Container Security, and Application Security into a cohesive platform. Essential attributes comprise anti-malware, intrusion prevention, firewall functionalities, integrity monitoring, and log analysis, all of which contribute to the protection of cloud workloads. Cloud One facilitates DevOps integration by embedding security policies and compliance checks into CI/CD pipelines. This guarantees that security is an ongoing effort throughout the application lifecycle (Trend Micro, 2024).

Check Point CloudGuard

Checkpoint CloudGuard offers sophisticated threat mitigation and cloud security posture management across many cloud environments. Utilizing artificial intelligence and machine learning, CloudGuard provides functionalities including automatic security policies, identity protection, and network traffic analysis. The integrated security administration facilitates uniform policy enforcement across AWS, Azure, and GCP, while the Identity Protection Engine detects too permissive roles and probable privilege escalations. CloudGuard's extensive threat intelligence assists enterprises in mitigating complex cyber risks while ensuring compliance with international regulatory standards (Check Point Software Technologies, 2023).

McAfee MVISION Cloud

McAfee MVISION Cloud is a Cloud Access Security Broker (CASB) platform intended to deliver visibility, data loss prevention (DLP), and threat protection across Software as a Service (SaaS), Platform as a Service (PaaS), and Infrastructure as a Service (IaaS) offerings. It facilitates real-time policy enforcement, user behavior analysis, and risk evaluation for apps operating in AWS, Azure, and GCP. A significant characteristic is its capacity to identify and manage Shadow IT, unauthorized cloud applications utilized within a business. Moreover, MVISION Cloud facilitates compliance initiatives by aligning risks with regulations including GDPR, HIPAA, and ISO 27001 (McAfee, 2023).

Microsoft Defender for Cloud

Microsoft Defender for Cloud offers integrated security management and sophisticated threat protection for hybrid and multi-cloud settings. It integrates Cloud Security Posture Management (CSPM) with Cloud Workload Protection (CWP) for resources deployed in Azure, AWS, and Google Cloud. Defender for Cloud provides secure score evaluations, ongoing compliance monitoring, and automated threat identification, facilitating proactive risk management. Its interaction with Microsoft Sentinel provides real-time security analytics and incident response functionalities. The platform's DevOps integration guarantees the incorporation of security tests inside deployment pipelines, fostering a shift-left security methodology (Microsoft, 2024).

These solutions collectively empower enterprises to efficiently identify, assess, and alleviate hazards in multi-cloud environments. Their capabilities exemplify a wider industry shift towards automation and intelligent security management, crucial for tackling the evolving issues presented by contemporary cloud systems.

RECENT ADVANCES IN MULTI-CLOUD THREAT DETECTION

Recent improvements in multi-cloud security have offered novel technologies and approaches to improve threat detection and response. Significant advancements encompass the implementation of Zero Trust frameworks, AI-powered analytics, and automation technologies.

Zero Trust Security Framework

The Zero Trust security approach requires ongoing verification of people and devices prior to allowing access to resources, guaranteeing that no entity is automatically trusted. This method is very efficacious in multi-cloud settings, as it aids in alleviating hazards linked to remote infrastructures. Implementing Zero Trust may entail the deployment of service meshes that authenticate and approve each request, thereby safeguarding cloud-native apps without requiring modifications to the application code. Research indicates that although this solution may elevate CPU and memory consumption, it concurrently diminishes delay variability in processing HTTP requests (Rodigari et al., 2021).

Artificial Intelligence-Driven Cloud Threat Intelligence

Artificial intelligence has emerged as a fundamental element in augmenting cloud threat intelligence. AI systems can now assess global cyber threats in real-time, enhancing proactive security strategies. Platforms like as Wiz's Cloud Threat Landscape provide extensive datasets on cloud security incidents, actors, tools, and strategies. This resource allows enterprises to obtain quick insights on new cloud-native attacker strategies, techniques, and processes, hence enhancing proactive threat detection and response (Cohen & Schindel, 2024).

Cloud Workload Protection Platforms (CWPPs)

Cloud Workload Protection Platforms are intended to oversee and safeguard workloads across various cloud environments. These platforms offer automated, real-time security solutions that safeguard workloads operating on both on-premises and hybrid cloud infrastructures, encompassing containerized and virtualized environments. Cloud Workload Protection Platforms (CWPPs) concentrate on safeguarding diverse cloud workloads, including virtual machines, containers, and serverless operations, by proactively screening for vulnerabilities before to deployment and providing ongoing runtime protection to mitigate emerging threats (Sysdig, n.d.).

RESEARCH METHODOLOGY

This chapter delineates the research strategy followed to investigate threat detection mechanisms for multi-cloud environments. Due to the futuristic nature of cloud technologies and the ever-changing threat environment, this research followed a qualitative, exploratory methodology, guided by a comprehensive literature review and case study analysis.

RESEARCH DESIGN

This study takes an exploratory qualitative method that is most suited for examining emergent challenges and security solutions in multi-cloud setups. The study concentrates on evaluating current security tools, methods, and threat detection systems present in cloud service situations, including Amazon Web Services (AWS), Microsoft Azure, and Google Cloud Platform (GCP).

A qualitative framework allows for a deep understanding of the present-day scenario of multi-cloud threat detection, as well as emphasizing emerging technologies and prevailing industry trends.

DATA COLLECTION METHOD

All the data for this research were collected from secondary sources only. Secondary data sources are:

- Peer-reviewed scholarly journals (from sites like IEEE Xplore, Elsevier, ACM Digital Library).
- Industry whitepapers and security reports issued by major cybersecurity companies (e.g., Palo Alto Networks, Cisco, IBM, and Splunk).
- Government reports and guidelines from organizations (e.g., NIST, CISA, ENISA).
- Wrote case studies of real-world multi-cloud security incidents and breaches (SolarWinds attack, Uber data breach, Capital One breach).

These data sources gave information about threat detection tools, attack scenarios, and risk reduction in multi-cloud environments.

SELECTION CRITERIA

The case studies and books were chosen on the basis of the following criteria:

- Relevance: Dedicated to materials related to threat detection, cloud security, and multi-cloud settings.
- Credibility: Utilization of peer-reviewed scholarly articles, reputable industry reports, and official government documents.
- Practical Applicability: Focus on case studies, security frameworks, and tools with real-world implementation proof. Technological Coverage: Coverage of traditional security mechanisms (IDS, firewalls) and modern approaches (AI/ML, XDR, UEBA, Zero Trust Architecture).

DATA LIMITATIONS AND CONSTRAINTS

However, the reliance on secondary data presents several limitations which are stated in table 4 below with its corresponding impact and possible mitigation strategies.

Table 4. Data limitations and constraints

Limitation	Impact	Mitigation Strategy
Data Currency	Threat landscapes evolve rapidly; older studies may not capture the latest attack vectors.	Priority was given to recent publications.
Selection Bias	Enterprise security reports may emphasize their own products or solutions.	Cross-referencing between multiple vendors and independent academic sources was employed.
Threat Intelligence Gaps	Dark web monitoring and proprietary threat feeds used in industry may not be publicly available.	Use of publicly available threat databases (MITRE ATT&CK, OWASP, CVE) to complement findings.

ETHICAL CONSIDERATIONS

The study maintained ethical integrity by:

- Citing all sources appropriately (APA 7th Edition)
- Avoiding data manipulation or biased reporting

- Ensuring the confidentiality of sensitive threat intelligence reports by relying on publicly available information
- Acknowledging limitations transparently

VALIDITY AND RELIABILITY CONSIDERATIONS

To enhance the validity of the findings:

- Triangulation was applied by synthesizing insights from multiple reputable sources.
- Cross-case comparisons (different cyberattacks) were conducted to identify recurring patterns.
- Industry standards (NIST SP 800-53, ISO 27001) provided baseline reliability for security recommendations.

THREAT DETECTION TECHNIQUES IN MULTI-CLOUD

Traditional Threat Detection Techniques

Multi-cloud environments present a number of security challenges, as was brought up in Section 2.4. This chapter focuses on various threat detection techniques that have been built to help mitigate the problem that is being presented. Within the realm of cybersecurity, Traditional Threat Detection Systems are instruments that assist in the protection of computer networks from cybercriminals and other forms of online danger. According to Stallings and Brown (2018), these systems consist of firewalls and Intrusion Detection Systems (IDS), which collaborate to ensure the security of networks using their combined capabilities.

Firewalls

A firewall is like a security guard for a computer network. It decides which data can enter or leave based on preset rules. This helps stop hackers from accessing private information while allowing safe communication (Scarfone & Hoffman, 2009).

Older firewalls only check basic information, such as the sender and receiver of data. However, Next-Generation Firewalls (NGFWs) are more advanced. They can analyze the content of data and understand the context of network activity, making them better at detecting cyber threats (Conti et al., 2016). Even though NGFWs provide stronger protection, they require regular updates and proper configuration, which can make them harder to manage (Sharma & Sahay, 2020).

Intrusion Detection Systems (IDS)

An Intrusion Detection System (IDS) acts like a security camera for a network. It watches for unusual activity and alerts administrators if something suspicious happens. There are two main types of IDS:

- **Network-Based IDS (NIDS):** This type monitors the whole network and looks for signs of cyberattacks by checking the flow of data (Bace & Mell, 2001).
- **Host-Based IDS (HIDS):** This type is installed on individual devices. It watches system files and logs for any suspicious changes (Mukherjee et al., 1994).

IDS uses different techniques to detect cyber threats:

- **Signature-Based Detection:** This approach functions similarly to an anti-virus software. It contrasts network activity with a repository of recognized risks. This strategy effectively identifies prevalent attacks but encounters difficulties with novel threats that remain unrecorded (Sommer & Paxson, 2010).
- **Anomaly-Based Detection:** This approach analyzes typical network behavior and notifies managers upon the occurrence of any irregularities. As it does not depend on a catalog of recognized threats, it is capable of identifying novel forms of attacks (Garcia-Teodoro et al., 2009).

Firewalls vs. IDS: How They Work Together

Even though firewalls and IDS both help protect networks, they serve different purposes. Firewalls prevent unauthorized access by blocking or allowing data based on rules (Scarfone & Hoffman, 2009). IDS, on the other hand, detects suspicious activity and alerts administrators when something unusual happens (Mukherjee et al., 1994).

Since cyberattacks are becoming more advanced, using both firewalls and IDS are some of the best ways to secure a network. Firewalls stop many attacks before they happen, while IDS identify and report threats that get past the firewall.

Traditional security systems like firewalls and IDS are important for keeping computer networks safe. Firewalls control what enters and leaves a network, while IDS watch for signs of danger. Using them together helps prevent both known and unknown cyber threats, ensuring better security for businesses, schools, and other organizations.

STATE-OF-THE-ART SECURITY (THREAT DETECTION) PRACTICES IN MULTI-CLOUD ENVIRONMENTS

As organizations increasingly adopt multi-cloud architectures, the need for advanced security practices has become more critical than ever. Traditional security measures often struggle to address the complexity, scalability, and distributed nature of multi-cloud environments. As a result, organizations must implement cutting-edge security strategies to ensure data protection, threat detection, and regulatory compliance across diverse cloud platforms.

State-of-the-art security practices in multi-cloud environments go beyond traditional firewalls and Intrusion Detection Systems (IDS) to include AI-driven threat detection, automated security orchestration, Zero Trust architectures, and proactive compliance management. These modern approaches leverage automation, real-time analytics, and adaptive security models to mitigate risks effectively.

Machine Learning and AI in Multi-Cloud Threat Detection

Artificial Intelligence (AI) and Machine Learning (ML) have become formidable instruments for threat identification, response, and mitigation in multi-cloud security.

Artificial Intelligence and Machine Learning augment security in multi-cloud systems by:

1. Automating Threat Detection

Conventional rule-based detection systems necessitate human programming and continual upgrades to maintain relevance. Conversely, AI-driven security systems examine extensive datasets to detect trends, anomalies, and suspicious behaviors that could signify possible attacks (Harris, 2024). This alleviates the workload of security analysts and enhances firms' ability to identify cyber threats more swiftly and accurately.

AI-driven threat detection can concurrently monitor logs from AWS, Google Cloud, and Microsoft Azure, detecting anomalous login attempts from many sites that may signify credential compromise or brute-force attacks.

2. Real-Time Analysis

In multi-cloud setups, security problems can occur within seconds, rendering real-time threat detection essential. Machine learning systems perpetually observe and assess security data streams, enabling the detection of anomalous activities in real-time (Harris, 2024). This real-time functionality guarantees the detection of threats, including DDoS attacks, unauthorized access, and data exfiltration, prior to the onset of substantial damage.

An ML system monitoring cloud storage access patterns may identify unusual data downloads from an unfamiliar IP address and classify it as a potential insider threat or the use of stolen credentials by an attacker.

3. Adaptive Learning

In contrast to static security solutions, machine learning-based detection systems perpetually enhance by assimilating knowledge from previous instances, novel attack methodologies, and advancing cyber threats (Harris, 2024). This functionality enables AI-powered security systems to:

- Modify detection algorithms in real-time to anticipate emerging cyber threats.
- Reduce false positives by differentiating between authentic activities and genuine dangers.
- Identify advanced persistent threats (APTs) employing stealth tactics to circumvent conventional security protocols.

An AI-driven intrusion detection system (IDS) used in a multi-cloud context may assess attack patterns from one cloud provider and autonomously implement preventive measures across other clouds in real-time, therefore mitigating the danger of cross-cloud attacks.

AI/ML in Multi-Cloud Security: Key Advantages

In the realm of multi-cloud security, the use of Artificial Intelligence (AI) has become imperative to tackle the complexities and issues associated with managing varied cloud systems. AI-driven solutions improve security by offering extensive visibility, sophisticated threat detection, and automated response systems.

1. Integrating Diverse Data Sources

Multi-cloud infrastructures integrate many cloud platforms, each with own logging systems, security protocols, and monitoring tools. This heterogeneity frequently results in fragmented data silos, limiting the detection and correlation of security risks across many contexts. AI-driven Security Information and Event Management (SIEM) systems have emerged as an essential approach to tackle this dilemma. These solutions consolidate and evaluate security logs from several cloud providers, providing a unified perspective of security incidents that improves situational awareness. Jones et al. (2025) assert that this integrated methodology facilitates the swift detection of coordinated attack efforts that may go unnoticed in segregated contexts. An AI-driven SIEM can concurrently gather and correlate logs from services like AWS CloudTrail, Google Cloud Security Command Center, and Azure Security Center, allowing security teams to identify patterns indicative of cross-cloud attacks, such as coordinated unauthorized access attempts or orchestrated exploitation of cloud APIs.

2. Enhancing Anomaly Detection

AI-driven security systems function by creating benchmarks of standard user and system behavior across several cloud services. When behavior diverges from these baselines—such as unanticipated logins from unfamiliar geographic regions, abrupt increases in data transfer volumes, or atypical API calls, the system recognizes these anomalies as potential security events, necessitating additional investigation. Tatineni (2023) elucidates that this AI-driven approach is especially proficient in detecting zero-day attacks, account takeovers, and data breaches prior to their escalation. In contrast to traditional rule-based detection systems, AI technologies may perpetually learn and adapt, hence improving their proficiency in identifying new and previously unrecognized dangers over time. An AI system overseeing cloud storage access may detect a substantial, unauthorized file transfer to an external IP address, indicating a possible insider threat or an effort at data exfiltration.

3. Automating Response Mechanisms

A primary benefit of employing AI in multi-cloud security is its capacity to automate incident response, therefore markedly decreasing the interval between threat discovery and action. AI-driven security solutions can implement prompt preventative actions autonomously, hence reducing response times and alleviating potential damage. Tatineni (2023) emphasizes that automated actions may involve isolating compromised user accounts or workloads following the detection of suspicious activity, blocking malicious IP addresses, or implementing real-time access limits. For instance, if an AI system detects malware deployment on a virtual machine in a cloud environment, it can promptly terminate the affected instance, sever its network connection to inhibit lateral movement, initiate a forensic investigation to ascertain the attack vector, and implement security patches to prevent future exploitation. This level of automation not only improves the effectiveness of security operations but also strengthens the organization's resistance against emerging cyber threats.

Comparison of Supervised vs. Unsupervised Learning for Anomaly Detection

In multi-cloud security, accurately detecting abnormalities is essential for protecting data and ensuring system integrity. Machine Learning (ML) provides effective methodology for anomaly identification, typically via supervised and unsupervised learning techniques.

Supervised Learning entails training models on labeled datasets, allowing them to identify known hazards by categorizing data into established classifications. This approach is proficient at identifying previously recognized assault patterns. However, its effectiveness declines when addressing fresh or zero-day threats not included in the training data. Salman et al. (2018) revealed that although supervised models attained excellent accuracy in identifying known anomalies, their efficacy diminished with unfamiliar attack types (Salman et al., 2017).

Conversely, Unsupervised Learning does not necessitate labeled data. Algorithms examine datasets to reveal intrinsic structures or patterns, enabling them to effectively discover anomalies from established standards. This expertise is essential for identifying emerging dangers without historical data. Studies demonstrate that unsupervised methods can proficiently detect previously unrecognized anomalies in cloud systems (Hagemann & Katsarou, 2020).

The choice between utilizing supervised or unsupervised learning depends on particular security goals and the characteristics of the accessible data. A hybrid approach that combines both strategies often produces a more effective defense, utilizing the strengths of both to improve overall threat detection capabilities. Strategies have been suggested to enhance the precision and flexibility of anomaly detection systems in dynamic cloud environments (Shahzad et al., 2022).

USER AND ENTITY BEHAVIOR ANALYTICS (UEBA) IN THREAT DETECTION

User and Entity Behavior Analytics (UEBA) is a cybersecurity approach that leverages machine learning and advanced analytics to monitor and analyze the behaviors of users and entitiessuch as devices, applications, and network components, within an organization's network. By establishing baselines of normal activity, UEBA systems can detect deviations that may indicate malicious actions, including insider threats, compromised accounts, or advanced persistent threats (APTs) (Sharma, 2021). This methodology enhances traditional security measures by providing a dynamic and context-aware layer of defense, making it a critical tool in modern multi-cloud security environments (Olaniyan & Rakshit, 2023).

Key Components of UEBA

- Data Aggregation

User and Entity Behavior Analytics (UEBA) solutions initiate the process by aggregating and synthesizing extensive security data from diverse sources, including system logs, network traffic, user access logs, and API activity records. This extensive data compilation facilitates the development of intricate behavioral profiles for both people and entities inside an organization. The efficacy of UEBA systems is largely contingent upon the quality, quantity, and diversity of the aggregated data, since these factors constitute the essential basis for precise behavioral modeling and dependable anomaly detection (Rengarajan & Babu, 2021).

- Behavioral Modeling:

Subsequent to data collection, UEBA systems employ machine learning methodologies and statistical models to delineate the parameters of typical behavior for users, apps, and network entities. Sharma (2021) indicates that these models evaluate behavioral characteristics including login times and locations, frequency of resource use, network activity patterns, and application usage histories. UEBA systems progressively enhance these models by perpetually learning from changing behavior patterns, hence augmenting their capacity to differentiate between benign and suspicious actions (Khaliq, Tariq, & Masood, 2020).

- Anomaly Detection

The principal aim of UEBA is to identify aberrant behavior by assessing current user activities against predefined baselines. Unauthorized access to sensitive resources, quick logins from numerous geographic locations, or sudden surges in data transfer volumes trigger security alarms. Khaliq et al. (2020) assert that this methodology is especially proficient in detecting sophisticated threats, such as zero-day exploits, account breaches, and insider assaults, which frequently evade conventional security measures.

Enhancing Cloud Security with UEBA

The integration of UEBA into an organization's cloud security infrastructure enhances its ability to detect and mitigate threats that traditional firewall and Intrusion Detection Systems (IDS) might overlook. Since multi-cloud environments involve dynamic workloads, decentralized access control, and scalable resources, behavior-based security monitoring plays a pivotal role in ensuring a proactive and adaptive defense strategy (Shelke & Hamalainen, 2024).

Case Studies on the Implementation of User and Entity Behavior Analytics (UEBA) in Cloud Security

The implementation of User and Entity Behavior Analytics (UEBA) in cloud security environments has proven to be a crucial mechanism for detecting and mitigating threats, including insider threats, compromised accounts, privilege abuse, and data exfiltration. The following case studies illustrate successful real-world applications of UEBA in different industries.

1. Insider Threat Detection in a Corporate Cloud Environment

A multinational firm deployed Netskope's Advanced User and Entity Behavior Analytics (UEBA) to oversee and evaluate employee behavior across its cloud storage systems. The system detected a user who downloaded more than 2,180 files from the corporate OneDrive account and later uploaded almost 2,400 files to a personal Google Drive account. Significantly, over 1,500 of these uploaded files activated Data Loss Prevention (DLP) policies, signifying the existence of sensitive information. Netskope

This activity markedly diverged from the user's usual behavioral norms. Netskope's UEBA system, utilizing machine learning-driven behavioral analysis, identified this as a potential data exfiltration attempt. The User Confidence Index (UCI), which evaluates risk levels based on behavioral anomalies, issued a diminished score to this user, necessitating rapid inquiry. The firm employed advanced analytics to proactively identify and address the illicit transfer of critical corporate data, therefore averting a potential data breach. (Netskope, 2023).

2. Compromised Account Detection in a Legal Firm

A large law firm, Winthrop & Weinstine, deployed UEBA to strengthen its cybersecurity posture. The firm needed a centralized log monitoring system to detect anomalies in user activities and prevent potential cyber threats.

The UEBA solution identified unusual login attempts from an external IP address at an irregular time. Additionally, the system flagged an increase in access to confidential client documents, indicating a potential account compromise. Security teams were alerted, and immediate action was taken to revoke access and mitigate the breach (AIMultiple, 2025).

3. Detection of Compromised Devices in a Healthcare Cloud Infrastructure

In a healthcare organization, UEBA detected unusual communication patterns between an internal hospital device and a malicious external IP address. The device, which normally interacted only with internal hospital systems, was flagged for anomalous behavior after it began exchanging encrypted data packets with an external entity.

Upon investigation, it was determined that the device had been compromised by malware, posing a risk of data exfiltration of patient records. Thanks to UEBA-driven early threat detection, the compromised device was isolated from the network, preventing data loss and unauthorized access (Gurucul, 2021).

4. Preventing Data Exfiltration in a Financial Institution

A leading financial institution used UEBA to prevent employees from transferring sensitive financial data to external cloud storage services. The system analyzed user activities and flagged an employee attempting to move large volumes of files to a personal cloud storage account.

The UEBA system identified the deviation from normal behavior, triggering an immediate alert to security teams, who then blocked the unauthorized transfer and conducted an internal review. This preventive action reduced the risk of financial data leaks and insider threats (Netskope, *2023).*

5. Identifying Privilege Abuse in a Government Cloud Network

A governmental organization deployed Gurucul's User and Entity Behavior Analytics (UEBA) solution to oversee privileged users and administrators who have access to sensitive information. The system identified an administrator attempting to access classified files beyond their designated jurisdiction and outside of business hours. This illicit conduct was deemed high risk, necessitating an immediate inquiry. Subsequent investigation indicated that the administrator had been endeavoring to exfiltrate classified government secrets. The agency averted a potential security attack by revoking unlawful access and implementing tougher permission limits (Gurucul, 2020).

Gurucul's UEBA solution utilizes machine learning to create behavioral baselines for users, facilitating the identification of anomalies that may signify insider threats or data exfiltration attempts. The system continuously analyzes user activity and assigns risk scores, delivering real-time alerts for questionable behaviors, enabling enterprises to proactively mitigate possible security incidents. Gurucul, 2021.

ZERO TRUST ARCHITECTURE (ZTA)) IN THREAT DETECTION

Implementing Zero Trust Architecture (ZTA) in multi-cloud environments has become essential for organizations aiming to enhance their security posture and improve their threat detection. ZTA operates on the principle of "never trust, always verify," ensuring that every access request is continuously authenticated, authorized, and validated, regardless of its origin within or outside the network (Rose et al., 2020).

Principles of Zero Trust Architecture (ZTA)

- **Continuous Verification:** Every access request is subject to real-time authentication and authorization, ensuring that users and devices are consistently validated (Rose et al., 2020).
- **Least-Privilege Access:** Access rights are minimized to the essential level required for users to perform their tasks, reducing potential attack surfaces (Kindervag, 2010).
- **Micro-Segmentation:** Networks are divided into smaller, isolated segments to prevent lateral movement of threats within the environment (Sarkar et al., 2022).
- **Assume Breach:** Organizations operate under the assumption that breaches are inevitable, focusing on early detection and response to mitigate potential damage (Rose et al., 2020).

Implementation of Least-Privilege Access

In multi-cloud environments, implementing least-privilege access necessitates the integration of Identity and Access Management (IAM) systems across all cloud platforms. This guarantees that users possess only the essential rights, with access allocated according to dynamic policies that take into account user roles, device health, and contextual elements (Bartakke & Kashyap, 2024).

Continuous Authentication and Monitoring

Continuous authentication in a multi-cloud environment necessitates the implementation of sophisticated security protocols, including multi-factor authentication (MFA) and real-time monitoring systems. These instruments assess user behavior and network traffic to identify anomalies, facilitating swift reactions to potential attacks (Chimakurthi, 2020).

Case Studies on Zero Trust Adoption in Multi-Cloud Security

- **Financial Services Sector:** A multinational bank implemented ZTA across its multi-cloud infrastructure to enhance data protection and regulatory compliance. By adopting micro-segmentation and continuous monitoring, the bank reduced unauthorized access incidents by 40% within a year (Adahman et al., 2022).
- **Healthcare Industry:** A healthcare provider transitioned to a Zero Trust model to secure patient data across multiple cloud platforms. The implementation of least-privilege access and continuous authentication mechanisms led to a significant decrease in data breaches, ensuring compliance with health data protection regulations (Ahmadi, 2024).

These examples underscore the effectiveness of Zero Trust Architecture in fortifying security within multi-cloud environments, highlighting its role in mitigating risks and enhancing organizational resilience.

CLOUD SECURITY POSTURE MANAGEMENT (CSPM)

As organizations shift toward multi-cloud infrastructures, ensuring threat detection and proactive security monitoring has become a critical challenge. Cloud Security Posture Management (CSPM) plays a pivotal role in identifying security misconfigurations, automating compliance enforcement, and reducing security risks across AWS, Azure, and Google Cloud (Verdet et al., 2023). Unlike traditional security measures, CSPM focuses on continuous cloud security assessments, risk mitigation, and compliance management to enhance threat detection and response.

Role of CSPM in Misconfiguration Detection for Threat Prevention

One of the most common causes of cloud security breaches is misconfiguration, accounting for nearly 65% of cloud-based cyberattacks (Banse et al., 2022). CSPM solutions enhance threat detection by continuously scanning cloud environments to identify security vulnerabilities caused by:

- Over-permissioned IAM roles that allow unauthorized data access
- Misconfigured storage buckets (e.g., public exposure of sensitive files)
- Unrestricted network access controls leading to unauthorized access

By leveraging automated security monitoring, CSPM solutions detect misconfigurations in real-time, alerting security teams to take corrective action before adversaries can exploit vulnerabilities. Threat detection capabilities within CSPM platforms analyze security configurations against industry best practices, such as CIS Benchmarks and NIST Security Frameworks, ensuring proactive threat mitigation (Jimmy, 2023)

Figure 4. Cloud security posture management (CSPM) diagram

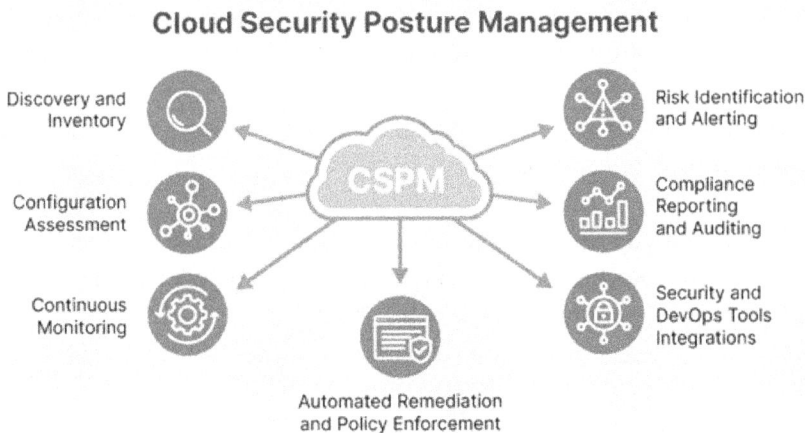

Cloud Security Posture Management

Discovery and Inventory

Configuration Assessment

Continuous Monitoring

CSPM

Automated Remediation and Policy Enforcement

Risk Identification and Alerting

Compliance Reporting and Auditing

Security and DevOps Tools Integrations

From Cloud Security Posture Management (CSPM), by AppOmni, n.d. (https://appomni.com/saas-glossary/cloud-security-posture-management-cspm/). Copyright by AppOmni.

Figure 4 above illustrates some key components of CSPM.

Automating Security Policy Enforcement for Threat Reduction

Manually managing security policies across multiple cloud providers can introduce security gaps that increase the risk of undetected threats. CSPM automates security policy enforcement by:

- Applying predefined security templates across multi-cloud resources
- Automatically remediating security risks by adjusting misconfigurations
- Enforcing compliance controls to detect unauthorized changes

By automating security enforcement, CSPM solutions significantly reduce the attack surface and improve threat detection by eliminating human error and ensuring that security policies remain consistent across AWS, Azure, and Google Cloud (Kadar, 2023).

Best Practices for Maintaining Compliance and Governance to Strengthen Threat Detection

Ensuring regulatory compliance in multi-cloud security environments is crucial for mitigating cyber risks. CSPM solutions enhance compliance management by:

- Conducting automated security assessments to detect configuration drift
- Providing real-time compliance scoring based on regulatory standards such as ISO 27001, NIST 800-53, GDPR, and HIPAA
- Generating security reports that help organizations demonstrate adherence to cloud security regulations

CSPM technologies facilitate real-time detection and response to security threats by consistently monitoring and enforcing compliance regulations, hence upholding a robust governance structure (Banse et al., 2022).

Extended Detection and Response (XDR) in Multi-Cloud Threat Detection

Extended Detection and Response (XDR) is a modern security approach that enhances multi-cloud threat detection by integrating security telemetry from multiple cloud platforms, correlating security events, and utilizing AI-driven analytics to improve detection accuracy and response efficiency (Freitas & Gharib, 2024). Unlike traditional security solutions, XDR provides centralized visibility and real-time threat intelligence, making it a crucial security framework for organizations operating in AWS, Microsoft Azure, and Google Cloud (Farzaan et al., 2024).

How XDR Integrates Telemetry from Different Cloud Providers

XDR aggregates security data from various cloud services, including network logs, endpoint detection, identity management systems, and API activity monitoring. This integration enables security teams to detect, analyze, and respond to cloud threats in a cohesive and efficient manner (Manzoor et al., 2022). Key components of XDR telemetry integration include:

- **Cloud Workload Protection Platforms (CWPP):** Detects security threats at the virtual machine and container level across AWS, Azure, and Google Cloud (Reece et al., 2023).
- **Identity and Access Management (IAM) Logs:** Identifies anomalous authentication patterns, helping prevent account takeovers and unauthorized access attempts (Farzaan et al., 2024).
- **Network Flow Logs and API Monitoring:** Tracks unusual data transfers or unauthorized API calls that could indicate data exfiltration or privilege escalation attacks (KuppingerCole Analysts, 2024).

By combining multiple data sources, XDR enhances threat detection across multi-cloud environments, reducing blind spots that attackers could exploit (S&P Global Market Intelligence, 2021).

Correlation of Security Events Across Cloud Environments

In multi-cloud security, correlating security incidents is a major challenge due to disparate logging formats, security policies, and infrastructure configurations (Reece et al., 2023). XDR solves this problem by:

- Standardizing and analyzing logs from different cloud providers in a centralized security framework.
- Detecting multi-stage attacks that involve multiple cloud services, such as an attacker gaining initial access in AWS, escalating privileges in Azure, and exfiltrating data via Google Cloud (CrowdStrike, 2024).
- Utilizing behavioral analytics to correlate user activity, system anomalies, and network deviations, enabling faster threat identification (Freitas & Gharib, 2024).

A cloud-native XDR solution may identify an unauthorized login attempt in Microsoft Azure, associate it with dubious data transfer in Google Cloud, and initiate automated incident response measures to thwart the attack before to data compromise (eSentire, 2024).

AI-Driven Threat Intelligence and Analytics

Artificial Intelligence (AI) and Machine Learning (ML) play a pivotal role in XDR-driven threat detection by analyzing massive security datasets and identifying attack patterns in real time (Farzaan et al., 2024). AI-powered XDR solutions enhance multi-cloud threat detection by:

- Detecting anomalies in cloud traffic and user behavior using predictive threat models (SentinelOne, 2024).
- Minimizing false positives by differentiating between legitimate user actions and potential security threats (Manzoor et al., 2022).
- Automating incident response workflows, ensuring faster containment of security breaches (KuppingerCole Analysts, 2024).

For instance, AI-driven XDR has been shown to reduce false positive alerts by 40%, enabling security teams to focus on real threats rather than spending time on false alarms (Palo Alto Networks, 2024).

Real-World Applications of XDR in Cyber Defense

Case Study 1: Preventing Cloud-Based Ransomware Attacks

A global financial institution implemented XDR across its AWS, Azure, and Google Cloud environments to prevent ransomware attacks. The system detected:

- Unusual spikes in file encryption requests on cloud storage.
- Simultaneous login attempts from different geographic locations.
- Automated privilege escalation events triggered by an attacker's script.

By correlating these events, the XDR system isolated compromised cloud workloads and blocked the attack before financial data could be encrypted (Farzaan et al., 2024).

Case Study 2: Detecting Insider Threats in Cloud Environments

A multi-national technology firm leveraged XDR to monitor user behavior in cloud applications and detected:

- Anomalous data downloads from a privileged account outside business hours.
- Unauthorized API access to modify security settings in Google Cloud.
- Attempts to exfiltrate intellectual property via external file-sharing platforms.

The AI-driven XDR system flagged the activity as high risk, and automated response mechanisms blocked data transfers and revoked access before sensitive information was leaked (Freitas & Gharib, 2024).

Extended Detection and Response (XDR) strengthens threat detection in multi-cloud environments by integrating security telemetry, correlating security events, and leveraging AI-driven analytics. By providing a centralized, automated security framework, XDR allows organizations to detect and mitigate advanced cyber threats in real time across AWS, Azure, and Google Cloud. As multi-cloud adoption grows, XDR is becoming an essential tool for cyber resilience, automated threat intelligence, and incident response optimization.

THREAT INTELLIGENCE AND DARK WEB MONITORING IN MULTI-CLOUD THREAT DETECTION

The integration of threat intelligence and dark web surveillance into multi-cloud security systems offers a proactive method for recognizing and alleviating future cyber threats. By utilizing these capabilities, businesses can improve situational awareness, identify emerging attack patterns, and bolster their entire security posture in response to evolving threats.

Role of Threat Intelligence in Multi-Cloud Security

Threat intelligence encompasses the methodical gathering and examination of material pertaining to existing and potential threats from diverse sources, including open-source platforms and covert forums. In multi-cloud architectures, threat intelligence offers essential insights into vulnerabilities and attack vectors unique to each cloud platform, allowing enterprises to adopt customized security solutions. The incorporation of threat intelligence feeds into cloud settings has demonstrated an improvement in an organization's cybersecurity capabilities by enabling early detection and proactive response against emerging threats (Agufenwa, 2023).

Importance of Dark Web Monitoring

The dark web functions as a marketplace for cybercriminals to exchange stolen data, credentials, and hacking tools. Surveillance of this hidden sector of the internet is essential for detecting compromised data pertaining to an organization's multi-cloud infrastructure. By identifying the sale of access credentials or conversations regarding vulnerabilities in certain cloud services, organizations can promptly act to avert illegal access and potential breaches. The BlackWidow system has been designed to surveil dark web services, gathering and analyzing cybersecurity data to detect potential risks (Kloft et al., 2019).

Integration of Threat Intelligence and Dark Web Monitoring

Combining threat intelligence with dark web monitoring offers a comprehensive approach to multi-cloud threat detection. This integration enables organizations to:

- **Identify and Mitigate Risks:** Proactively address vulnerabilities before they are exploited by adversaries.
- **Enhance Incident Response:** Utilize real-time data to inform and expedite security measures.
- **Strengthen Security Posture:** Continuously adapt defenses based on the evolving threat landscape.

Monitoring dark web forums can uncover talks regarding attacks aimed at certain cloud services, enabling enterprises to implement preemptive measures to prevent identified risks (Gopireddy, 2020).

Integrating threat intelligence and dark web surveillance into multi-cloud security architectures is vital for effective threat detection and proactive protection. This strategy allows enterprises to anticipate potential risks, safeguarding the security and integrity of their multi-cloud infrastructures.

IDENTITY AND ACCESS MANAGEMENT (IAM) IN MULTI-CLOUD THREAT INTELLIGENCE

In multi-cloud systems, Identity and Access Management (IAM) is crucial for enhancing security by overseeing user identities and controlling access to resources across various cloud platforms. Effective IAM deployment is crucial for improving threat intelligence and reducing potential security risks in complex infrastructures.

Challenges in Multi-Cloud IAM

Administering IAM in multi-cloud environments poses specific issues because to the diversity of cloud services and their individual security protocols. The variation in IAM configurations among platforms might result in inconsistencies, heightening the risk of misconfigurations and possible security breaches. An extensive examination of multi-cloud setups revealed authentication and architectural vulnerabilities as significant concerns, highlighting the necessity for strong IAM techniques to mitigate these risks (Reece et al., 2023).

AI-Driven Approaches to IAM

Integrating Artificial Intelligence (AI) has emerged as a promising way to handle the complexity of multi-cloud Identity and Access Management (IAM). AI-augmented IAM frameworks may scrutinize extensive access data to identify anomalies and anticipate potential attacks, ultimately fortifying the security posture of multi-cloud infrastructures. Recent studies indicate that AI-driven IAM solutions markedly diminish security threats and guarantee adherence to regulatory standards, establishing a new benchmark for access control in intricate cloud environments (Agufenwa, 2023).

Enhancing Threat Intelligence through IAM

Efficient IAM solutions boost threat intelligence by offering extensive visibility into user actions across all cloud platforms. This visibility allows for the identification of illegal access attempts and anomalous behavior patterns, enabling swift and informed reactions to possible threats. Furthermore, IAM systems are essential in the aggregation and examination of security data, facilitating the formulation of proactive threat mitigation methods (Pöhn & Hommel, 2024).

Implementing effective Identity and Access Management systems in multi-cloud environments is crucial for sustaining a solid security posture. By tackling fundamental difficulties with AI-driven strategies and improving threat intelligence capabilities, organizations may efficiently manage identities and access, thus protecting their resources from growing cyber threats.

CLOUD-NATIVE THREAT DETECTION TOOLS AND FRAMEWORKS

Cloud-native threat detection tools play a pivotal role in providing real-time threat monitoring, vulnerability management, and automated security responses. AWS GuardDuty, Azure Security Center, and Google Cloud Security Command Center (SCC) are among the most widely used cloud-native security solutions that enhance security operations (Sharma & Kumar, 2023).

Overview of AWS GuardDuty, Azure Security Center, and Google Cloud Security Command Center

Table 5 below gives some details about the comparison between threat detection tools that are native to each top three cloud provider in the world.

Table 5. Feature comparison of cloud-native security tools. Source: (Amazon Web Services, 2023; Microsoft, 2023; Google Cloud, 2023)

Feature	AWS GuardDuty	Azure Security Center	Google Cloud Security Command Center
Threat Detection Method	Machine Learning & Threat Intelligence	Behavioral Analytics & ML	Google Threat Intelligence Feeds
Security Log Analysis	CloudTrail, VPC Flow Logs, DNS Logs	Virtual Machines, Databases, Kubernetes	GCP Workloads & API Monitoring
Cross-Cloud Support	AWS only	Multi-cloud (AWS, Azure, GCP)	Primarily GCP, integrates with SIEM
Automated Remediation	Yes (via AWS Lambda)	Yes (via Microsoft Defender)	Yes (via Security Playbooks)
Compliance Management	FedRAMP, PCI DSS, HIPAA	CIS, GDPR, ISO 27001	SOC 2, PCI DSS, HIPAA

AWS GuardDuty

AWS GuardDuty is a managed threat detection service that leverages machine learning, anomaly detection, and threat intelligence to identify potential security risks across AWS accounts and workloads. It continuously analyzes AWS CloudTrail logs, VPC Flow Logs, and DNS query logs to detect threats such as unauthorized API calls, compromised credentials, and data exfiltration attempts (Amazon Web Services, 2023).

Azure Security Center

Microsoft's Azure Security Center is a unified cloud security posture management (CSPM) solution that helps organizations protect workloads, detect threats, and enforce security best practices. It uses advanced analytics, machine learning, and behavioral monitoring to detect threats across Azure, AWS, and Google Cloud environments (Microsoft, 2023).

Google Cloud Security Command Center (SCC)

Google Cloud SCC provides real-time security monitoring, threat intelligence, and risk assessment for Google Cloud resources. It integrates with Google Chronicle, an advanced security analytics platform, to enhance threat hunting and investigation capabilities (Google Cloud, 2023).

Strengths and Weaknesses of Cloud-Native vs. Third-Party Security Tools

Comparison of Cloud-Native vs. Third-Party Security Tools

Strengths of Cloud-Native Security Tools

Cloud-native security products are effortlessly included inside their respective cloud environments, guaranteeing native compatibility with current workloads (Sharma & Kumar, 2023). Moreover, these solutions frequently provide adaptable pricing structures, rendering them more economical than third-party security alternatives (Ali & Siddiqui, 2022). Additionally, they provide centralized security management through the cloud provider's dashboard, streamlining security administration and oversight (Microsoft, 2023).

Weaknesses of Cloud-Native Security Tools

Cloud-native security technologies predominantly concentrate on their specific cloud platforms, hence complicating multi-cloud security management (Patel et al., 2023). Moreover, these solutions may be deficient in advanced analytics, configurable rules, and extended detection and response (XDR) functionalities, which are frequently found in third-party security systems (Mohan et al., 2022).

Strengths of Third-Party Security Tools

Third-party security solutions offer multi-cloud compatibility, guaranteeing cross-cloud security visibility and centralized threat monitoring (Patel et al., 2023). They provide sophisticated threat detection via improved threat intelligence, AI-powered analytics, and tailored security setups (Ali & Siddiqui, 2022). Moreover, these solutions enhance flexibility by mitigating vendor lock-in risks and facilitating security management across AWS, Azure, and Google Cloud (Sharma & Kumar, 2023).

Weaknesses of Third-Party Security Tools

Implementing third-party security technologies frequently necessitates supplementary setups, hence augmenting operational complexity and requiring specialist skills (Patel et al., 2023). The license and maintenance costs for these tools might be considerably greater than those of cloud-native security solutions, rendering them a more expensive alternative (Mohan et al., 2022).

Cloud-native threat detection systems, such AWS GuardDuty, Azure Security Center, and Google Cloud Security Command Center, are crucial for safeguarding cloud workloads. Cloud-native technologies facilitate smooth integration and cost reduction, whilst third-party security solutions offer enhanced flexibility, cross-cloud visibility, and sophisticated threat analytics. Organizations must meticulously assess their security requirements, cloud infrastructure, and compliance obligations when selecting between cloud-native and third-party security solutions.

Table 6 presents a comparative analysis of the features of cloud-native tools (e.g., AWS GuardDuty, Azure Security Center, Google Cloud Security Command Center) in relation to third-party solutions. This matrix encompasses critical elements like as threat detection, compliance assistance, integration with cloud services, and cost-effectiveness.

Table 6. Comparison of cloud-native vs third-party security tools

Feature	AWS GuardDuty	Azure Security Center	Google Cloud Security Command Center	Third-Party Security Tools (e.g., CrowdStrike, Palo Alto Cortex XDR)
Threat Detection	AI-powered anomaly detection, continuous monitoring of AWS resources	Security alerts and recommendations for Azure services	Real-time threat detection with AI/ML, centralized alerts	Advanced threat detection, AI/ML-driven insights, broad cross-cloud coverage
Integration with Cloud Services	Seamless integration with AWS services (S3, EC2, Lambda, etc.)	Full integration with Azure services (VMs, databases, etc.)	Integration with Google Cloud resources, Kubernetes, etc.	Broad integration with multiple cloud providers (AWS, Azure, GCP, etc.)
Compliance Support	Built-in compliance checks for AWS (e.g., PCI DSS, HIPAA)	Compliance management for Azure (e.g., ISO 27001, SOC)	Compliance and auditing for Google Cloud services	Extensive support for various compliance frameworks (e.g., GDPR, SOC 2, PCI DSS)
Automation & Response	Automated responses to detected threats (via CloudWatch, Lambda)	Automated security policies and remediation actions	Automated actions to mitigate threats and respond to incidents	Comprehensive SOAR capabilities for automated incident response across environments
Data Privacy & Encryption	In-depth visibility into encryption and data privacy across AWS	Encryption policies, security posture, and identity protection	Centralized monitoring of data privacy settings and encryption	Strong encryption management and data privacy features across cloud platforms
Cost Efficiency	Cost-effective for AWS users, pay-as-you-go model	Free tier for basic services, pricing based on usage	Usage-based pricing, cost-effective for Google Cloud environments	Varies widely, often higher cost but more extensive cross-platform coverage
Customization & Flexibility	Limited to AWS ecosystem, flexible with AWS-specific features	Highly customizable within Azure services	Primarily tailored for Google Cloud, customizable alerts	Highly customizable across multiple cloud platforms, with integration into broader enterprise environments
Real-Time Monitoring & Alerts	24/7 monitoring and alerting for security threats	Continuous security monitoring and alerting for Azure environments	Real-time alerting and centralized view for Google Cloud	Comprehensive monitoring and alerting across all platforms with centralized dashboards

IMPLEMENTATION STRATEGIES AND FRAMEWORKS

Strong threat detection systems in multi-cloud settings call for a thorough strategy including architectural design, integration best practices, automation frameworks, hybrid tool use, and ongoing monitoring. This part explores these important elements and offers thorough advice for using efficient threat detection solutions in multi-cloud environments.

Threat Detection Architecture in Multi-Cloud

Creating a good threat detection architecture in multi-cloud settings calls for a unified security framework guaranteeing uniform monitoring and reaction across several cloud platforms. Important factors are:

- Centralized logging systems aggregating security logs from all cloud platforms into a single repository are crucial. This strategy improves the capacity to identify complex attacks by means of thorough analysis and correlation of security events. For example, Torkura et al. (2020) suggested CSBAuditor, a unique cloud security tool that continuously analyzes cloud infrastructure to find unauthorized changes and harmful behaviors (Torkura et al., 2020)
- Interoperable Security Policies: It is absolutely vital to create security policies that uniform and enforced across all cloud environments. Using policy-as-code techniques guarantees consistent application of security settings, hence lowering the possibility of misconfigurations. Emphasizing the need of consistent security policies, Reece et al. (2023) underlined the significance of a holistic security and vulnerability assessment approach relevant to multi-cloud environments.
- Secure Inter-Cloud Communication: Establishing secure communication channels between different cloud platforms using encryption protocols and virtual private networks (VPNs) safeguards data integrity and confidentiality during inter-cloud data transfers. Chauke et al. (2025) discussed the critical need to enhance security in multi-cloud environments by applying adaptive threat detection techniques powered by machine learning and software-defined networks

Figure 5. Unified threat detection architecture with SIEM integration

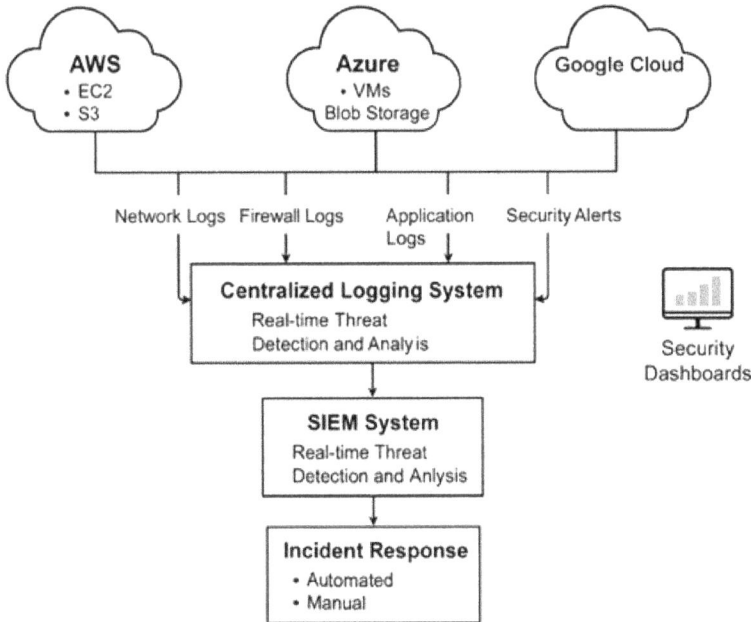

Figure 5 above illustrates the unified threat detection architecture, where security logs and events from AWS, Azure, and Google Cloud are aggregated into a single Security Information and Event Management (SIEM) system. This architecture enables real-time analysis, threat correlation, and incident response, ensuring a robust defense against evolving cyber threats across multiple cloud platforms.

CROSS-CLOUD INTEGRATION BEST PRACTICES

Effective cross-cloud integration is vital for maintaining a cohesive security posture across multiple cloud platforms. Best practices include:

- **Standardized Identity and Access Management (IAM):** Implementing a centralized IAM system that spans all cloud environments ensures consistent user authentication and authorization policies. The concept of Identity Threat Detection and Response (ITDR) enhances IAM by adding detection and response capabilities, providing visibility into potential credential misuse and abuse of privileges (Proofpoint, n.d.).
- **API Security Measures:** Securing APIs used for inter-cloud interactions by employing authentication tokens, rate limiting, and regular security assess-

159

ments prevents unauthorized access and abuse. Reece et al. (2023) emphasized the importance of addressing API vulnerabilities in their systemic risk and vulnerability analysis of multi-cloud environment

- **Data Consistency and Synchronization:** Utilizing data replication and synchronization techniques maintains data consistency across cloud platforms, ensuring data integrity and availability. Paul (2024) explored security challenges in multi-cloud environments and proposed comprehensive mitigation strategies to address these challenges, highlighting the importance of data consistency

Frameworks for Orchestration and Automation (e.g., SOAR in Multi-Cloud)

Security Orchestration, Automation, and Response (SOAR) frameworks play a pivotal role in automating threat detection and response processes in multi-cloud environments. Key strategies include:

- Creating automated playbooks that specify uniform response actions for different security issues helps to provide fast and consistent reactions across all cloud platforms. Olaoye (2022) suggested a low-cost, cloud-native SOAR solution meant to satisfy operational requirements, showing notable cost reductions while preserving scalability.
- Integration with Existing Tools: Ensuring that the SOAR platform integrates seamlessly with existing security tools and cloud services facilitates comprehensive threat detection and response. Kummarapurugu (2024) presents an architectural framework for integrating threat intelligence with SIEM and SOAR in hybrid cloud security environments, highlighting the importance of seamless integration
- Continuous Improvement: Regularly updating and refining automation scripts and playbooks based on emerging threats and lessons learned from past incidents enhances the effectiveness of the SOAR framework. Olaoye (2022) emphasize the need for continuous improvement in their study on integrating security orchestration in cloud environments.

Figure 6 below illustrates the automated incident response process within a SOAR framework. The diagram shows how the integration of cloud services and security tools feeds into the SOAR platform, which performs incident analysis and triage, initiates automated response actions, and concludes with incident resolution. This workflow significantly reduces the response time and enhances the efficiency of security operations

Figure 6. Automated incident response process within a SOAR framework

Automated Incident Response

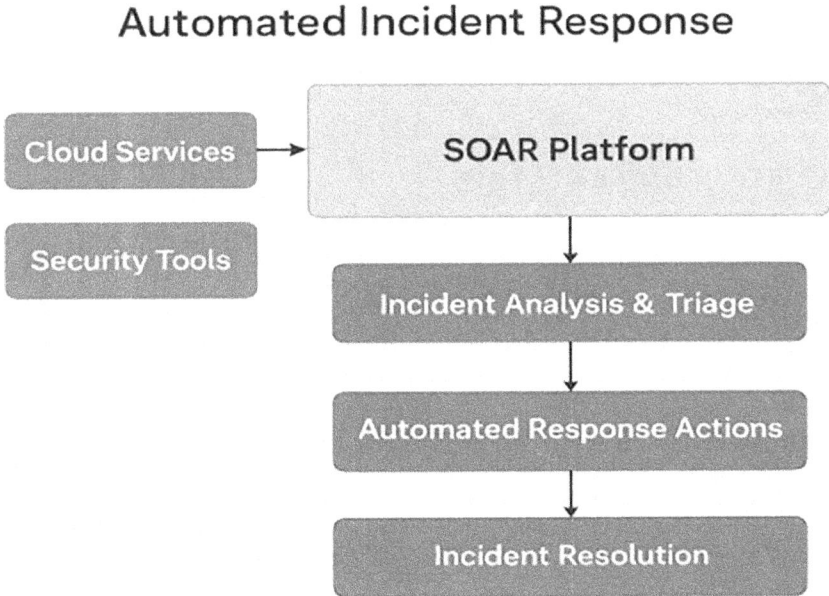

Figure 6. Automated incident response process within a SOAR framework

Hybrid Use of Native and Third-Party Tools

Leveraging both native cloud security tools and third-party solutions can enhance threat detection capabilities in multi-cloud environments. Considerations include:

- **Native Tool Utilization:** Employing security tools provided by cloud service providers for platform-specific threat detection and compliance monitoring is essential. An article on multi-cloud security challenges and best practices emphasizes the importance of utilizing native tools for platform-specific security measures (Sasovets, 2023).
- **Third-Party Tool Integration:** Incorporating third-party security solutions that offer advanced threat detection features and cross-platform compatibility addresses gaps not covered by native tools. Chauke et al. (2025) discussed the integration of adaptive threat detection techniques powered by machine learning and software-defined networks to enhance security in multi-cloud environments (Chauke et al., 2025).
- **Unified Management Console:** A centralized security management console gives visibility and control over both native and third-party security products,

hence enabling unified threat management. Torkura et al. (2020) underlined the efficiency of a unified management strategy in simplifying security operations, lowering reaction times, and improving an organization's capacity to identify and minimize complicated threats across hybrid settings.

EVALUATION AND ANALYSIS

Evaluating the effectiveness of threat detection mechanisms in multi-cloud environments is crucial for ensuring that security operations are proactive, accurate, and responsive. This section outlines essential metrics, KPIs, comparative tool assessments, and simulation methodologies to determine performance quality.

Metrics for Evaluating Threat Detection Effectiveness

Effective threat detection requires robust metrics that measure performance across precision, speed, and coverage. According to Bou Nassif, Talib, Nasir, and Dak Albab (2021), detection systems should be evaluated based on how accurately and quickly they identify malicious activity, while minimizing false positives.

- Detection Rate is the percentage of actual threats correctly identified. A higher rate indicates better system sensitivity to malicious behavior.
- False Positive Rate (FPR) is critical to operational efficiency, as an excess of false alarms can overwhelm analysts and obscure real threats (Kim et al., 2019).
- Time to Detect (TTD) and Time to Respond (TTR) measure how quickly the system detects and mitigates incidents. Ahmed, Mahmood, and Hu (2016) emphasized that shorter detection and response times are directly correlated with reduced damage.
- Coverage refers to the breadth of attack types a system can detect, from known malware to zero-day exploits.

Table 7 below compares essential metrics, such as Detection Rate, False Positive Rate (FPR), Time to Detect (TTD), Time to Respond (TTR), and Coverage, along with their definitions, formulas, and optimal ranges for multi-cloud environments. This comparison helps identify the ideal thresholds for effective threat detection while minimizing false alarms and response times.

Table 7. Comparison of key metrics for evaluating threat detection effectiveness in multi-cloud environments

Metric	Definition	Formula	Optimal Range
Detection Rate	Percentage of actual threats correctly identified by the system.	Detection Rate = (True Positives / (True Positives + False Negatives)) × 100	> 90% (higher is better)
False Positive Rate (FPR)	Percentage of non-malicious activities incorrectly identified as threats.	FPR = (False Positives / (False Positives + True Negatives)) × 100	< 5% (lower is better)
Time to Detect (TTD)	Time taken by the system to identify a threat from the moment it occurs.	Measured in seconds or minutes from detection onset.	< 5 minutes (faster detection is optimal)
Time to Respond (TTR)	Time taken from detecting the threat to mitigating the issue (response time).	Measured in seconds or minutes from detection to remediation action.	< 30 minutes (quick response is crucial)
Coverage	Breadth of attack types detected, including known and unknown threats.	Assessed by the number of attack types detected versus total attack types.	95%+ (higher coverage ensures better protection)

Benchmarks and Key Performance Indicators (KPIS)

Key Performance Indicators (KPIs) offer a standardized framework to track and benchmark threat detection performance.

Dasgupta, Akhtar, and Sen (2022) recommend using:

- **Mean Time to Detect (MTTD):** A lower MTTD suggests faster recognition of threats and is essential for damage control.
- **Mean Time to Respond (MTTR):** Reflects how efficiently a threat is neutralized after detection.
- **Precision and Recall:** Particularly for AI-driven systems, these indicators reflect how accurate the model is at identifying threats and avoiding false alarms.
- **Threat Detection Coverage:** An essential metric in multi-cloud deployments where varied services and configurations exist (Bou Nassif et al., 2021).

Monitoring these KPIs allows cloud security teams to identify gaps and continuously improve detection mechanisms (Ahmed et al., 2016).

COMPARATIVE ANALYSIS OF TOOL PERFORMANCE ACROSS CLOUDS

Each major cloud provider, AWS, Microsoft Azure, and Google Cloud—offers native threat detection tools that differ in architecture, integration, and analytical capabilities.

Kim et al. (2019) observed that Amazon GuardDuty emphasizes network flow anomaly detection, while Microsoft Defender offers superior compliance policy enforcement due to its Active Directory integration. Google Security Command Center integrates more deeply with GCP services and excels in threat aggregation.

Bou Nassif et al. (2021) suggest that multi-cloud environments benefit from hybrid detection frameworks that combine the strengths of each native tool, supported by a unifying SIEM or SOAR layer.

Simulation or Emulation-Based Testing (Using Labs or Testbeds)

Simulated environments allow organizations to safely test threat detection mechanisms under controlled attack scenarios. Red team–blue team exercises, emulated environments, and dataset-based machine learning simulations are all useful methods.

Kim et al. (2019) propose the use of labeled datasets like UNSW-NB15 or CICIDS for testing intrusion detection models before deployment. Madasu (2023) emphasizes the role of synthetic testing in evaluating access control and incident response systems under stress conditions.

Machine learning-based simulations further allow for testing against unknown threats, making them valuable in adaptive security settings (Dasgupta et al., 2022).

Figure 7 below illustrates the cycle of simulation-based testing, starting from setting up the test environment, followed by attack emulation, threat detection, incident logging, and analysis. This cycle ensures that the system is tested against both known and unknown threats, providing critical insights for improvement.

Figure 7. Simulation-based testing cycle for threat detection

Simulation or Emulation-Based Testing

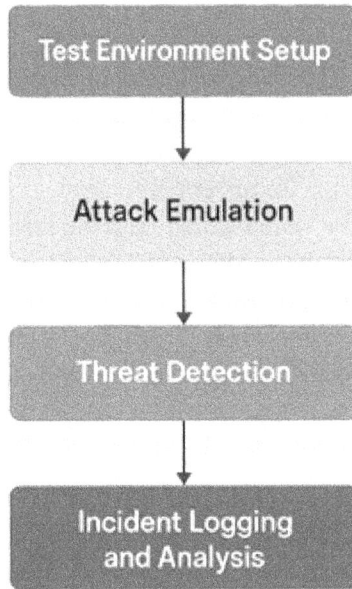

```
┌─────────────────────────────┐
│   Test Environment Setup    │
└─────────────────────────────┘
              │
              ▼
┌─────────────────────────────┐
│      Attack Emulation       │
└─────────────────────────────┘
              │
              ▼
┌─────────────────────────────┐
│      Threat Detection       │
└─────────────────────────────┘
              │
              ▼
┌─────────────────────────────┐
│     Incident Logging        │
│      and Analysis           │
└─────────────────────────────┘
```

ETHICAL, LEGAL, AND COMPLIANCE CONSIDERATIONS

Data moves across several jurisdictions in multi-cloud settings, each with different legal and ethical criteria. Organizations must grasp and handle these issues if they are to remain compliant and preserve ethical values.

Data Sovereignty Challenges

Data sovereignty denotes the notion that digital information is governed by the legal framework of the nation in which it is stored or processed. Multi-cloud configurations provide issues because data is located in various jurisdictions, each with own restrictions.

Key Challenges:
- **Jurisdictional Conflicts:** Data stored in one country may be subject to that nation's laws, potentially conflicting with the data owner's local regulations (Hummel et al., 2021).
- **Data Localization Requirements:** Some countries mandate that specific data types be stored within their borders, complicating global data management strategies (Mathew, 2024).
- **Compliance Management:** Ensuring adherence to diverse regional regulations requires robust governance frameworks and continuous monitoring (Mathew, 2024).

Figure 8 below illustrates the global landscape of data sovereignty regulations, highlighting regions with strict data localization requirements (green), countries with existing data sovereignty laws (yellow), and regions where data sovereignty laws are not well-defined (gray). This map provides a visual representation of the challenges organizations face in ensuring compliance with data regulations across multiple jurisdictions.

Figure 8. World map highlighting data sovereignty laws and regions with strict data localization requirements.

Data Sovereignty Laws

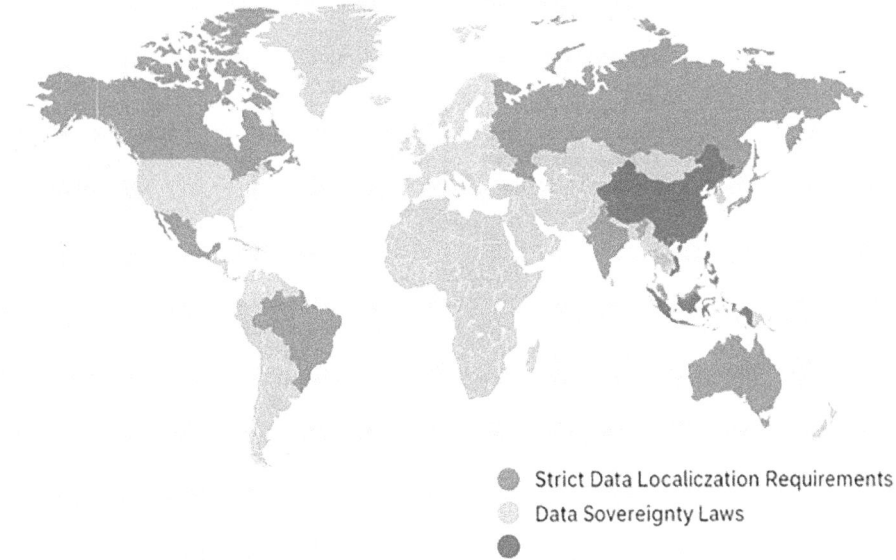

- Strict Data Localiczation Requirements
- Data Sovereignty Laws

GDPR, HIPAA, and Other Regulatory Impacts

Global enterprises must traverse an intricate network of legislation aimed at safeguarding data privacy and security. Two notable instances are the General Data Protection Regulation (GDPR) in the European Union and the Health Insurance Portability and Accountability Act (HIPAA) in the United States. The GDPR mandates rigorous regulations for data collection, processing, and transfer, with substantial fines for non-compliance. It impacts any entity processing the personal data of EU individuals, irrespective of the entity's geographical location. HIPAA establishes rules for safeguarding sensitive patient information within the healthcare industry, mandating measures to assure data confidentiality, integrity, and availability. Table 8 offers a comparative analysis of GDPR, HIPAA, and other significant data protection rules. This table delineates the principal provisions of each rule and its ramifications for cloud computing operations, emphasizing their influence on organizational strategies for compliance and data privacy safeguarding.

Table 8. Comparison of data protection regulations and their impact on cloud security practices

Regulation	Key Provisions	Impact on Multi-Cloud Environments	Cloud Compliance Strategies
GDPR (General Data Protection Regulation)	- Requires data protection by design and by default - Data subject consent and transparency - Right to access, rectification, and erasure - Strict data breach notification requirements	- Complicated by data transfers across multiple cloud providers and regions - Challenges in ensuring compliance when using third-party cloud services	- Implement data encryption at rest and in transit - Use cloud providers offering GDPR-compliant services - Establish strong access control policies and consent management processes
HIPAA (Health Insurance Portability and Accountability Act)	- Ensures the confidentiality, integrity, and availability of electronic protected health information (ePHI) - Requires business associate agreements (BAAs) with cloud providers handling ePHI - Mandates audit controls, access controls, and secure transmission of health data	- Cloud services must meet specific standards to store and process ePHI - Risks related to outsourcing ePHI processing and storage without a BAA in place	- Ensure cloud providers are HIPAA-compliant and sign a BAA - Implement encryption and strong authentication for ePHI - Conduct regular risk assessments and audits
CCPA (California Consumer Privacy Act)	- Provides California residents with the right to access, delete, and opt-out of data sales - Focuses on the right to be forgotten and data portability - Requires businesses to disclose data collection practices and share data requests	- Multi-cloud deployments increase the complexity of tracking data subject rights across regions - Data ownership and control must be clearly defined across platforms	- Implement a clear data governance framework for CCPA compliance - Use cloud tools that support rights management and data subject requests - Maintain detailed records of data processing activities

continued on following page

Table 8. Continued

Regulation	Key Provisions	Impact on Multi-Cloud Environments	Cloud Compliance Strategies
LGPD (Lei Geral de Proteção de Dados – Brazil)	- Similar to GDPR with requirements for data consent, processing transparency, and data protection rights - Mandates data breach reporting within 72 hours - Companies must appoint a Data Protection Officer (DPO)	- Multi-cloud environments may struggle with cross-border data transfers - Challenges in coordinating compliance for organizations operating in multiple jurisdictions	- Ensure compliance with both local and international regulations - Work with cloud providers that comply with LGPD data handling practices - Establish clear roles for DPO and cross-border data transfer agreements
PIPEDA (Personal Information Protection and Electronic Documents Act – Canada)	- Requires organizations to obtain consent for the collection, use, and disclosure of personal information - Individuals have the right to access and correct their personal information - Businesses must safeguard personal information and notify of breaches	- Multi-cloud setups can complicate compliance with cross-border data flow regulations - Special considerations for cloud storage of personal data	- Cloud providers should have robust data protection measures - Create mechanisms for tracking consent and handling breach notifications - Perform regular privacy audits

Implications for Multi-Cloud Environments

- **Data Transfer Restrictions:** Regulations may limit the transfer of data across borders, necessitating data localization or specific contractual agreements (Mathew, 2024).
- **Vendor Compliance:** Organizations must ensure that all cloud service providers comply with relevant regulations, requiring thorough due diligence and continuous oversight (ISACA, 2024).
- **Incident Reporting:** Timely breach notification requirements demand robust monitoring and reporting mechanisms across all cloud platforms.

Ethical Implications of AI in Threat Detection

Artificial Intelligence (AI) has become essential to contemporary threat detection systems, providing improved capabilities for spotting and responding to security problems. Nonetheless, its implementation presents numerous ethical dilemmas.

Concerns:

- Bias and discrimination: Artificial intelligence systems have the potential to unintentionally strengthen biases that are already present in training data, which can result in unjust treatment of particular populations (Cowls et al., 2023).
- Transparency and Accountability: The "black box" nature of some AI models makes it challenging to understand their decision-making processes, complicating accountability (Taddeo & Floridi, 2018).
- Misuse of AI Capabilities: There is a risk that AI tools could be exploited for malicious purposes, such as developing sophisticated cyber-attacks (Gupta et al., 2023).

Managing Privacy vs. Security Trade-Offs

Balancing privacy and security is a perennial challenge in multi-cloud environments. Enhanced security measures can sometimes infringe on individual privacy, necessitating careful consideration.

Strategies for Balance:

- **Data Minimization:** Collect only necessary data to reduce privacy risks while maintaining security effectiveness.
- **Anonymization Techniques:** Employ data anonymization to protect individual identities without compromising security analysis (Taddeo & Floridi, 2018).
- **Transparent Policies:** Maintain open policies on data use to foster confidence and guarantee adherence to privacy regulations.

DISCUSSION AND ANALYSIS

This chapter offers a critical examination of the data obtained from the literature research and case study analysis, focusing on the assessment of contemporary developments in threat detection within multi-cloud setups. It examines the intricacies and security problems associated with multi-cloud methods, while highlighting deficiencies in current threat detection methodologies. The discourse emphasizes novel tactics and cutting-edge technologies that are influencing the future of multi-cloud security. As organizations progressively embrace multi-cloud environments for their adaptability and scalability, it is essential to comprehend how these infrastructures present new vulnerabilities and how emerging technologies, such as AI-driven analytics and Zero Trust frameworks, can assist in mitigating these risks.

This chapter seeks to integrate these insights, provide pathways for advancement, and provide recommendations for enhancing threat detection systems in multi-cloud environments.

Key Trends in Multi-Cloud Threat Detection

The study shows a notable change in the way companies handle threat detection in multi-cloud environments. Among the main industry trends are:

- **Enhanced Utilization of AI/ML-Driven Threat Detection for Immediate Anomaly Recognition and Behavioral Assessment:** Artificial Intelligence (AI) and Machine Learning (ML) are integral to contemporary threat detection. These technologies provide real-time analysis of extensive data streams from many cloud providers, detecting irregularities that may signify cyber dangers. Organizations can utilize AI/ML algorithms to autonomously identify patterns that diverge from typical behavior, including atypical user activity or irregular access to cloud resources. This method enhances both the velocity and precision of threat detection while facilitating dynamic, adaptive responses to new threats, so assisting enterprises in outpacing sophisticated attackers.

- **Adoption of Zero Trust Architecture to Minimize Trust Boundaries in Distributed Environments:** The Zero Trust security approach, founded on the premise of "never trust, always verify," is increasingly used in multi-cloud setups. This methodology reduces the inherent confidence afforded to users and devices within the network, necessitating ongoing authentication and authorization for each access request, irrespective of location. In multi-cloud environments, where data and applications are distributed across many cloud platforms, Zero Trust mitigates security vulnerabilities arising from disparate access regulations. It guarantees that every user and device within the cloud undergoes stringent validation prior to accessing sensitive resources, hence minimizing the possible attack surface..

- **Implementation of Extended Detection and Response (XDR) for Unified Threat Visibility Across AWS, Azure, and GCP:** Extended Detection and Response (XDR) is emerging as a holistic solution for threat detection and response inside multi-cloud infrastructures. XDR consolidates security data from many cloud platforms, including AWS, Azure, and Google Cloud Platform (GCP), into a consolidated framework, offering a cohesive perspective of the security environment. This consolidated view allows security teams to detect, evaluate, and respond to threats across cloud environments with greater efficiency. XDR improves situational awareness by integrating data from several sources, facilitating the rapid identification of multi-cloud assaults and minimizing response times.

- **Automating Security Processes Through Cloud Security Posture Management (CSPM) to Mitigate Configuration Errors:** As cloud environments grow increasingly intricate, maintaining uniform security configurations across many cloud platforms poses significant challenges. Cloud Security Posture Management (CSPM) automates security configuration and compliance monitoring processes. CSPM technologies consistently examine cloud settings for misconfigurations, like excessively permissive access rules or publicly accessible storage buckets, and notify security teams of potential threats. Through the automation of these operations, CSPM diminishes the probability of human error and guarantees the uniform application of security regulations throughout the entire multi-cloud infrastructure.
- **Increasing Significance of Threat Intelligence and Dark Web Surveillance for Proactive Detection of Emerging Attacks:** Threat intelligence and dark web surveillance are becoming essential for proactive threat identification. By persistently collecting and scrutinizing data from various threat intelligence sources, including global security feeds and dark web forums, companies can obtain early insights into new threats, attack strategies, and potential vulnerabilities. Surveillance of the dark web for pilfered passwords, compromised information, and dialogues regarding cloud platform vulnerabilities allows security teams to recognize and alleviate dangers prior to their escalation into significant attacks. This proactive strategy not only bolsters an organization's capacity to counter threats but also elevates its overall security posture by anticipating the dynamic cyber threat landscape.

These trends collectively reflect a shift in the cybersecurity industry from traditional, reactive measures to more proactive, automated, and integrated strategies. As organizations continue to embrace multi-cloud environments, the adoption of these advanced security measures is essential for maintaining robust protection against emerging and increasingly sophisticated cyber threats.

GAPS AND CHALLENGES IDENTIFIED

Despite the availability of advanced security tools, several critical challenges persist: Table 9 offers some challenges and their Implications

Table 9. Persisting challenges and their implications

Challenge	Implication
Lack of Unified Cross-Cloud Threat Detection Frameworks	Organizations struggle to integrate telemetry from multiple providers.
Complexity of Multi-Cloud IAM	Managing identities and access across different platforms remains difficult.
Limited Threat Correlation	Tools may fail to detect multi-stage attacks spanning different cloud environments.
Compliance and Data Sovereignty Issues	Adherence to regulatory standards is complicated across global cloud regions.
Insider Threats	Behavioral detection remains a weak point for many organizations.

These gaps suggest the need for improved interoperability between cloud providers and more advanced security orchestration tools.

PRACTICAL IMPLICATIONS FOR ORGANIZATIONS

Organizations operating in multi-cloud environments must take several strategic actions to ensure robust security across their distributed infrastructures. These actions are essential for building resilient, scalable, and proactive security frameworks capable of addressing the unique challenges of multi-cloud architectures. Key strategies include:

- **Establish Vendor-Neutral Security Frameworks to Ensure Consistency Across Platforms:** Given that multi-cloud settings typically encompass many cloud service providers (e.g., AWS, Azure, GCP), each possessing distinct tools and security procedures, the formulation of vendor-neutral security frameworks is essential. These frameworks ought to standardize security procedures, controls, and policies applicable uniformly across all platforms, irrespective of the cloud provider. By implementing standardized security baselines and governance frameworks, organizations can guarantee that security measures are not reliant on any singular provider, hence minimizing complexity and fostering a unified security stance across their multi-cloud environment.
- **Invest in XDR and SIEM Solutions Featuring Multi-Cloud Support:** Extended Detection and Response (XDR) and Security Information and Event Management (SIEM) solutions with multi-cloud capabilities are essential for improving visibility and monitoring across cloud environments. These solutions consolidate security data from all cloud platforms, offering

centralized analytics that empower security teams to identify, correlate, and address risks in real-time. Utilizing XDR and SIEM systems designed for multi-cloud environments enables enterprises to combine and analyze security events from diverse cloud services, enhancing their capacity to detect coordinated assaults and mitigate risks prior to escalation.

- **Implement IAM Governance Frameworks Incorporating AI-Driven Analytics:** Governance of Identity and Access Management (IAM) is essential for multi-cloud security. Implementing IAM models that use AI-driven analytics can improve access control and mitigate the risk of unwanted access across platforms. Organizations can utilize AI and machine learning to evaluate user behavior and access patterns, enabling the preemptive detection of anomalous behaviors and the enforcement of dynamic access controls in real time. This AI-driven methodology for Identity and Access Management (IAM) guarantees that users and devices receive just the minimal privileges necessary for their functions, thereby enhancing overall security and reducing human error and misconfigurations.

- **Automate Regulatory Oversight Employing Cloud Security Posture Management and Policy-as-Code Methodologies:** Maintaining ongoing compliance across many cloud environments can be intricate and labor-intensive. Automating compliance monitoring with Cloud Security Posture Management (CSPM) and Policy-as-Code methodologies enables enterprises to enhance the efficiency of identifying and rectifying security misconfigurations. CSPM tools autonomously examine multi-cloud infrastructures for policy infractions and non-compliance with industry standards, whereas Policy-as-Code facilitates the direct implementation of security policies within the cloud architecture. These automated solutions diminish the human labor required for compliance management, enhance precision, and guarantee that firms can perpetually uphold security and regulatory standards.

- **Implement Ongoing Threat Intelligence Exchange with Industry Collaborators:** Efficient threat intelligence dissemination is crucial for preempting emerging attacks within a multi-cloud environment. Through the constant exchange of threat intelligence with industry partners, enterprises can obtain early insights into cyber threats and vulnerabilities that may impact their multi-cloud architecture. Collaboration with other organizations, governmental bodies, and cybersecurity corporations facilitates the exchange of current information regarding attack methodologies, vulnerabilities, and mitigation tactics. This proactive strategy enhances an organization's capacity to identify emerging threats promptly, react swiftly, and establish more resilient defenses throughout their cloud environments.

These strategic actions, when implemented collectively, are critical to building resilient and scalable multi-cloud security infrastructures. By developing standardized, automated, and adaptive security measures, organizations can ensure that they are well-equipped to handle the evolving security landscape in multi-cloud environments while minimizing vulnerabilities and operational risks.

CONCLUSION AND FUTURE WORK

Conclusion

This research has provided a comprehensive exploration into the evolving landscape of threat detection in multi-cloud environments. The study examined the increasing complexity of cloud computing architectures, driven by the widespread adoption of multi-cloud strategies across diverse industries. While the multi-cloud model offers significant operational benefits, including flexibility, scalability, and avoidance of vendor lock-in, it also introduces an expanded attack surface and new cybersecurity challenges that organizations must address.

The research critically analyzed various threat detection techniques, tools, and frameworks utilized across leading cloud platforms such as AWS, Microsoft Azure, and Google Cloud. Traditional security mechanisms - including firewalls and intrusion detection systems — were found to have significant limitations when applied within distributed, dynamic cloud infrastructures. In response, organizations have increasingly adopted state-of-the-art security practices, leveraging artificial intelligence (AI), machine learning (ML), user and entity behavior analytics (UEBA), Zero Trust Architecture (ZTA), and Extended Detection and Response (XDR) solutions.

The study also assessed critical challenges in multi-cloud security, such as the lack of unified threat detection frameworks, fragmented security policies, cloud misconfigurations, identity and access management (IAM) complexities, and regulatory compliance issues. Furthermore, real-world case studies, including the SolarWinds attack and the Uber data breach, underscored the urgency of adopting proactive, intelligent, and automated threat detection systems.

Equally important, this research explored ethical, legal, and compliance considerations within multi-cloud security, emphasizing the need for organizations to align their security practices with regulatory standards such as GDPR, HIPAA, and data sovereignty laws to ensure the protection of sensitive information, maintain user privacy, and uphold legal accountability across different jurisdictions.

The study further highlighted that ethical deployment of AI-driven threat detection systems must ensure transparency, fairness, and user privacy while preventing unintended biases or misuse of sensitive data.

Ultimately, the research confirms that securing multi-cloud environments requires more than isolated technical controls, it necessitates an integrated, holistic approach that combines technical innovation, operational governance, regulatory compliance, and ethical best practices. Organizations that adopt a proactive, adaptive, and intelligence-driven security strategy will be best positioned to mitigate evolving cyber threats in multi-cloud ecosystems.

Research Contributions

This study contributes to the growing body of cybersecurity literature and provides practical insights for industry practitioners. The key contributions of this research include:

- Providing a comprehensive analysis of threat detection techniques tailored to multi-cloud environments.
- Proposing a vendor-neutral *Multi-Cloud Threat Detection Architecture* illustrating the integration of security tools across AWS, Azure, and Google Cloud.
- Identifying critical challenges and limitations in existing security tools when applied to distributed cloud infrastructures.
- Offering strategic recommendations for organizations on implementing AI-driven security systems, Zero Trust Architecture, and centralized threat monitoring platforms.
- Addressing ethical, legal, and compliance considerations that inform responsible security practices within multi-cloud settings.

Immediate Application for Organizations

For organizations currently operating or transitioning to multi-cloud environments, there are several immediate steps that can be taken to apply the findings of this research:

- **Adopt AI and Machine Learning-Based Threat Detection Tools:** Organizations should prioritize the integration of AI-powered threat detection tools to enhance their ability to detect anomalies and respond in real-time. AI and machine learning systems are particularly effective in multi-cloud environments, where diverse data streams need to be analyzed continuously.
- **Implement Zero Trust Architecture (ZTA):** Adopting a Zero Trust security model should be an immediate priority for any organization aiming to fortify its cloud security posture. Zero Trust focuses on the continuous validation of

all users, devices, and applications, regardless of whether they are inside or outside the organization's perimeter. This architecture helps mitigate risks arising from identity theft and privilege escalation, which are common in multi-cloud environments.

- **Prioritize Security Posture Management Tools (CSPM):** With multi-cloud environments increasing the likelihood of misconfigurations, Cloud Security Posture Management (CSPM) tools must be integrated into the security strategy. These tools help continuously monitor cloud resources for misconfigurations that could lead to vulnerabilities. Automated remediation can be implemented to reduce human errors in policy management, making CSPM a crucial part of proactive security.
- **Enhance Incident Response Capabilities with XDR:** Extended Detection and Response (XDR) solutions provide centralized visibility and automated response capabilities across multiple cloud platforms. By implementing XDR, organizations can not only detect complex threats but also take swift action across the entire cloud environment. This system correlates security data from cloud resources, endpoint devices, and network traffic, enabling a coordinated defense against cyber threat

Future Work

While this research provides significant insights, several opportunities for future work exist to build upon its findings:

- Cross-Cloud Threat Detection Frameworks: One of the primary challenges in multi-cloud environments is the integration and correlation of security data across different cloud platforms. Future research should focus on developing cross-cloud threat detection frameworks that standardize security telemetry across platforms like AWS, Azure, and GCP. This would enable security teams to get a unified view of security events and improve response times.
- Exploration of *Cross-Cloud Forensic Investigation* techniques that facilitate post-incident analysis, evidence preservation, and legal compliance in multi-cloud breaches.
- Quantum-Resistant Cryptography for Cloud Security: As quantum computing advances, traditional cryptographic methods may become obsolete. Research should explore the development of quantum-resistant cryptographic techniques to safeguard multi-cloud environments against potential future threats posed by quantum computing capabilities.

- Investigation of *Blockchain-based Security Solutions* for decentralized identity management, secure logging, and trustless security event verification in multi-cloud ecosystems.

- Integration of Insider Threat Detection and Behavioral Analytics: Insider threats remain one of the most challenging risks in multi-cloud environments. Future research could explore integrating User and Entity Behavior Analytics (UEBA) with AI and Zero Trust architectures to enhance insider threat detection. By leveraging behavioral analytics, organizations can identify abnormal user activities, potentially preventing insider attacks before they escalate.

- Explainable AI (XAI) for Threat Detection: The adoption of Explainable AI (XAI) can significantly improve the transparency of machine learning models used in threat detection. By focusing on creating interpretable AI systems, organizations can gain better insights into why certain threats are flagged and how mitigation decisions are made. This will foster trust in AI-driven security tools, especially in critical infrastructures.

- Establishment of *Global Cyber Threat Intelligence Sharing Standards* and collaborative platforms for real-time threat detection across different organizations, cloud providers, and jurisdictions.

- Compliance-Driven Security Models for Multi-Cloud: As data protection regulations continue to evolve globally, future work should explore compliance-driven security models that integrate regulatory requirements directly into security systems. This approach will help organizations automate compliance management and ensure they meet legal obligations across various jurisdictions and cloud platforms.

This research serves as a foundational guide for security professionals, cloud architects, and researchers dedicated to enhancing security posture within multi-cloud environments. The findings not only address current cybersecurity challenges but also provide a roadmap for developing more resilient, intelligent, and ethically grounded security solutions in an increasingly complex cloud-driven world.

REFERENCES

Adahman, Z., Malik, A. W., & Anwar, Z. (2022). An analysis of zero-trust architecture and its cost-effectiveness for organizational security. *Computers & Security*, *122*, 102911. DOI: 10.1016/j.cose.2022.102911

Agufenwa, O. J. (2023, December 20). *The crucial role of threat intelligence feeds integration in cloud security*. ResearchGate. https://doi.org/DOI: 10.13140/RG.2.2.19905.33123

Ahmadi, S. (2024). Zero trust architecture in cloud networks: Application, challenges and future opportunities. *Journal of Engineering Research and Reports*, *26*(2), 215–228. DOI: 10.9734/jerr/2024/v26i21083

Ahmed, M., Mahmood, A. N., & Hu, J. (2016). A survey of network anomaly detection t echniques. *Journal of Network and Computer Applications, 60*, 19–31. h t tps://DOI: 10.1016/j.jnca.2015.11.016

AIMultiple. (2025). *Top 15 UEBA Use Cases for Today's SOCs in 2025.* h t tps:// research.aimultiple.com/ueba-use-cases

Ali, M., & Siddiqui, N. (2022). A comparative study of cloud-native and third-party security tools for multi-cloud environments. *Journal of Cybersecurity Research*, *14*(3), 45–61.

Alotaibi, F. G., Clarke, N., & Furnell, S. M. (2021). A novel approach for improving information security management and awareness for home environments. *Information and Computer Security*, *29*(1), 25–48. DOI: 10.1108/ICS-05-2020-0073

Alotaibi, M., et al. (2023). *Enhancing Threat Detection in Multi-Cloud Environments.* C ybersecurity Journal, 7(2), 35-51.

Amazon Web Services. (2023). *AWS GuardDuty: Threat detection and continuous monitoring.* https://aws.amazon.com/guardduty/

AppOmni. (n.d.). Cloud Security Posture Management (CSPM) diagram. A p p Omni. https://appomni.com/saas-glossary/cloud-security-posture-management-cspm/

Asthana, K. (2024, November 26). *Top 8 cloud vulnerabilities*. CrowdStrike. h t tps://www.crowdstrike.com/en-us/cybersecurity-101/cloud-security/cloud- v u lnerabilities/

Asthana, N. (2024). Multi-cloud security challenges and best practices. *Cloud Computing Journal, 15*(2), 45–58.

Asthana, P. (2024). Security Risks and Threat Detection in Multi-Cloud Environments. *Journal of Cloud Security*, *12*(4), 87–102.

Babu, S., & Irudhayaraj, R. (2019, March). User-Entity Behavior Analytics (UEBA) – A S ystematic Review of Literatures. In *9th Annual International Conference on Industrial Engineering and Operations Management*, https://doi.org/DOI: 10.46254/AN09.20190828

Bace, R. G., & Mell, P. (2001). *Intrusion detection systems*. National Institute of Standards and Technology. DOI: 10.6028/NIST.SP.800-31

Banse, C., Kunz, I., Schneider, A., & Weiss, K. (2021, September). Cloud property graph: Connecting cloud security assessments with static code analysis. In *2021 IEEE 14th International Conference on Cloud Computing (CLOUD)* (pp. 13-19). IEEE.

Bartakke, J., & Kashyap, R. (2024). The Usage of Clouds in Zero-Trust Security Strategy: An Evolving Paradigm. *Journal of Information and Organizational Sciences*, *48*(1), 149–165. DOI: 10.31341/jios.48.1.8

Basan, M. (2023, October 26). *What is multi-cloud security? Everything to know*. eSecurity Planet. https://www.esecurityplanet.com/cloud/multi-cloud-security/

Chauke, K. O., Muchenje, T., & Makondo, N. (2025). *Enhancing network security in multi-cloud environments through adaptive threat detection.*

Check Point Software Technologies. (2023). *CloudGuard: Prevention-first cloud security*. https://www.checkpoint.com/cloudguard/

Chhabra, S., & Singh, A. K. (2022). A Comprehensive Vision on Cloud Computing Environment: Emerging Challenges and Future Research Directions. *arXiv preprint*

Chhabra, S., & Singh, M. (2022). A comprehensive survey on multi-cloud computing: Challenges and future directions. *Journal of Network and Computer Applications*, *213*, 103966.

Chimakurthi, V. N. S. S. (2020). The challenge of achieving zero trust remote access in multi- cloud environment. *ABC Journal of Advanced Research*, *9*(2), 89–102. DOI: 10.18034/abcjar.v9i2.608

CISA. (2021). *Mitigating Supply Chain Risks in Cloud Services*. Retrieved from h t tps://www.cisa.gov/supply-chain-security

Cisco Systems. (2022). *Cisco Secure Cloud Analytics Data Sheet*. Retrieved from h ttps://www.cisco.com/c/dam/en/us/products/collateral/security/stealthwatch/secure-cloud- analytics-ds.pdf

Cloudwithease. (n.d.). *Cloud security comparison: AWS vs Azure vs GCP.* Cloudwithease. https://cloudwithease.com/cloud-security-comparison-aws-vs-azure-vs-gcp/

Cloudwithease. (n.d.). Comparison of security services between GCP, AWS and Azure [Image]. Retrieved February 22, 2025, from [URL omitted]

Cohen, A., & Schindel, A. (2024, January 24). *Introducing the Cloud Threat Landscape, a new TI resource for cloud defenders.* Wiz. https://www.wiz.io/blog/introducing-the-cloud- threat-landscape

Conti, M., Kumar, E. S., Lal, C., & Ruj, S. (2018). A survey on security and privacy issues of bitcoin. *IEEE Communications Surveys and Tutorials*, *20*(4), 3416–3452. DOI: 10.1109/COMST.2018.2842460

Cowls, J., Tsamados, A., Taddeo, M., & Floridi, L. (2023). The AI gambit: Leveraging artificial intelligence to combat climate change—Opportunities, challenges, and recommendations. *AI & Society*, *38*(1), 1–25. DOI: 10.1007/s00146-021-01294-x PMID: 34690449

Creswell, J. W. (2014). *Research design: Qualitative, quantitative, and mixed methods a pproaches* (4th ed.). Sage Publications.

CrowdStrike. (2024). *What is XDR? Extended Detection & Response.* h t tps://www .crowdstrike.com/en-us/cybersecurity-101/endpoint-security/extended- d etection-and-response-xdr/

Dasgupta, D., Akhtar, Z., & Sen, S. (2022). Machine learning in cybersecurity: A comprehensive survey. *The Journal of Defense Modeling and Simulation*, *19*(1), 57–106. DOI: 10.1177/1548512920951275

eSentire. (2024). *What is Extended Detection and Response (XDR)?* h t tps://www .esentire.com/cybersecurity-fundamentals-defined/glossary/what-is-extended - detection-and-response-xdr

Farzaan, M. A. M., Ghanem, M. C., El-Hajjar, A., & Ratnayake, D. N. (2024). Ai-enabled system for efficient and effective cyber incident detection and response in cloud e nvironments. *arXiv preprint arXiv:2404.05602.*

Farzaan, M. A. M., Ghanem, M. C., El-Hajjar, A., & Ratnayake, D. N. (2024). *AI-Enabled System for Efficient and Effective Cyber Incident Detection and Response in Cloud Environments.* arXiv preprint, arXiv:2404.05602.

Farzaan, R.. (2024). AI-Driven Cybersecurity: Enhancing Threat Detection in Cloud Environments. *Information Security Research*, *8*(1), 22–38.

Farzaan, S., Sarkar, B., & Chowdhury, F. (2024). AI-enabled automated cyber incident detection and response in cloud environments. *IEEE Transactions on Cloud Computing, 12*(3), 789–801.

Freitas, S., & Gharib, A. (2024). *GraphWeaver: Billion-Scale Cybersecurity Incident Correlation.* arXiv preprint, arXiv:2406.01842. DOI: 10.1145/3627673.3680057

Garcia-Teodoro, P., Diaz-Verdejo, J., Maciá-Fernández, G., & Vázquez, E. (2009). Anomaly- based network intrusion detection: Techniques, systems and challenges. *computers & security, 28*(1-2), 18-28.

Google Cloud. (2023). *Security Command Center: Real-time threat intelligence for cloud security.* https://cloud.google.com/security-command-center/docs

Gopireddy, R. R. (2020). Dark Web Monitoring: Extracting and Analyzing Threat Intelligence.

Gupta, M., Akiri, C., Aryal, K., Parker, E., & Praharaj, L. (2023). From ChatGPT to ThreatGPT: Impact of generative AI in cybersecurity and privacy. *IEEE Access : Practical Innovations, Open Solutions, 11*, 80218–80245. DOI: 10.1109/ACCESS.2023.3300381

Gurucul. (2020, May 20). *ABCs of UEBA: P is for PRIVILEGE.* https://gurucul.com/blog/abcs- of-ueba-p-is-for-privilege/

Gurucul. (2021, August 2). *ABCs of UEBA: X is for eXfiltration.* https://gurucul.com/blog/abcs- of-ueba-x-is-for-exfiltration/

Gurucul. (2021). *Top UEBA Use Cases to Fuel Modern, Next-Gen Security Operations.* https://gurucul.com/blog/top-ueba-use-cases

Hagemann, T., & Katsarou, K. (2020, December). A systematic review on anomaly detection for cloud computing environments. In *Proceedings of the 2020 3rd Artificial Intelligence and Cloud Computing Conference* (pp. 83-96). DOI: 10.1145/3442536.3442550

Horev, R. (2024, February 13). *Multi-cloud security challenges: A best practice guide.* Vulcan. https://vulcan.io/blog/multi-cloud-security-challenges-a-best-practice -guide/

Horev, R. (2024). Advanced DDoS mitigation strategies for multi-cloud environments. *Network Security, 2024*(2), 8–13.

Horev, R. (2024). Supply Chain Vulnerabilities in Multi-Cloud. *Cybersecurity & Digital Trust, 11*(2), 18–35.

Hummel, P., Braun, M., Tretter, M., & Dabrock, P. (2021). Data sovereignty: A review. *Big Data & Society*, *8*(1), 2053951720982012. DOI: 10.1177/2053951720982012

IBM. (2022, August 10). *What is user and entity behavior analytics (UEBA)?* IBM. h ttps://www.ibm.com/think/topics/ueba

IBM. (n.d.). *IBM Cloud Pak for Security.* cloud.ibm.com/catalog/content/ibm-cp -security%3A%3A1-b25bd169-0fbd-4cf3-a8ea-

Ibrahim, A. S., Hamlyn-Harris, J., & Grundy, J. (2016). Emerging security challenges of cloud virtual infrastructure. *arXiv preprint arXiv:1612.09059.*

Jimmy, F. N. U. (2023). Cloud security posture management: tools and techniques. Journal of Knowledge Learning and Science Technology ISSN: 2959-6386 (online), 2(3).

Jones, J., Smith, B., Micheal, O., Barnes, M., & Adebayo, H. (2025). Revolutionizing Cybersecurity with AI-Driven SIEM: Optimizing Threat Detection in Multi-Cloud. *Environments.*

Kadar, A. (2023, December 20). *Enhancing cloud security: Posture management tools and approaches.* ResearchGate.

Khaliq, S., Tariq, Z. U. A., & Masood, A. (2020, October). Role of user and entity behavior analytics in detecting insider attacks. In *2020 International Conference on Cyber Warfare and Security (ICCWS)* (pp. 1-6). IEEE.

Kim, H., Kim, J., Kim, Y., Kim, I., & Kim, K. J. (2019). Design of network threat detection and classification based on machine learning on cloud computing. *Cluster Computing*, *22*(S1), 2341–2350. DOI: 10.1007/s10586-018-1841-8

Kindervag, J., & Balaouras, S. (2010). No more chewy centers: Introducing the zero trust model of information security. *Forestry Research*, *3*, 56682.

Kumari, S. (2022). Cybersecurity in Digital Transformation: Using AI to Automate Threat Detection and Response in Multi-Cloud Infrastructures. *Journal of Computational Intelligence and Robotics*, *2*(2), 9–27.

Kumari, S. (2022). Machine learning approaches for threat detection in cloud computing: A systematic review. *Information Sciences*, *580*, 340–366.

Kummarapurugu, R. (2024). An architectural framework for threat intelligence integration in hybrid cloud using SIEM and SOAR. [IJIRCT]. *International Journal of Innovative Research in Computer Technology*, *10*(1), 133–138. https://www.ijirct .org/download.php? a_pid=2411031

KuppingerCole Analysts. (2024). *Leadership Compass: eXtended Detection and Response (XDR)*. Retrieved from https://www.kuppingercole.com/research/lc80923/ extended- detection-and-response-xdr

Madasu, S. (2023). Access control models and technologies for big data processing and m anagement. *European Chemical Bulletin, 12*, 6886–6902.

Manzoor, A., Hussain, M., & Mehrban, S. (2022). Security and privacy issues in cloud c omputing: A comprehensive survey. *Journal of Systems Architecture, 128*, 102533.

Manzoor, S., Gouglidis, A., Bradbury, M., & Suri, N. (2022). ThreatPro: Multi-Layer Threat Analysis in the Cloud. *arXiv preprint arXiv:2209.14795*.

Manzoor, S., Gouglidis, A., Bradbury, M., & Suri, N. (2022). *ThreatPro: Multi-Layer Threat Analysis in the Cloud*. arXiv preprint, arXiv:2209.14795.

Mathew, A. (2024, November 18). *Cloud data sovereignty: Governance and risk implications of cross-border cloud storage*. ISACA. https://www.isaca.org/resources/ news-and-t rends/industry-news/2024/cloud-data-sovereignty-governance-and-risk-implications-of- cross-border-cloud-storage

McAfee. (2023). *MVISION Cloud: Cloud access security broker (CASB)*. ht t p s:// www.mcafee.com/enterprise/en-us/products/mvision-cloud.html

Microsoft. (2021). *Microsoft Security Report: Exchange Attacks & Mitigation Strategies*. https://www.microsoft.com/security

Microsoft. (2024). *Microsoft Defender for Cloud overview*. https://learn.microsoft .com/en- us/azure/defender-for-cloud/defender-for-cloud-introduction

Mohan, T., Kumar, V., & Sharma, R. (2022). *Advancements in cloud security monitoring: Analyzing the efficiency of AWS, Azure, and Google Cloud security frameworks*. I nternational Journal of Information Security and Cyber Forensics, 11(2), 88-104.

Morrow, T. (2018, March 5). *12 risks, threats, and vulnerabilities in moving to the cloud*. S oftware Engineering Institute. https://insights.sei.cmu.edu/blog/12-risks -threats- v ulnerabilities-in-moving-to-the-cloud/

Morrow, T. (2018). Best Practices for Cloud Security. *Journal of IT Security, 3*(2), 50–65.

Mukherjee, B., Heberlein, L. T., & Levitt, K. N. (1994). Network intrusion detection. *IEEE Network, 8*(3), 26–41. DOI: 10.1109/65.283931

Nassif, A. B., Talib, M. A., Nasir, Q., Albadani, H., & Dakalbab, F. M. (2021). Machine learning for cloud security: A systematic review. *IEEE Access : Practical Innovations, Open Solutions, 9*, 20717–20735. DOI: 10.1109/ACCESS.2021.3054129

National Institute of Standards and Technology (NIST). (2011). *The NIST Definition of Cloud Computing*. Special Publication 800-145.

Netskope. (2023, March 22). *Operationalizing advanced UEBA: Detection scenarios and UCI alerts.*

Olaniyan, R., Rakshit, S., & Vajjhala, N. R. (2023). Application of user and entity behavioral analytics (UEBA) in the detection of cyber threats and vulnerabilities management. In *Computational Intelligence for Engineering and Management Applications: Select Proceedings of CIEMA 2022* (pp. 419-426). Singapore: Springer Nature Singapore.

Olaoye, A. O. (2022). *Multi-cloud architecture for cloud computing.* Iowa State University. https://dr.lib.iastate.edu/server/api/core/bitstreams/48ab713c-0f6c-472a -9564- 9 d6bee1b394b/content

Palo Alto Networks. (2024). *Prisma Cloud: At a Glance.* Retrieved from h t t ps://www.paloaltonetworks.com/resources/datasheets/prisma-cloud-at-a-glance

Palo Alto Networks. (2024). *What Is Extended Detection and Response (XDR)?* https://www.paloaltonetworks.com/cyberpedia/what-is-extended-detection-response -XDR

Patel, S., Gupta, R., & Ahmed, Z. (2023). A review of cloud security tools: Native vs third-party solutions. *Cybersecurity and Cloud Computing Journal, 17*(1), 121–136.

Paul, A. (2024). *Security challenges and solutions in multi-cloud environments.* ResearchGate.

Piskorz, P. (n.d.). *Top 5 challenges of protecting multi-cloud environments.* Storware. h ttps://storware.eu/blog/top-5-challenges-of-protecting-multi-cloud-environment/

Pöhn, D., & Hommel, W. (2023). Towards an improved taxonomy of attacks related to digital identities and identity management systems. *Security and Communication Networks, 2023*(1), 5573310. DOI: 10.1155/2023/5573310

Proofpoint. (n.d.). *Identity Threat Detection and Response (ITDR).* www.proofpoint .com/us/threat-reference/identity-threat-detection-and-response-itdr

Quist, A. (2023). Insider Threats in Multi-Cloud: Risk Analysis & Prevention Strategies. *Journal of Cloud Computing Security, 6*(3), 55–72.

Quist, N. (2023, May 15). *10 cloud security risks*. Palo Alto Networks. h t t ps://www.paloaltonetworks.com/blog/prisma-cloud/10-cloud-security-risks/

Reece, J., Jones, A., & Williams, P. (2023). Comprehensive security assessment frameworks for multi-cloud environments. *IEEE Security and Privacy, 21*(4), 28–36.

Reece, M., Lander, T., Jr., Mittal, S., Rastogi, N., Dykstra, J., & Sampson, A. (2023). Emergent (In) Security of Multi-Cloud Environments. *arXiv preprint arXiv:2311.01247.*

Reece, M., Lander, T., Jr., Mittal, S., Rastogi, N., Dykstra, J., & Sampson, A. (2023). *E mergent (In)Security of Multi-Cloud Environments*. arXiv preprint, arXiv:2311.01247.

Reece, M., Lander, T. E., Jr., Stoffolano, M., Sampson, A., Dykstra, J., Mittal, S., & Rastogi, N. (2023). Systemic Risk and Vulnerability Analysis of Multi-Cloud Environments. *arXiv preprint arXiv:2306.01862.*

Reece, M., Lander, T. E., Jr., Stoffolano, M., Sampson, A., Dykstra, J., Mittal, S., & Rastogi, N. (2023). Systemic risk and vulnerability analysis of multi-cloud environments. *arXiv preprint arXiv:2306.01862.*

Reece, M., Lander, T. E.Jr, Stoffolano, M., Sampson, A., Dykstra, J., Mittal, S., & Rastogi, N. (2023). *Systemic risk and vulnerability analysis of multi-cloud environments* (a rXiv:2306.01862). arXiv. https://arxiv.org/abs/2306.01862

Rengarajan, R., & Babu, S. (2021, March). Anomaly detection using user entity behavior a nalytics and data visualization. In *2021 8th International Conference on Computing for Sustainable Global Development (INDIACom)* (pp. 842-847). IEEE.

Rodigari, S., O'Shea, D., McCarthy, P., McCarry, M., & McSweeney, S. (2021). *Performance Analysis of Zero-Trust Multi-Cloud*. arXiv. https://arxiv.org/abs/2105.02334 DOI: 10.1109/CLOUD53861.2021.00097

Rose, S., Borchert, O., Mitchell, S., & Connelly, S. (2020). Zero trust architecture. *NIST Special Publication 800-207*. National Institute of Standards and Technology.

Salman, T., Bhamare, D., Erbad, A., Jain, R., & Samaka, M. (2017, June). Machine learning for anomaly detection and categorization in multi-cloud environments. In *2017 IEEE 4th international conference on cyber security and cloud computing (CSCloud)* (pp. 97-103). IEEE. DOI: 10.1109/CSCloud.2017.15

Sarkar, S., Choudhary, G., Shandilya, S. K., Hussain, A., & Kim, H. (2022). Security of zero trust networks in cloud computing: A comparative review. *Sustainability (Basel), 14*(18), 11213. DOI: 10.3390/su141811213

Sasovets, I. (2023, August 1). *Multi-cloud security: Benefits, challenges, and best practices*. TechMagic. https://www.techmagic.co/blog/multi-cloud-security

Sasovets, I. (2024, October 1). *What is multi-cloud security? Challenges and best practices*. TechMagic. https://www.techmagic.co/blog/multi-cloud-security

Sasovets, I. (2024). The Evolution of Cloud Threats: Lessons from Recent Attacks. *Cyber Risk Review*, *14*(1), 78–93.

Sasovets, Y. (2024). Unified visibility in multi-cloud security: Challenges and solutions. *International Journal of Cloud Applications and Computing*, *14*(1), 1–15.

Saxena, A., Gupta, S., & Singh, Y. K. (2021). A survey on multi-cloud computing: Benefits and research directions. *Journal of Parallel and Distributed Computing*, *157*, 34–51.

Saxena, D., Gupta, R., & Singh, A. K. (2021). A Survey and Comparative Study on Multi-Cloud Architectures: Emerging Issues and Challenges for Cloud Federation. *arXiv preprint arXiv:2108.12831*.

Scarfone, K., & Hoffman, P. Guidelines on firewalls and firewall policy: Recommendations of the National Institute of Standards and Technology. *NIST Special Publication*, 800-41.

Schäfer, M., Fuchs, M., Strohmeier, M., Engel, M., Liechti, M., & Lenders, V. (2019, May). BlackWidow: Monitoring the dark web for cyber security information. In *2019 11th International Conference on Cyber Conflict (CyCon)* (Vol. 900, pp. 1-21). IEEE.

SentinelOne. (2024, July 31). *Cloud security vulnerabilities*. SentinelOne. h t t ps://www.sentinelone.com/cybersecurity-101/cloud-security/cloud-security- v ulnerabilities/

SentinelOne. (2024). *Cyber Threat Report: Multi-Cloud Attacks on the Rise*. Retrieved from https://www.sentinelone.com/research

SentinelOne. (2024). *Benefits of XDR (Extended Detection and Response)*. h ttps:// www.sentinelone.com/cybersecurity-101/endpoint-security/benefits-of-xdr/

Shahzad, F., Mannan, A., Javed, A. R., Almadhor, A. S., Baker, T., & Al-Jumeily, O. B. E. (2022). Cloud-based multiclass anomaly detection and categorization using ensemble learning. *Journal of Cloud Computing (Heidelberg, Germany)*, *11*(1), 74. DOI: 10.1186/s13677-022-00329-y

Sharma, A., & Sahay, S. K. (2020). Evolution and adoption of AI in cybersecurity. *Computers & Security*, *98*, 101935.

Sharma, H. (2021). Behavioral Analytics and Zero Trust. *International Journal of Computer Engineering and Technology, 12*(1), 63–84.

Sharma, R., & Kumar, A. (2023). Threat intelligence and security analytics in multi-cloud environments. *Computers & Security, 132*, 104317.

Sharma, S., & Modi, C. (2021). A review of service mesh in cloud-native applications: A rchitecture and security. *IEEE Access*, 9, 23487-23500. ht t p s ://DOI: 10.1109/ACCESS.2021.3056014

Shelke, P., & Hämäläinen, T. (2024). Analysing multidimensional strategies for cyber threat detection in security monitoring. In *Proceedings of the European Conference on Cyber Warfare and Security* (Vol. 23, No. 1). Academic Conferences International Ltd. DOI: 10.34190/eccws.23.1.2123

Sommer, R., & Paxson, V. (2010, May). *Outside the closed world: On using machine learning for network intrusion detection. In 2010 IEEE symposium on security and privacy*. IEEE.

S&P Global Market Intelligence. (2021). *The Rise of Extended Detection and Response*. https://www.spglobal.com/marketintelligence/en/documents/the-rise-of - extended- detection-and-response.pdf

Spiceworks (2025). MultiCloud Infrastructure [Photograph]. What Is Multicloud Infrastructure? https://www.spiceworks.com/tech/cloud/articles/what-is-multicloud -infrastructure/

Spiceworks Editorial Team. (2025). What is multicloud infrastructure? [Image]. Spiceworks. https://www.spiceworks.com/tech/cloud/articles/what-is-multicloud -infrastructure/

Splunk Inc. (2020). *Splunk Security Cloud Product Brief.* www.splunk.com/pdfs/ product-briefs/splunk-security-cloud.pdf

Stallings, W., & Brown, L. (2015). *Computer security: principles and practice.* Pearson.

Süß, F., Freimuth, M., Aßmuth, A., Weir, G. R. S., & Duncan, B. (2024). *Cloud Security and Security Challenges Revisited.* arXiv preprint arXiv:2405.11350

Sysdig. (n.d.). AWS vs. Azure vs. Google Cloud: Security Comparison. https://sysdig .com/learn- cloud-native/threat-detection-in-the-cloud-defender-vs-guardduty-vs-security-command- center/

Sysdig. (n.d.). Cloud-based threat detection: A comparative analysis. Retrieved February 22, 2025

Sysdig. (n.d.). *What is a Cloud Workload Protection Platform (CWPP)?.* sysdig .com/learn-cloud-native/what-is-a-cloud-workload-protection-platform-cwpp/

Taddeo, M., & Floridi, L. (2018). Regulate artificial intelligence to avert cyber arms race. *Nature, 556*(7701), 296–29. DOI: 10.1038/d41586-018-04602-6 PMID: 29662138

Tatineni, S. (2023). AI-infused threat detection and incident response in cloud security. *I nternational Journal of Science and Research (IJSR), 12*(11), 998-1004.

Torkura, K., Sukmana, M. I. H., Cheng, F., & Meinel, C. (2020). *Continuous auditing & threat detection in multi-cloud infrastructure.* https://doi.org/DOI: 10.36227/ techrxiv.13108313

Trend Micro. (2024). *Trend Micro Cloud One: Workload security overview.* https:// cloudone.trendmicro.com/docs/workload-security/protection-modules/

Tuyishime, E., Balan, T. C., Cotfas, P. A., Cotfas, D. T., & Rekeraho, A. (2023). Enhancing cloud security—Proactive threat monitoring and detection using a siem-based approach. *Applied Sciences (Basel, Switzerland), 13*(22), 12359. DOI: 10.3390/app132212359

Verdet, A. (2023). *Exploring security practices in infrastructure as code: An empirical study* (Master's thesis, Ecole Polytechnique, Montreal (Canada)).

Wiz. (2024). *Wiz cloud security platform.* https://www.wiz.io/platform

Wiz Experts Team. (2024, October 11). *Multi-cloud security.* Wiz. www.wiz.io/ academy/multi-cloud-security

Wiz Experts Team. (2024). Securing APIs in Multi-Cloud Environments. *Cloud Security Best Practices, 9*(1), 11–27.

Wiz Experts Team. (2024). Best practices for IAM configuration in multi-cloud setups. *Cloud Security Insights, 7*(3), 112–125.

Zhao, H., Benomar, Z., Pfandzelter, T., & Georgantas, N. (2022). Supporting multi-cloud in serverless computing. *In 2022 IEEE/ACM 15th International Conference on Utility and Cloud Computing (UCC)* (pp. 285–290). IEEE.

Zhao, L., Benomar, O., Pfandzelter, T., & Georgantas, N. (2022). Multi-cloud orchestration: Current practices and challenges. *IEEE Software, 39*(5), 53–59.

Chapter 5
The Role of English Language and AI in Scientific Writing:
Ethical and Academic Implications

Hewa Majeed Zangana
https://orcid.org/0000-0001-7909-254X
Duhok Polytechnic University, Iraq

Pratiwi Amelia
https://orcid.org/0000-0001-9778-6772
Universitas Muhammadiyah Bangka Belitung, Indonesia

Firas Mahmood Mustafa
Duhok Polytechnic University, Iraq

Shuai Li
University of Oulu, Finland

ABSTRACT

In the contemporary era of digital transformation, English remains the lingua franca of scientific discourse, serving as a gateway to global recognition and academic collaboration. Simultaneously, artificial intelligence (AI) tools—ranging from grammar checkers to large language models—are reshaping the landscape of scientific writing. This chapter critically explores the intersection of English language proficiency and AI-powered writing assistance in science, technology, and medical (STM) research. It discusses the ethical implications of AI-generated content, authorship attribution, and linguistic biases, while considering the academic pressures faced by non-native English-speaking researchers. By examining both opportunities and challenges, this

DOI: 10.4018/979-8-3373-4252-8.ch005

chapter emphasizes the need for equitable language practices and responsible AI use to maintain the integrity and inclusivity of scientific communication.

1. INTRODUCTION

The landscape of academic and scientific writing has undergone a dramatic transformation in recent years, driven by two converging forces: the global predominance of English as the primary language of scholarly communication and the rapid rise of artificial intelligence (AI) tools in the writing process. In science, technology, and medical (STM) fields, the expectation that researchers publish in English-dominant journals remains a significant barrier for non-native English-speaking scholars, who often face challenges in producing linguistically polished and academically acceptable manuscripts (Li et al., 2024; Fedoriv et al., 2024).

Simultaneously, the integration of AI-powered writing assistants—such as grammar correctors, citation generators, summarization tools, and large language models (LLMs) like ChatGPT—has revolutionized the writing process, offering unprecedented support in drafting, revising, and even conceptualizing scientific manuscripts (Khalifa & Albadawy, 2024; Salvagno et al., 2023). These tools are increasingly employed to bridge linguistic gaps and improve the overall readability and structure of research outputs, thus enhancing the visibility and acceptance of non-native speakers' work in high-impact journals (Li et al., 2024; Nurchurifiani et al., 2025).

However, this technological shift is not without ethical and academic concerns. The use of AI in scientific writing raises profound questions about authorship, originality, accountability, and academic integrity (Lund et al., 2023; Chetwynd, 2024). Scholars and educators are debating whether AI-generated content undermines the educational value of writing, promotes plagiarism, or creates a dependency that could weaken the development of critical language and research skills (Aljuaid, 2024; Molligan & Pérez-López, 2024; Casal & Kessler, 2023).

Another dimension of the ethical debate is the issue of transparency and disclosure. Researchers often fail to disclose the use of AI tools in manuscript preparation, raising questions about intellectual honesty and compliance with publishing guidelines (Aslam & Nisar, 2024; Harati, 2024). Furthermore, disparities in access to advanced AI tools may exacerbate global inequities in scientific publishing, disproportionately benefiting those in well-resourced institutions while marginalizing scholars in low-income regions (Guleria et al., 2023; Omodan & Marongwe, 2024).

Pedagogically, AI tools have been both lauded and criticized in English as a Foreign Language (EFL) contexts. On one hand, they offer scaffolded support to learners struggling with academic conventions; on the other, they may lead to overreliance and inhibit independent learning (Alghamdy, 2023; Khup & Bantugan, 2025). Educators must thus strike a balance between embracing technological tools and maintaining the pedagogical principles that foster authentic skill development.

Given these evolving dynamics, this chapter seeks to provide a comprehensive analysis of the dual role played by the English language and AI in scientific writing. It will explore how AI tools are reshaping the writing practices of researchers worldwide, the implications for non-native English speakers, and the ethical considerations that must be addressed to uphold academic standards. Drawing from a wide array of recent literature and international perspectives, this chapter contributes to ongoing discussions about equity, authorship, and ethical responsibility in an AI-augmented academic environment.

In the sections that follow, we will examine:

1. The centrality of English in STM publishing and its implications for global equity.
2. The functionalities and limitations of AI-assisted writing tools.
3. The ethical concerns surrounding authorship, integrity, and the misuse of AI.
4. Pedagogical implications and institutional responsibilities.
5. Recommendations for transparent and responsible AI integration in academic writing.

This exploration is not merely theoretical; it responds to an urgent need for guidelines, best practices, and ethical standards in an era where AI is no longer optional but integral to the research enterprise (Ajiye & Omokhabi, 2025; AlSamhori & Alnaimat, 2024; Eberlin, 2024).

2. AI TOOLS IN SCIENTIFIC WRITING AND EDITING

The integration of artificial intelligence (AI) into scientific writing and editing has transformed the landscape of academic research. AI tools offer a range of functionalities, from grammar and style enhancements to literature synthesis and data analysis. However, their adoption also raises ethical considerations that necessitate careful deliberation.

2.1 Enhancing the Writing Process with AI

AI tools have become invaluable in streamlining various stages of scientific writing. They assist in:

- **Idea Generation and Content Structuring**: AI can help researchers brainstorm and organize their thoughts, facilitating a more coherent narrative.
- **Literature Synthesis**: By analyzing vast amounts of data, AI tools can summarize existing literature, aiding in comprehensive reviews.
- **Grammar and Style Editing**: Tools like Grammarly and Hemingway Editor enhance clarity and readability by correcting grammatical errors and suggesting stylistic improvements.
- **Reference Management**: AI can automate citation formatting and manage bibliographies, reducing manual workload.

These functionalities not only improve the quality of manuscripts but also expedite the writing process, allowing researchers to focus more on their core scientific contributions.

To visualize how AI tools contribute at different stages of the scientific writing process, the following flowchart maps key integration points, from idea generation to citation management.

Figure 1. AI integration points in the scientific writing workflow

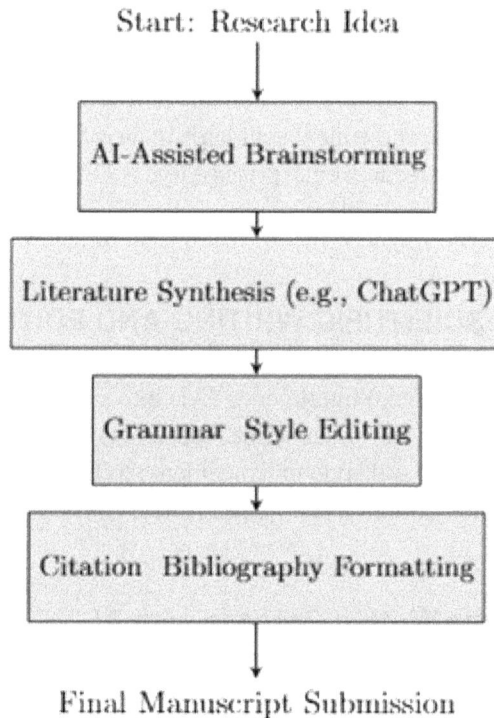

2.2 Ethical Considerations in AI-Assisted Writing

While AI offers numerous benefits, its use in scientific writing is accompanied by ethical challenges:

- **Plagiarism and Authorship**: AI-generated content may inadvertently replicate existing works, raising concerns about originality and proper attribution.
- **Bias and Accuracy**: AI models trained on biased data can perpetuate existing prejudices, and their outputs may contain inaccuracies if not properly supervised.
- **Transparency**: The extent of AI involvement in manuscript preparation should be disclosed to maintain transparency and uphold academic integrity.
- **Overreliance on AI**: Excessive dependence on AI tools may hinder the development of critical thinking and writing skills among researchers.

Addressing these ethical concerns requires the establishment of clear guidelines and the promotion of responsible AI usage in academia.

2.3 Best Practices for Integrating AI in Scientific Writing

To harness the benefits of AI while mitigating ethical risks, researchers should consider the following best practices:

1. **Use AI as a Complementary Tool**: Employ AI to assist with specific tasks, such as grammar checking or literature summarization, rather than relying on it for entire manuscript generation.
2. **Maintain Human Oversight**: Always review and edit AI-generated content to ensure accuracy, coherence, and alignment with research objectives.
3. **Disclose AI Usage**: Clearly state the extent of AI involvement in the writing process to maintain transparency with readers and reviewers.
4. **Stay Informed About Ethical Guidelines**: Familiarize oneself with institutional and journal policies regarding AI use in academic writing.

By adhering to these practices, researchers can effectively integrate AI tools into their workflows while upholding the standards of scientific integrity.

2.4 Summary

AI tools have undeniably enhanced the efficiency and quality of scientific writing and editing. However, their integration into academic research necessitates a balanced approach that considers both the advantages and the ethical implications. Through responsible usage and adherence to established guidelines, AI can serve as a valuable ally in the pursuit of scientific advancement.

3. ETHICAL IMPLICATIONS OF AI USE IN ACADEMIC WRITING

The integration of artificial intelligence (AI), particularly large language models (LLMs) such as ChatGPT, into academic writing has spurred significant ethical debate across global educational and research communities. While these tools offer undeniable benefits in enhancing productivity, refining language, and supporting non-native English speakers, their use raises substantial ethical concerns related to authorship, academic integrity, accountability, data privacy, and equity in education.

3.1. Academic Integrity and Authorship

One of the most pressing ethical dilemmas centers on authorship and academic honesty. The ability of AI tools to generate coherent, often sophisticated text blurs the lines between human and machine authorship (Casal & Kessler, 2023; Eberlin, 2024). When AI-generated content is submitted without transparency, it may constitute plagiarism or academic fraud (Perkins, 2023; Chetwynd, 2024). Concerns regarding 'ghostwriting by machines' have led some researchers to argue for mandatory disclosures when AI tools are used in manuscript preparation (BaHammam, 2023; Lund et al., 2023).

3.2. Misuse and Overreliance on AI Tools

The overuse or uncritical dependence on AI-powered writing assistance can lead to superficial learning, diminished critical thinking, and erosion of original scholarly contributions (Khup & Bantugan, 2025; Guleria et al., 2023). Aslam and Nisar (2024) emphasize that students and early-career researchers may substitute AI assistance for genuine engagement with academic material, thereby undermining educational objectives. Similarly, Aljuaid (2024) warns that AI tools could shift academic writing from a reflective, cognitive process to a mechanical activity, diminishing intellectual rigor.

3.3. Bias, Inaccuracy, and Hallucination

AI-generated content is not immune to error. LLMs have been found to "hallucinate" references, fabricate facts, or reflect the biases embedded in their training data (Kirov, 2023; Liu et al., 2025; Salvagno et al., 2023). Such inaccuracies can compromise the reliability of academic texts and mislead readers or reviewers (Harati, 2024; Nam & Bai, 2023). Miao et al. (2023) advocate for stringent peer-review frameworks to detect and mitigate such risks.

3.4. Transparency, Accountability, and Disclosure

Transparency in AI-assisted writing is pivotal to maintaining academic ethics. Espino et al. (2024) and Ocampo et al. (2023) suggest the development of standardized disclosure protocols to inform readers and editors of AI involvement. Dinçer (2024) further proposes institutional guidelines and audit mechanisms to ensure accountability and responsible AI use. These protocols would not only uphold research integrity but also foster trust in AI-enhanced publications.

3.5. Equity and Access

While AI tools can support non-native English speakers and students from under-resourced academic environments (Li et al., 2024; Fedoriv et al., 2024), unequal access to premium AI tools risks widening existing academic disparities (Ajiye & Omokhabi, 2025; Khalifa & Albadawy, 2024). This digital divide creates ethical concerns around fairness and inclusivity in academic knowledge production (Omodan & Marongwe, 2024).

3.6. Pedagogical and Institutional Ethics

Educators face a paradox: whether to ban, restrict, or incorporate AI writing tools into curricula. Alghamdy (2023) and Nurchurifiani et al. (2025) advocate for integrating AI literacy into academic instruction, enabling students to use these tools ethically and responsibly. Educational policies must strike a balance between harnessing innovation and preserving pedagogical integrity (AlSamhori & Alnaimat, 2024; Alahdab, 2024).

3.7. Evolving Norms in Research and Publishing

The academic publishing industry is also adapting to the ethical challenges posed by AI-generated content. Granjeiro et al. (2025) and Molligan & Pérez-López (2024) suggest that journals and conferences establish explicit policies regarding AI usage in manuscript preparation. Meanwhile, Ersöz and Engin (2024) stress the importance of a consensus-based approach among researchers, publishers, and institutions to manage evolving norms around AI authorship and ethics.

3.8. Best Practices and Ethical Frameworks

Multiple scholars have proposed ethical frameworks and practical recommendations for responsible AI use in academic writing. These include ensuring human oversight, disclosing AI involvement, validating content accuracy, and prioritizing originality (Shofiah et al., 2023; Shofiah, Putera, & Solichah, 2023). For instance, Subaveerapandiyan et al. (2025) found that while students perceive AI tools as beneficial, they express concern about overreliance and misapplication. Similarly, Espino et al. (2024) recommend a stakeholder-driven approach to co-develop guidelines that uphold academic values.

3.9 Summary

The ethical implications of AI in academic writing are multifaceted and evolving. While these tools offer efficiency and accessibility, their use necessitates careful navigation of ethical terrain, from authorship attribution to equity in access. Institutions, educators, and researchers must collaboratively construct and uphold ethical guidelines that preserve academic integrity, promote transparency, and ensure equitable and responsible use of AI technologies in scholarly communication.

The chart below categorizes and compares the most commonly cited ethical concerns associated with AI use in academic writing, based on discussions in this chapter.

Figure 2. Key ethical concerns of AI in academic writing

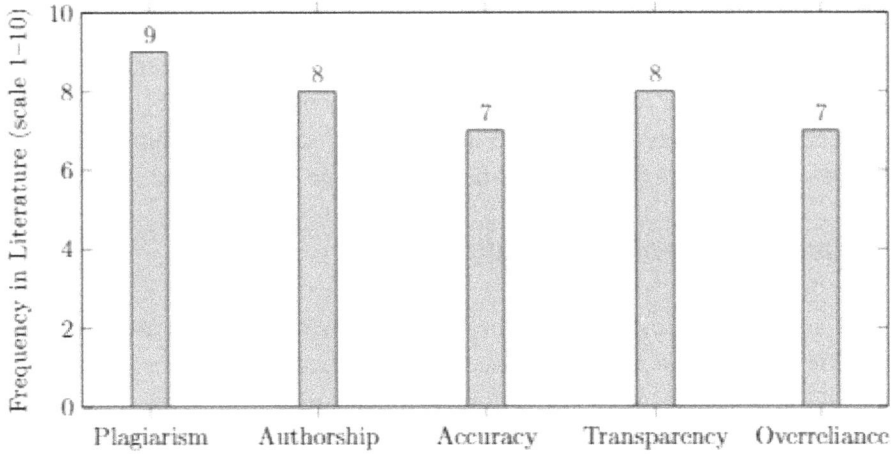

4. AI AND LINGUISTIC BIAS IN SCIENTIFIC COMMUNICATION

The integration of artificial intelligence (AI), particularly large language models (LLMs) like ChatGPT, into scientific writing has revolutionized academic communication. These tools offer unprecedented assistance in drafting, editing, and translating scholarly texts. However, their widespread adoption has also surfaced concerns regarding linguistic biases that may inadvertently perpetuate inequities in scientific discourse. This section delves into the multifaceted nature of AI-induced linguistic biases in scientific communication, exploring their origins, manifestations, and implications.

4.1 Origins of Linguistic Bias in AI Models

AI language models are trained on vast datasets sourced predominantly from the internet, encompassing a plethora of texts that reflect existing societal biases. Consequently, these models may internalize and reproduce linguistic patterns that favor dominant languages and dialects, particularly standardized English, while marginalizing others (Chetwynd, 2024). For instance, non-native English speakers often find their writing styles misclassified as AI-generated by detection tools, highlighting an inherent bias in these systems (Liang et al., 2023).

Moreover, the underrepresentation of minority languages and dialects in training data exacerbates this issue, leading to AI outputs that lack cultural nuance and fail to capture the richness of diverse linguistic expressions (Khan, 2024). This skewed representation not only hampers the inclusivity of scientific communication but also risks reinforcing existing disparities in academic visibility and recognition.

4.2 Manifestations in Scientific Writing

The linguistic biases embedded in AI models manifest in various ways within scientific writing. One prominent issue is the homogenization of academic prose, where AI-generated texts tend to conform to a standardized linguistic style that may not align with the author's cultural or disciplinary context (Jain & Jain, 2023). This uniformity can dilute the authenticity of scholarly voices, particularly those from underrepresented backgrounds.

Additionally, AI tools may inadvertently perpetuate stereotypes or biased narratives present in their training data. For example, when generating content related to specific regions or communities, AI models might reproduce outdated or prejudiced perspectives, thereby compromising the objectivity and integrity of scientific discourse (Mitchell, 2025).

The following pie chart illustrates the types of linguistic bias that may be embedded in AI-generated scientific content, including regional, syntactic, and cultural factors.

Figure 3. Types of linguistic bias in AI-generated scientific writing

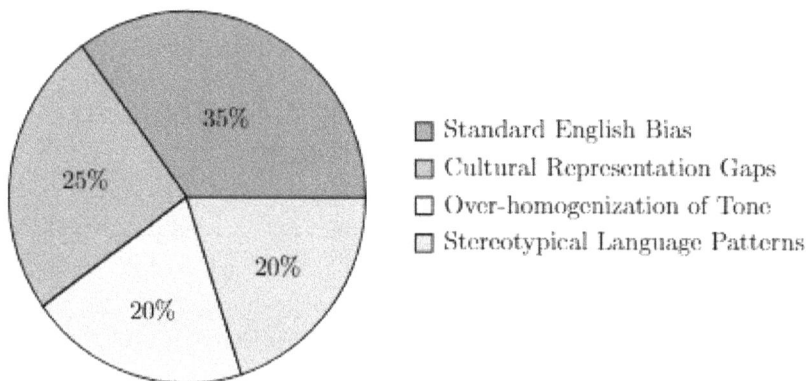

- Standard English Bias
- Cultural Representation Gaps
- Over-homogenization of Tone
- Stereotypical Language Patterns

35%
25%
20%
20%

4.3 Implications for Academic Equity

The linguistic biases inherent in AI tools have significant implications for academic equity. Non-native English speakers, in particular, may face challenges in ensuring their work is accurately represented and fairly evaluated. The misclassification of their writing as AI-generated not only undermines their credibility but also poses barriers to publication and dissemination (Liang et al., 2023).

Furthermore, the reliance on AI-generated content in peer review and editorial processes raises concerns about the perpetuation of existing biases. If unchecked, these tools could reinforce dominant linguistic norms, marginalizing diverse scholarly contributions and hindering the progress toward a more inclusive academic landscape (Chetwynd, 2024).

4.4 Mitigating Linguistic Bias in AI-Assisted Scientific Communication

Addressing linguistic bias in AI-assisted scientific communication necessitates a multifaceted approach:

1. **Diversifying Training Data**: Incorporating a broader range of languages, dialects, and cultural contexts into AI training datasets can enhance the models' ability to generate more inclusive and representative content (Khan, 2024).
2. **Enhancing Transparency**: Clear disclosure of AI assistance in scientific writing allows for critical evaluation of potential biases and fosters accountability in scholarly communication (Chetwynd, 2024).
3. **Developing Inclusive AI Tools**: Designing AI systems with built-in mechanisms to detect and mitigate linguistic biases can promote fairness and equity in academic publishing (Jain & Jain, 2023).
4. **Promoting Human Oversight**: Ensuring that human editors and reviewers critically assess AI-generated content can help identify and correct biases, preserving the integrity of scientific literature (Mitchell, 2025).

4.5 Summary

While AI offers valuable tools for enhancing scientific communication, it is imperative to recognize and address the linguistic biases these technologies may introduce. By fostering diversity in training data, enhancing transparency, developing inclusive AI tools, and promoting vigilant human oversight, the academic community can harness the benefits of AI while safeguarding the equity and integrity of scientific discourse.

5. LANGUAGE PRECISION, AMBIGUITY, AND MISINTERPRETATION

The integration of large language models (LLMs) into academic writing and scientific communication has brought to the forefront significant concerns surrounding **language precision, ambiguity, and the potential for misinterpretation**. While LLMs like ChatGPT have demonstrated remarkable capabilities in generating coherent and structured text, their outputs often lack the semantic precision and contextual awareness essential in scholarly discourse (Casal & Kessler, 2023; Dinçer, 2024). This section explores these linguistic limitations and the ethical challenges they pose, especially in high-stakes academic environments.

5.1. Linguistic Fluency vs. Semantic Accuracy

LLMs are designed to mimic human language patterns based on probabilistic predictions, often prioritizing fluency over factual or contextual precision (Liu et al., 2025; Lund et al., 2023). This trade-off may result in seemingly accurate text that subtly misrepresents concepts, data, or arguments. For instance, Ajiye and Omokhabi (2025) emphasize that AI-generated text may appear academically polished but often contains nuanced errors that can mislead readers, especially in complex fields like medicine or law.

This phenomenon, sometimes termed **"semantic hallucination"** (Harati, 2024), reflects the model's tendency to fill knowledge gaps with plausible-sounding content. In research writing, where precise terminology and methodological clarity are paramount, such hallucinations can lead to the dissemination of misinformation (BaHammam, 2023; Eberlin, 2024).

5.2. Ambiguity and Interpretative Variance

One of the major ethical challenges in using AI-generated academic text lies in its potential to introduce ambiguity that is not immediately apparent. Aslam and Nisar (2024) point out that students and researchers may rely on outputs that contain syntactic or semantic ambiguity, which can affect peer interpretation and academic assessment. Similarly, Espino et al. (2024) warn that such ambiguities are particularly detrimental in cross-cultural or multilingual academic contexts, where meaning can easily be distorted.

In EFL (English as a Foreign Language) settings, Alghamdy (2023) and Fedoriv et al. (2024) note that AI tools may amplify confusion by generating contextually inappropriate phrases or idiomatic expressions, hindering learner comprehension. This becomes especially critical when LLMs are used to assist non-native speakers, as misinterpretations may not be evident to users lacking advanced language proficiency (Li et al., 2024).

5.3. Misrepresentation of Authorial Intent

Misinterpretation can also occur at the level of **authorial voice and intent**. Chetwynd (2024) and Ersöz and Engin (2024) argue that when LLMs are tasked with summarizing or rewriting user content, they may inadvertently shift the original emphasis or argument. This is particularly concerning in disciplines that demand rhetorical nuance, such as philosophy, ethics, or the social sciences (Nam & Bai, 2023).

Khup and Bantugan (2025) observed that in secondary education, students using AI to complete writing tasks sometimes submitted work that did not reflect their original ideas or intent, complicating assessments of learning outcomes. Similarly, in higher education, Khalifa and Albadawy (2024) highlight that even minor lexical changes by LLMs can alter the tone or epistemological stance of a paper, leading to unintended academic consequences.

5.4. Decontextualized Content Generation

A recurring concern is the **decontextualized nature of AI-generated content**, which can strip academic writing of its disciplinary specificity and socio-cultural context. As noted by Alahdab (2024) and Granjeiro et al. (2025), LLMs often lack an understanding of discipline-specific conventions, such as citation styles, hedging language, or argumentation structures. This can result in writing that, while syntactically correct, does not meet the expectations of scholarly communication in specific fields.

This issue is further exacerbated in multilingual academic contexts. Omodan and Marongwe (2024) argue that AI-generated writing often imposes a Westernized epistemology, thereby marginalizing indigenous or localized academic voices. In decolonizing academic writing practices, the blind adoption of AI-generated content may perpetuate linguistic and cultural homogenization (Perkins, 2023).

5.5. Ethical and Pedagogical Implications

The lack of precision and clarity in AI-generated academic text raises pressing **ethical questions** related to authorship, accountability, and student learning (Miao et al., 2023; Guleria et al., 2023). For educators, these tools can serve as a double-edged sword. On one hand, they provide scaffolding for learners; on the other, they risk undermining the development of critical writing and thinking skills if used uncritically (Aljuaid, 2024; Nurchurifiani et al., 2025).

Moreover, ambiguities introduced by AI can lead to **academic integrity violations**. If a model generates content with unclear sources or misrepresented arguments, students may unknowingly plagiarize or misattribute information (Molligan & Pérez-López, 2024; Shofiah et al., 2023). Educators are thus tasked with fostering ethical literacy and critical engagement with AI tools in the classroom (Ocampo et al., 2023; AlSamhori & Alnaimat, 2024).

6. POLICIES AND GUIDELINES FOR RESPONSIBLE USE

The advent of artificial intelligence (AI) tools, particularly large language models (LLMs), has transformed academic writing and research across disciplines. However, this transformation demands clearly defined policies and ethical guidelines to ensure the responsible use of such technologies. The responsible integration of AI in academia must balance innovation, integrity, transparency, and equity (BaHammam, 2023; Casal & Kessler, 2023).

6.1. Transparency and Disclosure Requirements

Academic institutions and publishers must require authors to disclose any use of AI tools such as ChatGPT, Grammarly, or other LLM-powered platforms in the writing or analysis phases of research (Casal & Kessler, 2023; Alahdab, 2024). Disclosure should include the extent and purpose of the tool's usage, including whether it assisted in language editing, data interpretation, or content generation (Miao et al., 2023; Eberlin, 2024). This transparency maintains the trustworthiness of scholarly communication (Lund et al., 2023).

6.2. Academic Integrity and Authorship

AI-generated content should not be credited with authorship, as these systems do not meet criteria related to accountability, intellectual contribution, or conflict-of-interest disclosure (Harati, 2024; Salvagno et al., 2023). Clear guidelines must be in place to delineate the human authors' responsibilities, even when AI is used as a co-writing or support tool (AlSamhori & Alnaimat, 2024; Espino et al., 2024).

Additionally, excessive dependence on AI may challenge the authenticity of students' and researchers' work, especially in contexts where original thought and critical analysis are central (Aslam & Nisar, 2024; Perkins, 2023; Ajiye & Omokhabi, 2025).

6.3. Ethical Training and Digital Literacy

Educators and researchers must receive training in digital ethics, with curricula that integrate AI literacy and awareness of its limitations and biases (Alghamdy, 2023; Omodan & Marongwe, 2024). Such training should prepare users to evaluate AI outputs critically and avoid over-reliance on these tools for knowledge production (Guleria et al., 2023; Fedoriv et al., 2024; Liu et al., 2025).

Institutions should emphasize that AI tools are assistive rather than authoritative, ensuring that human judgment remains central to academic evaluation and publication (Dinçer, 2024; Aljuaid, 2024).

6.4. Plagiarism and Misuse Prevention

The line between legitimate assistance and unethical appropriation of content can blur with AI tools. Institutions must update plagiarism detection policies to account for AI-generated text, which may be original in form but derivative in intent (Chetwynd, 2024; Nam & Bai, 2023).

There should also be institutional mechanisms to monitor and detect misuse, especially regarding ghostwriting, paraphrasing without understanding, and the fabrication of references or data (Ocampo et al., 2023; Subaveerapandiyan et al., 2025).

6.5. Context-Specific Guidelines

Policies must be sensitive to disciplinary differences and levels of education. For instance, while AI-assisted grammar correction may be acceptable for non-native English-speaking scholars (Li et al., 2024), content generation for undergraduate essays may breach academic conduct (Khup & Bantugan, 2025; Shofiah et al., 2023).

Furthermore, the cultural, linguistic, and infrastructural diversity in global academia necessitates adaptable frameworks. For example, faculty perspectives in Indonesia highlight the need for flexible, context-aware policies that respect local pedagogical norms (Nurchurifiani et al., 2025).

6.6. Promoting Inclusive and Ethical Use

The development of policies must involve diverse stakeholders—faculty, students, policymakers, and technologists—to ensure equitable access to AI tools and prevent knowledge disparities (Granjeiro et al., 2025; Molligan & Pérez-López, 2024). Such collaborative policy-making can help decolonize academic writing, giving marginalized communities agency in knowledge creation (Omodan & Marongwe, 2024).

Efforts should also be made to ensure AI tools used in academic contexts are designed ethically, avoiding biases and promoting responsible knowledge dissemination (Espino et al., 2024; Khalifa & Albadawy, 2024).

6.7. Journal and Publisher Guidelines

Publishers play a crucial role in regulating AI use in research publication. Many journals are now issuing guidelines on when and how AI tools can be used, often requiring authors to declare the presence of any machine-generated text (Chetwynd, 2024; Miao et al., 2023). Others are creating AI-review protocols to evaluate whether AI assistance contributed to scientific misrepresentation or plagiarism (Casal & Kessler, 2023).

6.8. Collaborative Oversight and Policy Evolution

Finally, given the rapid evolution of AI technologies, guidelines must remain dynamic. Institutions should develop AI ethics committees or working groups to periodically review and update policies (Kirov, 2023; AlSamhori & Alnaimat, 2024). These bodies can serve as a hub for assessing new tools, addressing emerging risks, and providing community-wide recommendations (Ersöz & Engin, 2024).

To promote responsible and equitable use of AI in academic writing, the following framework highlights key pillars of ethical AI integration: disclosure, training, oversight, and inclusivity.

Figure 4. Framework for ethical use of AI in academic writing

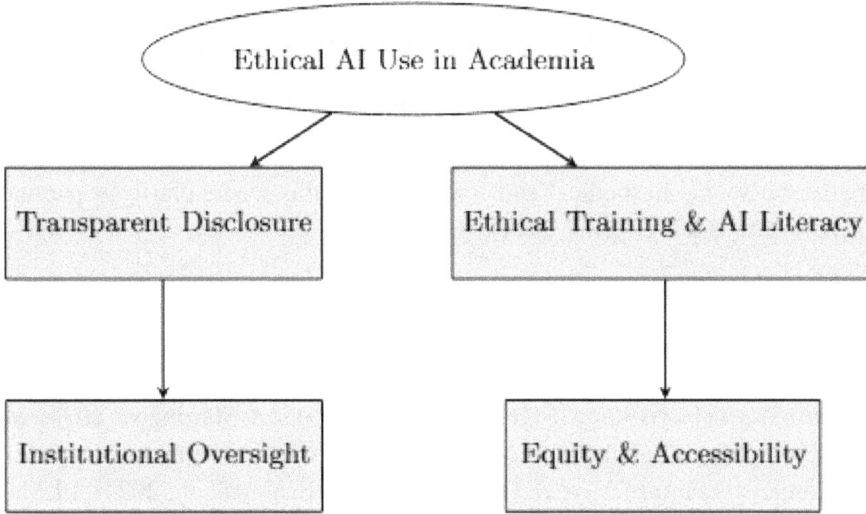

7. SUPPORTING NON-NATIVE ENGLISH-SPEAKING RESEARCHERS

The proliferation of Artificial Intelligence (AI)-powered writing tools, particularly large language models (LLMs), offers transformative potential for non-native English-speaking (NNES) researchers. These tools not only democratize academic writing but also provide linguistic scaffolding, grammar correction, content organization, and discipline-specific vocabulary assistance (Li et al., 2024; Khalifa & Albadawy, 2024; Salvagno et al., 2023). By narrowing the language proficiency gap, AI technologies can significantly enhance the global inclusivity of scholarly communication (Nurchurifiani et al., 2025; Khup & Bantugan, 2025).

7.1 Enhancing Academic Productivity and Confidence

For NNES scholars, academic writing often poses a dual burden: articulating complex ideas and ensuring linguistic accuracy. LLMs such as ChatGPT offer an interactive platform to address both concerns, enabling users to refine manuscripts, understand stylistic conventions, and simulate peer-review feedback (Alahdab, 2024;

Ocampo et al., 2023). This assistance fosters not only improved outcomes but also increased confidence among researchers (Liu et al., 2025).

AI tools can thus serve as "writing collaborators," offering suggestions on coherence, transitions, and argument structure (Aslam & Nisar, 2024; Guleria et al., 2023). For example, NNES users can ask LLMs to rephrase dense sentences or simplify jargon without compromising academic rigor (Harati, 2024). This capability is particularly vital in medical and scientific domains where clarity is paramount (BaHammam, 2023; Eberlin, 2024).

7.2 Promoting Educational Equity

By reducing language barriers, AI can help decolonize academic writing and promote more inclusive knowledge production (Omodan & Marongwe, 2024). Many scholars from the Global South face marginalization in international publishing due to linguistic biases (Ajiye & Omokhabi, 2025; Espino et al., 2024). LLMs can mitigate such inequalities by providing equitable support for manuscript preparation, formatting, and citation consistency (Miao et al., 2023).

These tools also support multilingual education strategies and empower English as a Foreign Language (EFL) educators to integrate writing technologies into classroom instruction (Alghamdy, 2023; Aljuaid, 2024). ChatGPT, in particular, has been applied to enhance writing curricula in both secondary and tertiary education (Khup & Bantugan, 2025), helping students bridge the gap between their ideas and academic expectations.

7.3 Ethical Considerations and Academic Integrity

Despite their benefits, AI tools must be used with caution to maintain ethical integrity in research. Issues such as over-reliance, misattribution of authorship, and lack of transparency in AI-assisted writing are prevalent concerns (Casal & Kessler, 2023; Dinçer, 2024; Molligan & Pérez-López, 2024). Researchers should disclose AI assistance in manuscripts and avoid delegating critical thinking or data interpretation to automated systems (AlSamhori & Alnaimat, 2024; Lund et al., 2023).

Furthermore, the inability of current LLMs to verify factual accuracy or cite credible sources necessitates careful human oversight (Kirov, 2023; Nam & Bai, 2023). Ethical frameworks and institutional policies are essential to guide responsible use (Perkins, 2023; Ersöz & Engin, 2024), particularly among novice researchers who may unknowingly blur the line between support and authorship (Shofiah et al., 2023).

7.4 Best Practices and Institutional Support

Promoting ethical and effective use of AI tools among NNES researchers requires structured guidance. Universities and publishers should offer training on AI literacy, transparency, and ethical citation practices (Chetwynd, 2024; Granjeiro et al., 2025). Moreover, departments can develop tool-specific guidelines to prevent misuse while maximizing pedagogical potential (Fedoriv et al., 2024; Subaveera-pandiyan et al., 2025).

Faculty members play a crucial role in modeling best practices by integrating AI-supported writing tasks with peer review, reflective critique, and human editing (Molligan & Pérez-López, 2024; Miao et al., 2023). Peer collaboration can further ensure that AI remains a complement to—not a substitute for—critical academic skills (Omodan & Marongwe, 2024; Guleria et al., 2023).

The following flowchart maps the institutional and technological pathways available to support non-native English-speaking (NNES) researchers in ethical and effective AI-supported writing.

Figure 5. Support pathways for NNES researchers using AI

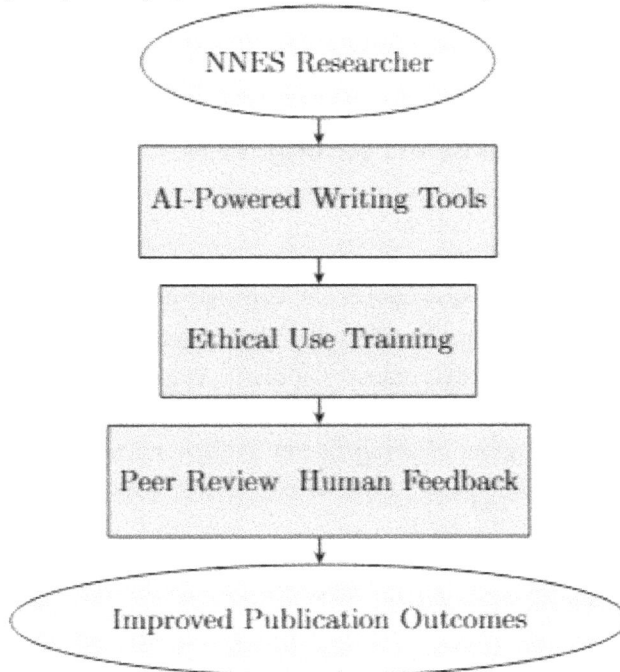

209

8. FUTURE TRENDS IN AI AND ACADEMIC ENGLISH

As artificial intelligence (AI) continues to evolve rapidly, its integration into academic English and scholarly communication is becoming both inevitable and transformative. The future of AI in academic English encompasses pedagogical innovation, enhanced inclusivity, ethical evolution, and increasingly sophisticated AI-human collaboration.

8.1. Personalized and Adaptive Learning Environments

One of the most promising trends is the deployment of AI to create personalized learning experiences tailored to the linguistic proficiency and academic goals of individual learners. AI-powered tools like ChatGPT can provide real-time feedback, identify writing weaknesses, and offer contextualized grammar or vocabulary suggestions. This supports second-language learners in academic settings, where precision and tone are critical (Li et al., 2024; Nurchurifiani et al., 2025; Alghamdy, 2023).

Moreover, by leveraging natural language processing (NLP), AI can analyze a student's historical performance to predict writing challenges and recommend targeted resources (Khup & Bantugan, 2025). Such systems could adapt writing instruction dynamically, moving from a one-size-fits-all model to personalized development pathways.

8.2. AI-Enhanced Academic Writing for Non-Native Speakers

AI is expected to play an increasingly significant role in supporting non-native English speakers, especially in writing research papers and dissertations. Tools like GPT-4 not only correct grammar and syntax but also aid in formulating discipline-specific expressions and complex academic arguments (Li et al., 2024; Khalifa & Albadawy, 2024). These developments can democratize academic publishing by reducing linguistic barriers (Omodan & Marongwe, 2024).

However, concerns about over-reliance on AI-generated text persist, as it may lead to the erosion of independent academic voice and identity (Ersöz & Engin, 2024; Kirov, 2023).

8.3. Ethical Frameworks and Governance

With AI increasingly involved in content generation, ethical considerations are becoming central to academic discourse. Key concerns include authorship attribution, data provenance, plagiarism, and transparency (BaHammam, 2023; Casal & Kessler, 2023; AlSamhori & Alnaimat, 2024). Emerging ethical frameworks are expected to focus on responsible AI use, informed consent, and clear disclosure of AI assistance (Aslam & Nisar, 2024; Dinçer, 2024; Eberlin, 2024).

Journals and universities are beginning to adopt AI ethics policies that delineate acceptable levels of AI use in research and writing (Miao et al., 2023; Perkins, 2023). These policies are likely to become more standardized and globally coordinated in the near future.

8.4. Interdisciplinary Collaborations and Research Innovation

AI is enabling new forms of interdisciplinary collaboration by streamlining the technical aspects of writing and allowing researchers to focus on ideation and critical analysis. From literature reviews to data interpretation, AI can now assist across stages of the research lifecycle, fostering innovation (Granjeiro et al., 2025; Espino et al., 2024; Chetwynd, 2024).

In fields such as medicine, engineering, and social sciences, where English writing skills may be unevenly distributed, AI acts as an equalizer (Harati, 2024; Salvagno et al., 2023; Molligan & Pérez-López, 2024).

8.5. The Role of AI in Academic Integrity and Publishing

Academic institutions are already grappling with questions about AI's influence on academic integrity. Distinguishing between human- and AI-generated content is a growing challenge (Casal & Kessler, 2023; Lund et al., 2023; Nam & Bai, 2023). Future trends suggest the development of advanced detection tools and the redefinition of what constitutes plagiarism in the AI age (Guleria et al., 2023; Shofiah et al., 2023).

Additionally, peer review models may be reimagined to account for AI-assisted writing, possibly incorporating machine-generated summaries or ethical declarations (Alahdab, 2024; Ajiye & Omokhabi, 2025).

8.6. Decolonizing and Democratizing Academic English

AI offers potential in decolonizing academic discourse by enabling multilingual and culturally contextualized writing support (Omodan & Marongwe, 2024). Tools trained on diverse linguistic datasets can help students frame their research narratives within non-Western paradigms, thus contributing to more inclusive knowledge production.

Efforts are already underway to customize AI tools for regional dialects and academic registers outside native English norms, ensuring that academic voice remains locally grounded yet globally communicable (Fedoriv et al., 2024).

8.7. Best Practices and Pedagogical Integration

Looking ahead, AI is likely to be embedded more systematically into academic curricula through workshops, writing labs, and institutional guidelines. Faculty development programs may train educators on leveraging AI while maintaining academic rigor (Aljuaid, 2024; Ocampo et al., 2023; Subaveerapandiyan et al., 2025). Institutions must balance empowerment with caution, encouraging ethical creativity over rote automation.

As highlighted by AlSamhori and Alnaimat (2024), fostering awareness about ethical AI usage from early academic stages is key to long-term integrity and innovation.

8.8 Summary

The future of AI in academic English is filled with promise and complexity. From empowering non-native speakers and promoting inclusive knowledge production to challenging existing notions of authorship and integrity, AI's role will continue to expand. However, this expansion must be guided by robust ethical frameworks, responsible policies, and a commitment to preserving the human elements of academic inquiry—curiosity, critical thinking, and creativity.

9. CONCLUSION

The integration of Artificial Intelligence (AI), particularly large language models (LLMs) like ChatGPT, into academic English and scholarly writing is reshaping the educational landscape. This transformation brings significant advantages, including enhanced productivity, improved linguistic precision, and broader accessibility for non-native English speakers. Researchers such as Khalifa and Albadawy (2024) and Li et al. (2024) have demonstrated how AI-powered tools can serve as essential aids in producing higher-quality academic texts, supporting both novice and experienced scholars alike.

However, this technological evolution is accompanied by pressing ethical, pedagogical, and practical considerations. Multiple scholars (e.g., Casal & Kessler, 2023; Dinçer, 2024; Chetwynd, 2024) emphasize the importance of maintaining academic integrity, transparency, and authorship accountability when employing AI tools. The risk of over-reliance on AI may undermine critical thinking, originality, and the learning process itself, especially in language education (Alghamdy, 2023; Ersöz & Engin, 2024).

Moreover, the growing capability of AI to emulate human writing prompts on-going debate over the boundaries of acceptable use in scholarly publishing. Ethical frameworks and institutional policies, as highlighted by Miao et al. (2023), Espino et al. (2024), and Aslam & Nisar (2024), are urgently needed to ensure responsible usage and prevent misuse, such as ghostwriting or unacknowledged AI-generated content.

Despite these challenges, the future of AI in academic English is promising. As proposed by Omodan and Marongwe (2024), AI can even contribute to decol-onizing academic discourse by democratizing access to scholarly communication and supporting inclusive knowledge production. Future developments will likely focus on establishing standardized guidelines (BaHammam, 2023; Granjeiro et al., 2025), refining AI tools for educational contexts (Nurchurifiani et al., 2025; Khup & Bantugan, 2025), and fostering AI literacy among students and educators alike.

In sum, while AI technologies hold remarkable potential to support and revolu-tionize academic English, their application must be tempered with ethical vigilance, pedagogical awareness, and a commitment to fostering human intellectual growth. As we advance, collaboration among educators, technologists, ethicists, and policymakers will be vital to navigating this evolving terrain with both innovation and integrity.

REFERENCES

Ajiye, O. T., & Omokhabi, A. A. (2025). The Potential And Ethical Issues Of Artificial Intelligence In Improving Academic Writing. *ShodhAI: Journal of Artificial Intelligence*, 2(1), 1–9. DOI: 10.29121/shodhai.v2.i1.2025.24

Alahdab, F. (2024). Potential impact of large language models on academic writing. *BMJ Evidence-Based Medicine*, 29(3), 201–202. DOI: 10.1136/bmjebm-2023-112429 PMID: 37620013

Alghamdy, R. Z. (2023). Pedagogical and ethical implications of artificial intelligence in EFL context: A review study. *English Language Teaching*, 16(10), 87–98. DOI: 10.5539/elt.v16n10p87

Aljuaid, H. (2024). The impact of artificial intelligence tools on academic writing instruction in higher education: A systematic review. *Arab World English Journal (AWEJ) Special Issue on ChatGPT.*

AlSamhori, A. F., & Alnaimat, F. (2024). Artificial intelligence in writing and research: Ethical implications and best practices. *Central Asian Journal of Medical Hypotheses and Ethics = Central'noaziatskij Zurnal Medicinskich Gipotez i Etiki = Medicinalyk Gipoteza Men Ètikanyn Orta Aziâlyk Zurnaly*, 5(4), 259–268. DOI: 10.47316/cajmhe.2024.5.4.02

Aslam, M. S., & Nisar, S. (2024). Ethical Considerations for Artificial Intelligence Tools in Academic Research and Manuscript Preparation: A Web Content Analysis. In *Digital Transformation in Higher Education, Part B: Cases, Examples and Good Practices* (pp. 155-196). Emerald Publishing Limited. DOI: 10.1108/978-1-83608-424-220241007

BaHammam, A. S. (2023). Balancing innovation and integrity: The role of AI in research and scientific writing. *Nature and Science of Sleep*, ●●●, 1153–1156. PMID: 38170140

Casal, J. E., & Kessler, M. (2023). Can linguists distinguish between ChatGPT/AI and human writing?: A study of research ethics and academic publishing. *Research Methods in Applied Linguistics*, 2(3), 100068. DOI: 10.1016/j.rmal.2023.100068

Chetwynd, E. (2024). Ethical use of artificial intelligence for scientific writing: Current trends. *Journal of Human Lactation*, 40(2), 211–215. DOI: 10.1177/08903344241235160 PMID: 38482810

Dinçer, S. (2024). The use and ethical implications of artificial intelligence in scientific research and academic writing. *Educational Research & Implementation*, *1*(2), 139–144. DOI: 10.14527/edure.2024.10

Eberlin, M. N. (2024). The Art of Scientific Writing and Ethical Use of Artificial Intelligence. *Journal of the Brazilian Chemical Society*, *35*(1), e-20230121.

Ersöz, A. R., & Engin, M. (2024). Exploring Ethical Dilemmas in the Use of Artificial Intelligence in Academic Writing: Perspectives of Researchers. *Journal of Uludag University Faculty of Education*, *37*(3), 1190–1208. DOI: 10.19171/uefad.1514323

Espino, A. R. C., Esto, M. R. A., Manalo, C. G. S., Santiago, C. S., & Pragacha, R. N. (2024, June). Ethical Implications of Using Assistive Writing Tools in the Academe: A Literature Review. In *International Conference on Frontiers of Intelligent Computing: Theory and Applications* (pp. 133-145). Singapore: Springer Nature Singapore.

Fedoriv, Y., Shuhai, A., & Pirozhenko, I. (2024). Foundations of Ethical Use of AI in EFL Academic Writing.

Granjeiro, J. M., Cury, A. A. D. B., Cury, J. A., Bueno, M., Sousa-Neto, M. D., & Estrela, C. (2025). The Future of Scientific Writing: AI Tools, Benefits, and Ethical Implications. *Brazilian Dental Journal*, *36*, e25–e6471. DOI: 10.1590/0103-644020256471 PMID: 40197923

Guleria, A., Krishan, K., Sharma, V., & Kanchan, T. (2023). ChatGPT: Ethical concerns and challenges in academics and research. *Journal of Infection in Developing Countries*, *17*(09), 1292–1299. DOI: 10.3855/jidc.18738 PMID: 37824352

Harati, K. (2024). ChatGPT and AI-Powered Writing Tools: Unveiling Risks and Ethical Challenges in Scientific Writing. *Journal of Research in Medical Sciences : the Official Journal of Isfahan University of Medical Sciences*, *4*(1), 1–6.

Khalifa, M., & Albadawy, M. (2024). Using artificial intelligence in academic writing and research: An essential productivity tool. *Computer Methods and Programs in Biomedicine Update*, *5*, 100145. DOI: 10.1016/j.cmpbup.2024.100145

Khup, V. K., & Bantugan, B. (2025). Exploring the Impact and Ethical Implications of Integrating AI-Powered Writing Tools in Junior High School English Instruction: Enhancing Creativity, Proficiency, and Academic Outcomes. *International Journal of Research and Innovation in Social Science*, *9*(IIIS, 3s), 361–378. DOI: 10.47772/ IJRISS.2025.903SEDU0022

Kirov, B. (2023, September). Artificial Intelligence in Creation of Scientific Written Works: Weighing the Benefits and Ethical Dilemmas-Should We Use It? In *2023 International Scientific Conference on Computer Science (COMSCI)* (pp. 1-5). IEEE. DOI: 10.1109/COMSCI59259.2023.10315821

Li, J., Zong, H., Wu, E., Wu, R., Peng, Z., Zhao, J., Yang, L., Xie, H., & Shen, B. (2024). Exploring the potential of artificial intelligence to enhance the writing of english academic papers by non-native english-speaking medical students-the educational application of ChatGPT. *BMC Medical Education*, *24*(1), 736. DOI: 10.1186/s12909-024-05738-y PMID: 38982429

Liu, Y., Kong, W., & Merve, K. (2025). ChatGPT applications in academic writing: A review of potential, limitations, and ethical challenges. *Arquivos Brasileiros de Oftalmologia*, *88*(3), e2024–e0269. DOI: 10.5935/0004-2749.2024-0269 PMID: 39879415

Lund, B. D., Wang, T., Mannuru, N. R., Nie, B., Shimray, S., & Wang, Z. (2023). ChatGPT and a new academic reality: Artificial Intelligence-written research papers and the ethics of the large language models in scholarly publishing. *Journal of the Association for Information Science and Technology*, *74*(5), 570–581. DOI: 10.1002/asi.24750

Miao, J., Thongprayoon, C., Suppadungsuk, S., Garcia Valencia, O. A., Qureshi, F., & Cheungpasitporn, W. (2023). Ethical dilemmas in using AI for academic writing and an example framework for peer review in nephrology academia: A narrative review. *Clinics and Practice*, *14*(1), 89–105. DOI: 10.3390/clinpract14010008 PMID: 38248432

Molligan, J., & Pérez-López, E. (2024). Artificial intelligence in academia: Opportunities, challenges, and ethical considerations. *Biochemistry and Cell Biology*, *103*, 1–3. DOI: 10.1139/bcb-2024-0216 PMID: 39611424

Nam, B. H., & Bai, Q. (2023). ChatGPT and its ethical implications for STEM research and higher education: A media discourse analysis. *International Journal of STEM Education*, *10*(1), 66. DOI: 10.1186/s40594-023-00452-5

Nurchurifiani, E., Maximilian, A., Ajeng, G. D., Wiratno, P., Hastomo, T., & Wicaksono, A. (2025). Leveraging AI-Powered Tools in Academic Writing and Research: Insights from English Faculty Members in Indonesia. *International Journal of Information and Education Technology (IJIET)*, *15*(2), 312–322. DOI: 10.18178/ijiet.2025.15.2.2244

Ocampo, T. S. C., Silva, T. P., Alencar-Palha, C., Haiter-Neto, F., & Oliveira, M. L. (2023). ChatGPT and scientific writing: A reflection on the ethical boundaries. *Imaging Science in Dentistry*, *53*(2), 175. DOI: 10.5624/isd.20230085 PMID: 37405199

Omodan, B. I., & Marongwe, N. (2024). The role of artificial intelligence in decolonising academic writing for inclusive knowledge production. *Interdisciplinary Journal of Education Research*, *6*(s1), 1–14. DOI: 10.38140/ijer-2024.vol6.s1.06

Perkins, M. (2023). Academic Integrity considerations of AI Large Language Models in the post-pandemic era: ChatGPT and beyond. *Journal of University Teaching & Learning Practice*, *20*(2), 1–24. DOI: 10.53761/1.20.02.07

Salvagno, M., Taccone, F. S., & Gerli, A. G. (2023). Can artificial intelligence help for scientific writing? *Critical Care*, *27*(1), 75. DOI: 10.1186/s13054-023-04380-2 PMID: 36841840

Shofiah, N., Putera, Z. F., & Solichah, N. (2023, December). Challenges and opportunities in the use of artificial intelligence in education for academic writing: A scoping review. In *Conference Psychology and Flourishing Humanity (PFH 2023)* (pp. 174-193). Atlantis Press. DOI: 10.2991/978-2-38476-188-3_20

Subaveerapandiyan, A., Kalbande, D., & Ahmad, N. (2025). Perceptions of effectiveness and ethical use of AI tools in academic writing: A study Among PhD scholars in India. *Information Development*, •••, 02666669251314840. DOI: 10.1177/02666669251314840

Tai, A. M. Y., Meyer, M., Varidel, M., Prodan, A., Vogel, M., Iorfino, F., & Krausz, R. M. (2023). Exploring the potential and limitations of ChatGPT for academic peer-reviewed writing: Addressing linguistic injustice and ethical concerns. *Journal of Academic Language and Learning*, *17*(1), T16–T30.

Ugwu, N. F., Igbinlade, A. S., Ochiaka, R. E., Ezeani, U. D., Okorie, N. C., Opele, J. K., Onayinka, T. S., Iroegbu, O., Onyekwere, O. K., Adams, A. B., Aigbona, P., & Ojobola, F. B. (2024). Clarifying Ethical Dilemmas in Using Artificial Intelligence in Research Writing: A Rapid Review. *Higher Learning Research Communications*, *14*(2), 29–47. DOI: 10.18870/hlrc.v142.1549

Wiwanitmkit, S., & Wiwanitkit, V. (2024). Artificial Intelligence, Academic Publishing, Scientific Writing, Peer Review, and Ethics. *Brazilian Journal of Cardiovascular Surgery*, *39*(4), e20230377. DOI: 10.21470/1678-9741-2023-0377 PMID: 39038191

Ya'u, M. S., & Mohammed, M. S. (2025). AI-Assisted Writing and Academic Literacy: Investigating the Dual Impact of Language Models on Writing Proficiency and Ethical Concerns in Nigerian Higher Education. *International Journal of Education and Literacy Studies*, *13*(2), 593–604. DOI: 10.7575/aiac.ijels.v.13n.2p.593

Zohouri, M., Sabzali, M., & Golmohammadi, A. (2024). Ethical considerations of ChatGPT-assisted article writing. *Synesis (ISSN 1984-6754), 16*(1), 94-113.

Chapter 6
Open Science vs. Responsible Science:
Balancing Transparency and Security

Syed Q. Raza
https://orcid.org/0009-0003-6952-2667
Lindsey Wilcosn College, USA

ABSTRACT

The integration of Artificial Intelligence (AI) into scientific, technological, and medical(STM) research raises critical questions about balancing open science and responsible science. Open science promotes transparency, accessibility, and knowledge sharing to foster collaboration, innovation, and reproducibility. However, as sensitive data and proprietary technologies become central to research, security and intellectual property protection are significant concerns. This chapter explores the tension between these objectives and provides a framework for balancing transparency with security in AI-driven research. It discusses the ethical, legal, and technical challenges faced by researchers and institutions in adopting practices that promote both open access and secure data management. The chapter offers insights into reconciling open science initiatives with responsible science practices that prioritize privacy, confidentiality, and safety, highlighting the importance of policies and strategies for secure and ethical research practices.

DOI: 10.4018/979-8-3373-4252-8.ch006

1. INTRODUCTION

The rapid advancement of artificial intelligence (AI) has significantly impacted various industries, including cybersecurity, healthcare, business, and communication systems. As organizations and individuals increasingly rely on AI-powered technologies to enhance operational efficiency, the need for robust security and privacy mechanisms has become more critical. AI offers promising solutions to mitigate cybersecurity risks and protect sensitive data across multiple domains. However, as AI systems evolve, they also raise complex challenges related to data privacy, security, and ethical considerations. The integration of AI into cybersecurity has led to the development of innovative techniques, such as threat detection, anomaly identification, and encryption, to address these challenges (Abbas & Qazi, 2024; Akhtar & Rawol, 2024).

This chapter explores the emerging trends and applications of AI-powered technologies in enhancing privacy and security across various sectors. The increasing reliance on AI has given rise to a new era of cybersecurity, where AI tools are used to detect vulnerabilities, prevent malicious attacks, and safeguard data in real time (Gholami & Omar, 2023). While AI's potential to improve security is immense, it is not without its challenges. Issues such as data privacy concerns, ethical implications, and the risk of AI misuse must be addressed to ensure that AI-powered security systems are both effective and responsible (Alhitmi et al., 2024; Patel & Desai, 2024). Moreover, with the introduction of AI-powered solutions in diverse fields such as healthcare (Arefin, 2024), finance (Brightwood & Jame, 2024), and the Internet of Things (IoT) (Arya et al., 2024), there is a growing need to explore the intersection of AI and data protection strategies.

In the healthcare industry, AI-powered systems have revolutionized data management, diagnosis, and treatment planning. However, these systems also generate vast amounts of sensitive patient data, raising concerns about data breaches and unauthorized access (Bhamidipaty et al., 2025). Similarly, AI-driven IoT devices offer immense convenience but also create new vulnerabilities that could be exploited by cybercriminals (Farea et al., 2024). These challenges highlight the need for strong AI-driven security frameworks capable of protecting data privacy while ensuring the safe and ethical use of these technologies.

This chapter will also delve into the ethical implications of AI-powered security systems and explore potential regulatory frameworks to manage the risks associated with their deployment (Sato, 2023). Ethical concerns such as bias in AI algorithms, transparency, and accountability in decision-making processes will be discussed, along with strategies to mitigate these risks (Pandey et al., 2024). Additionally, the chapter will highlight case studies and real-world applications where AI has been successfully integrated into cybersecurity strategies, showcasing its role in preventing cyber threats and protecting sensitive data (Jones et al., 2023; Huff et al., 2023).

By examining the current state and future directions of AI-powered security, this chapter aims to provide insights into how AI can continue to shape the future of cybersecurity and data protection. Key topics include advancements in AI-based encryption methods, the role of machine learning in threat detection, and the importance of ensuring privacy in AI-driven systems (Gopireddy, 2021; Anidjar et al., 2023). Moreover, the chapter will explore innovative AI techniques being employed across sectors such as finance, healthcare, and IoT to enhance security and ensure data privacy (Gholami & Omar, 2024; Kumar et al., 2024).

Ultimately, this introduction sets the stage for an in-depth exploration of AI-powered security and privacy configurations in modern systems, emphasizing the need for continual innovation and ethical oversight to navigate the complexities of an increasingly interconnected world.

2. THE PHILOSOPHY AND PRINCIPLES OF OPEN SCIENCE

Open Science is a transformative approach that seeks to democratize the process of scientific research, ensuring that scientific knowledge, data, and methodologies are freely available, accessible, and usable by all. The philosophy of Open Science is rooted in the belief that transparency, collaboration, and the open sharing of knowledge can accelerate innovation, improve research quality, and make science more inclusive. By making scientific outputs freely available, Open Science strives to create a more equitable and efficient research ecosystem.

2.1 Core Principles of Open Science

1. **Open Access to Research Outputs**

Open access is perhaps the most visible aspect of Open Science. It emphasizes the availability of scientific publications without paywalls, ensuring that research can be read, shared, and reused by anyone, anywhere. This principle advocates for researchers to publish their findings in open access journals or repositories, allowing the global community to benefit from their work (Abbas & Qazi, 2024; Sato, 2023). Open access also enables data reuse, where other researchers can build upon previously published results to further scientific advancements (Abolaji & Akinwande, 2024).

2. **Open Data**

Open Science promotes the idea that research data should be made publicly available so others can access, verify, and reuse the data. This principle enhances the reproducibility of research and enables others to perform secondary analyses or apply new methods to the data. Open data is essential for advancing fields such as AI and genomics, where large datasets are required to train models and uncover new insights (Farea et al., 2024; Gawankar et al., 2024). Ensuring privacy and security when making such data open remains a significant challenge, especially with sensitive data such as healthcare records (Gholami & Omar, 2023; Bhamidipaty et al., 2025).

3. **Open Methods and Tools**

Open Science encourages the development and dissemination of open-source software and methods, enabling researchers to access and contribute to the tools and algorithms used in scientific discovery. By providing transparent methods, the scientific community can build on each other's work and verify the results more effectively (Awad et al., 2024; Thirunagalingam, 2024). In fields like AI and machine learning, the sharing of models, code, and algorithms is crucial for fostering innovation and collaboration (Gholami & Omar, 2024).

4. **Collaboration and Inclusivity**

Open Science encourages the sharing of knowledge across disciplines, countries, and communities. This principle emphasizes that science should not be an isolated endeavor but a collective one. Collaboration among researchers, industry, policymakers, and the public enhances the quality and applicability of research. Inclusivity in Open Science also extends to underrepresented groups, ensuring that diverse perspectives are incorporated into scientific discovery (Anidjar et al., 2023; Jones et al., 2023). It also ensures that marginalized populations are considered when developing AI tools and technologies (Alhitmi et al., 2024).

5. **Transparency and Reproducibility**

Open Science promotes transparency in all aspects of the research process, including data collection, analysis, and reporting. Ensuring reproducibility means that others can replicate findings using the same methods and data. This principle is especially relevant in the context of AI and machine learning, where the ability to reproduce results is crucial for verifying the reliability and accuracy of models (Rehan, 2023; Mbah & Evelyn, 2024). By making methodologies and data open, researchers can enhance the credibility of their findings and build trust within the scientific community (Kumar et al., 2024; Gopireddy, 2021).

6. Ethical Considerations

As Open Science grows, so do the ethical challenges, especially in areas where privacy, data security, and personal information are involved. For instance, AI-driven research can raise concerns about the protection of personal data, such as in healthcare or genomics, where sensitive information must be handled with care. Ethical guidelines are crucial in Open Science to ensure that researchers are accountable for the use of data, the impact of their findings, and the potential harms their research could cause (Patel & Desai, 2024; Brightwood & Jame, 2024). The balance between openness and privacy is a delicate one, particularly in fields like healthcare where patient confidentiality is paramount (Gawankar et al., 2024).

7. Open Peer Review

Open Science calls for the adoption of open peer review, where the process of reviewing research papers is transparent. This approach helps to reduce biases in the review process and provides a platform for constructive feedback. Open peer review can also improve the quality of research by making reviewers' comments and authors' responses publicly accessible (Jones & Omar, 2024). Furthermore, it fosters collaboration and trust between researchers and reviewers.

To provide a quantitative overview of the emphasis placed on each principle of Open Science in recent literature, the following chart shows their relative importance based on frequency in scholarly discussions.

Figure 1. Relative emphasis on open science principles

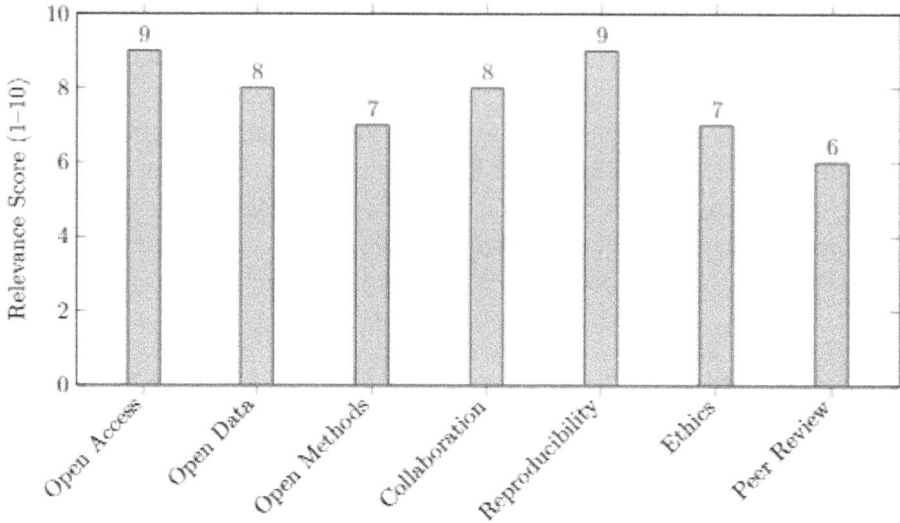

2.2 The Role of AI in Open Science

Artificial Intelligence (AI) plays a crucial role in advancing Open Science, particularly in areas such as data analysis, security, and privacy. AI can accelerate the process of data interpretation and enhance the reproducibility of results by automating tasks like data cleaning, analysis, and visualization. Furthermore, AI-powered systems can be used to secure sensitive data, ensuring that privacy is maintained even as data is made open (Gholami & Omar, 2024; Akhtar & Rawol, 2024).

In the context of AI-powered systems, ensuring privacy and security becomes a top priority, especially when dealing with sensitive data in fields like healthcare (Arefin, 2024; Gholami & Omar, 2023). AI models must be trained with privacy-preserving mechanisms, such as differential privacy or federated learning, to ensure that the data shared does not expose personal information (Bhamidipaty et al., 2025). Additionally, AI has the potential to detect security threats in open data repositories, thereby safeguarding against malicious activities that could exploit vulnerabilities in shared datasets (Awad et al., 2024; Arya et al., 2023).

2.3 Summary

The philosophy of Open Science advocates for openness, transparency, and collaboration, ensuring that scientific knowledge benefits everyone, not just a select few. By embracing Open Science, the research community can accelerate innovation, improve scientific integrity, and make significant strides in addressing global challenges. However, as Open Science continues to evolve, the integration of AI must be approached with careful consideration of privacy, security, and ethical implications. Only by addressing these challenges can we fully realize the potential of Open Science in the digital age.

This vision of Open Science is not just about making data and publications open but also about creating a collaborative, inclusive, and transparent research environment where AI plays a central role in advancing scientific discovery while safeguarding privacy and security (Abbas & Qazi, 2024; Sato, 2023).

3. RESPONSIBLE SCIENCE AND THE IMPERATIVE FOR CAUTION

As artificial intelligence (AI) becomes increasingly integrated into various sectors, the ethical, security, and privacy concerns surrounding its applications demand a cautious and responsible approach. The rapid pace of AI adoption, particularly in sensitive areas like healthcare, finance, and marketing, has brought attention to the need for robust frameworks that address the potential risks associated with AI technologies. This section explores the imperative for caution in AI-driven innovations and the responsibility of scientists, developers, and organizations to ensure that their work upholds ethical standards while mitigating risks.

3.1 Ethical Considerations in AI Development

AI technologies have the power to revolutionize industries and improve the quality of life across various domains. However, this transformation is not without its ethical implications. Issues such as data privacy, algorithmic bias, transparency, and accountability must be addressed to ensure that AI is used responsibly. As noted by Abbas and Qazi (2024), AI-powered systems, particularly in social media platforms, must be configured to prioritize user security and privacy, tailoring settings to meet individual needs while respecting privacy rights. Such considerations are vital to ensuring that AI innovations do not inadvertently perpetuate inequality or harm marginalized groups.

The ethical responsibility of AI developers is particularly pressing in sectors such as healthcare and business, where the stakes of mishandling sensitive data are high. Alhitmi et al. (2024) emphasize the need for AI-driven marketing systems to balance innovation with privacy protections, suggesting that businesses must design AI solutions that align with ethical principles while addressing data security concerns. AI-powered health systems, as highlighted by Almeida and Barr (2025), need to navigate complex ethical, legal, and technological challenges to ensure that innovations in healthcare are both secure and respectful of patient autonomy.

3.2 Security and Privacy in AI Applications

The integration of AI in systems that handle personal or sensitive information raises significant privacy and security concerns. Akhtar and Rawol (2024) stress the importance of AI-powered security mechanisms in enhancing cybersecurity efforts, particularly in protecting data from breaches and malicious attacks. AI's ability to predict and identify potential security threats is a game-changer, but it must be accompanied by robust encryption and security protocols to safeguard the data it processes.

In the context of healthcare, where the misuse of data can lead to catastrophic consequences, Arefin (2024) advocates for the use of AI-powered threat detection systems to protect sensitive health data. Such systems must be designed with a strong emphasis on privacy preservation, as personal health information is among the most vulnerable to misuse. The use of AI in genomics and precision medicine further underscores the need for careful consideration of data privacy, as highlighted by Daraf and Badi (2023), who argue that AI's power to analyze large datasets can advance medical research but must be coupled with strict safeguards to prevent unauthorized access to patient data.

Moreover, the rise of AI-powered biometrics, as discussed by Awad et al. (2024), introduces new challenges in terms of data protection and identity theft. These systems can offer significant advancements in security, but they also create new avenues for misuse if not carefully managed. The importance of integrating privacy-preserving mechanisms, such as anonymization and secure storage, cannot be overstated in this context.

3.3 AI in the Metaverse: Privacy Challenges

The burgeoning AI-powered Metaverse introduces an entirely new dimension to the debate around responsible AI use. Anidjar et al. (2023) highlight the potential risks of data infrastructure in the Metaverse, where users' personal and behavioral data can be exploited without proper safeguards. In such environments, the line between data sharing for convenience and privacy invasion can blur, and it is essential for developers to establish clear protocols to ensure that user consent is always obtained and data is handled transparently.

Furthermore, Sato (2023) points out the challenges of ensuring data privacy in the digital landscape, where AI's capabilities to aggregate and analyze personal data are becoming more sophisticated. In the Metaverse, this challenge is amplified, as the decentralized and immersive nature of virtual worlds can make it difficult to track and control data usage. Therefore, AI systems in these contexts must be designed to respect user privacy by default, implementing strong encryption and anonymization techniques to protect user identities.

3.4 Regulatory and Compliance Frameworks for AI

Given the growing concerns over privacy and security in AI applications, there is an increasing call for comprehensive regulatory frameworks that govern AI development and deployment. Almeida and Barr (2025) argue that innovations in health data protection, especially in AI-powered diagnostic systems, must be guided by clear ethical and legal frameworks to ensure patient data is protected and used responsibly. Similarly, Pandey et al. (2024) advocate for the establishment of stringent regulations to prevent AI-powered systems from becoming tools for unethical practices, particularly in communication tools that may inadvertently violate user privacy.

The role of regulatory bodies is crucial in ensuring that AI technologies are developed and deployed in a manner that prioritizes public safety and ethical considerations. As Farea et al. (2024) suggest, policymakers need to collaborate with AI researchers and practitioners to design regulations that address not only the technical capabilities of AI but also the social, ethical, and legal challenges that these technologies introduce. These regulations must also evolve as AI systems continue to advance, ensuring that privacy and security are always safeguarded.

3.5 The Future of Responsible AI

Looking to the future, responsible AI development requires a collective effort to balance innovation with caution. Gholami and Omar (2024) emphasize that AI's potential to enhance efficiency and security must not overshadow the need for responsible data handling and privacy preservation. As AI systems become more pervasive, there is a growing need for continuous research into AI's societal impacts, particularly in terms of its ability to perpetuate or mitigate inequality and discrimination.

Innovations in AI-powered healthcare, finance, and security must always be evaluated through a lens of ethical responsibility. As Bhamidipaty et al. (2025) note, AI's potential to revolutionize healthcare is immense, but it must be guided by ethical principles that prioritize patient privacy and security. Similarly, Kumar et al. (2024) stress the need for adaptive data protection mechanisms that can respond to the evolving risks posed by AI technologies. In the business and finance sectors, Brightwood and Jame (2024) highlight the importance of establishing clear ethical guidelines for AI-powered financial systems to protect consumer data and prevent exploitation.

3.6 Summary

In conclusion, while AI offers transformative potential across industries, its deployment must be approached with caution and responsibility. Developers, organizations, and regulators must work collaboratively to address the ethical, security, and privacy challenges posed by AI technologies. The imperative for responsible science is clear: AI must be used to enhance societal welfare while safeguarding individual rights and freedoms. By integrating robust privacy protections, ethical guidelines, and regulatory frameworks, we can ensure that AI continues to benefit society without compromising trust or security.

The following flowchart illustrates the dynamic interaction between the goals of Open Science and the ethical safeguards of Responsible Science, emphasizing areas of convergence and conflict.

Figure 2. Balancing open science with responsible science

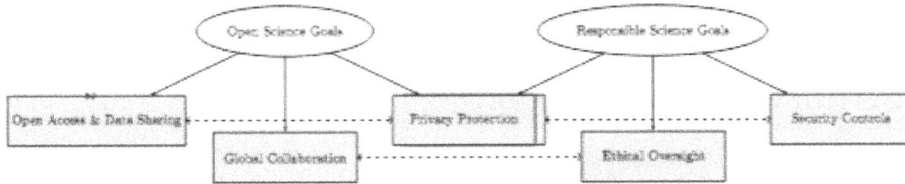

4. TENSIONS BETWEEN OPENNESS AND SECURITY IN AI AND STM FIELDS

The integration of artificial intelligence (AI) with various technological fields has sparked both optimism and caution. AI offers transformative potential across industries, but it also brings with it inherent tensions between openness and security. These concerns are particularly evident in the fields of AI, machine learning (ML), and smart technologies management (STM), where the accessibility of data and models can conflict with the need for robust security and privacy protection.

4.1 The Promise of Openness in AI

AI thrives on access to large datasets and open-source tools. Open access to AI models, codebases, and datasets fosters innovation, enabling researchers and developers to build on each other's work. Open-source AI frameworks like Tensor-Flow and PyTorch have accelerated the development of machine learning models across a wide range of applications. Similarly, the development of large language models (LLMs) and their open availability for research purposes has resulted in breakthrough advancements in NLP (Gholami & Omar, 2023; Gholami & Omar, 2024). This openness contributes to rapid innovation in fields such as healthcare, business, and cybersecurity, where AI solutions are deployed to solve complex problems (Omar & Zangana, 2024a).

However, the very same openness that fuels AI's growth also raises significant concerns regarding data security and privacy. With AI models and tools being widely accessible, malicious actors can exploit vulnerabilities within these systems. The use of open-source AI in sensitive areas such as healthcare, finance, and government presents a paradox: while openness accelerates progress, it exposes critical systems to higher risks of data breaches and misuse (Almeida & Barr, 2025; Brightwood

& Jame, 2024). This dilemma is particularly evident in sectors where patient data, financial transactions, and personal information are involved, leading to the necessity of balancing openness with security measures.

4.2 Security Risks in AI-Powered Systems

AI-powered systems, especially those integrated with Internet of Things (IoT) devices, cloud environments, and smart infrastructure, are highly vulnerable to security risks. AI algorithms rely on vast amounts of data, which can be intercepted, altered, or misused if proper safeguards are not in place. Studies have shown that AI models, including those used for cybersecurity, can themselves become targets of attacks, such as adversarial attacks that manipulate model outputs (Jones & Omar, 2024; Gholami & Omar, 2024). The use of AI in cybersecurity is a double-edged sword: while AI can enhance threat detection and response, it can also be weaponized by cybercriminals to bypass traditional security systems (Akhtar & Rawol, 2024).

Moreover, the growing deployment of AI-powered biometric systems for access control and surveillance further compounds the tension between openness and security. These systems, designed to improve security through facial recognition or fingerprint scanning, can inadvertently compromise user privacy, especially when personal data is stored or shared without adequate protection (Awad et al., 2024). In healthcare, for example, the collection of sensitive patient data for AI-driven diagnoses could be vulnerable to breaches if data privacy measures are not properly implemented (Arefin, 2024).

To clarify the distribution of security threats in AI-driven environments, the following pie chart highlights the key risks faced by AI-integrated systems.

Figure 3. Distribution of security risks in AI-powered systems

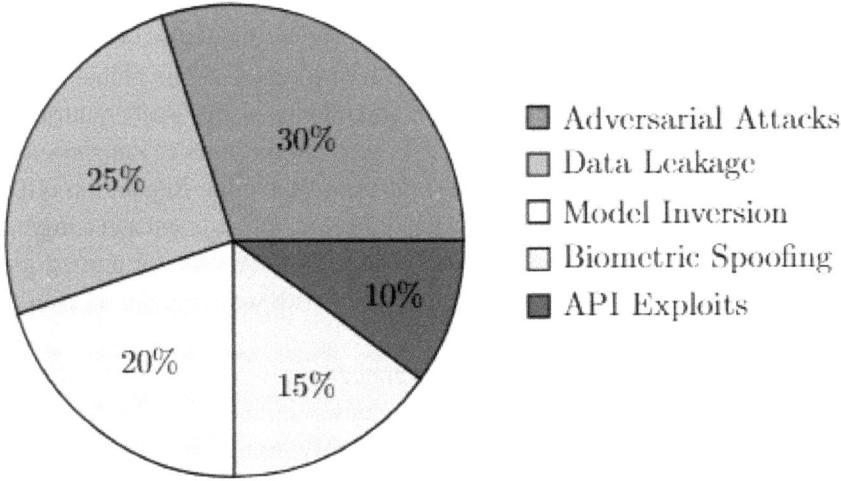

4.3 Ethical and Privacy Considerations

Alongside the technical challenges, ethical concerns also emerge as a major issue in the tension between openness and security. The increasing collection and analysis of personal data by AI systems raise profound questions about consent, transparency, and accountability. AI models, especially those used for marketing, financial analysis, or healthcare, often rely on data gathered without explicit user consent, which may infringe on individual privacy rights (Alhitmi et al., 2024). The introduction of AI-powered marketing systems, for instance, has led to growing concerns about the commodification of personal data and the potential exploitation of users (Alhitmi et al., 2024; Sato, 2023).

The use of AI in sensitive fields like genomics and precision medicine also underscores the ethical dilemma of balancing openness and privacy. AI-powered genomic analysis has the potential to revolutionize healthcare by enabling personalized treatment plans (Rehan, 2023), but the storage and sharing of genetic data pose significant risks if not managed securely. Ensuring that these data remain confidential and are used ethically remains a significant challenge for both AI developers and healthcare providers (Daraf & Badi, 2023; Dhinakaran et al., 2025).

4.4 Regulatory Frameworks and AI Security

As AI technologies continue to evolve, regulatory frameworks are struggling to keep pace with the need for privacy and security measures. The global nature of AI development presents an added layer of complexity, as different countries have varying laws and regulations governing data protection (Omar & Zangana, 2025). For instance, the European Union's General Data Protection Regulation (GDPR) has established strict guidelines for data handling, but enforcement and compliance vary widely across jurisdictions (Pandey et al., 2024). Without a unified global standard, AI systems can easily operate in regions with weak regulatory oversight, leaving sensitive data unprotected.

In response to these challenges, experts are calling for more robust frameworks that integrate security, privacy, and ethical considerations into the development and deployment of AI systems. This includes developing AI-powered solutions that ensure privacy preservation while allowing for innovation. For example, AI-based encryption and privacy-preserving techniques are being explored to protect data integrity while maintaining the functionality of AI models (Farea et al., 2024; Gopireddy, 2021).

Furthermore, businesses and organizations leveraging AI for decision-making must adopt comprehensive data governance policies that not only comply with regulations but also address the ethical implications of AI usage. Customizable AI-powered security and privacy configurations for specific sectors, such as social media or healthcare, are essential to ensure that both the openness of these platforms and the security of user data are maintained (Abbas & Qazi, 2024; Abolaji & Akinwande, 2024).

4.5 Future Directions: Striking the Balance

The future of AI security lies in developing systems that offer flexibility without sacrificing privacy or security. One promising avenue is the application of advanced encryption methods, such as homomorphic encryption, which allows data to be processed without being decrypted, ensuring both security and privacy (Abolaji & Akinwande, 2024). Another potential solution involves integrating AI models with secure hardware, such as trusted execution environments (TEEs), to provide a secure space for data processing (Gholami & Omar, 2024). Additionally, AI-driven security systems must be designed with transparency in mind, enabling users to understand how their data is being used and giving them control over its privacy settings (Kumar et al., 2024; Sato, 2023).

In conclusion, while AI's openness has the potential to drive substantial advancements in fields like healthcare, finance, and cybersecurity, it also introduces significant risks to privacy and security. As AI continues to evolve, it will be essential to navigate these tensions through the development of secure, transparent, and ethically responsible AI systems.

5. DATA GOVERNANCE, INTELLECTUAL PROPERTY, AND PRIVACY

In the context of AI, data governance, intellectual property (IP), and privacy are intricately connected, forming the backbone of responsible AI implementation and deployment. Data governance refers to the policies, procedures, and standards that dictate how data is managed, accessed, and used, ensuring its integrity, security, and privacy. Intellectual property, in turn, concerns the protection of creations and innovations, including AI algorithms, datasets, and software. Privacy relates to the safeguarding of personal data against unauthorized access and misuse, with a focus on transparency, accountability, and control for individuals whose data is being utilized. In AI systems, these elements need to coexist in a manner that promotes innovation while mitigating the risk of exploitation or unethical use.

5.1 Data Governance in AI

AI systems depend heavily on vast datasets, making data governance a critical concern. Effective data governance ensures that data is accurate, consistent, secure, and accessible when needed. It establishes frameworks to handle sensitive information, ensuring compliance with regulations such as the General Data Protection Regulation (GDPR) and the California Consumer Privacy Act (CCPA).

AI models, particularly those driven by machine learning, require continuous access to high-quality data to improve over time. However, data often resides in different departments, systems, or external sources, each with varying privacy and security measures. Abbas and Qazi (2024) highlight the need for customized AI-powered security and privacy configurations for platforms, such as social media websites, to ensure that sensitive data is well-governed and protected. Their work suggests that without robust data governance, the security of AI-driven systems can be compromised, leading to potential data breaches or misuse.

The integration of AI in healthcare and business systems further exacerbates these challenges. Alhitmi et al. (2024) explore the data security and privacy concerns of AI-driven marketing in the context of economics and business. They argue that while AI has the potential to improve business outcomes, without solid governance frameworks in place, companies risk breaching ethical guidelines and compromising consumer privacy. To mitigate these concerns, organizations must ensure that data management policies are not only comprehensive but also adaptable to the ever-changing landscape of AI technology.

5.2 Intellectual Property (IP) and AI

The intersection of IP and AI is an area of increasing concern as AI technologies become more sophisticated. AI-driven innovations, including algorithms, software, and even datasets, often push the boundaries of traditional IP laws. One challenge in the AI domain is the issue of ownership. Who owns the intellectual property generated by an AI model? Is it the developer, the organization that implemented the AI, or the AI itself? Abolaji and Akinwande (2024) emphasize that AI-powered privacy protection mechanisms must be designed to not only secure data but also respect IP rights. They argue that establishing clear ownership rights is essential for fostering innovation while protecting creators' and businesses' intellectual properties.

A significant concern in IP law with respect to AI is the protection of datasets. AI models are trained on data, and the datasets themselves can be considered valuable intellectual property. In this context, Akhtar and Rawol (2024) assert that the security of AI training datasets must be prioritized to prevent unauthorized access and exploitation. Intellectual property protections for datasets are often underdeveloped, posing risks of intellectual theft, especially in industries like healthcare and finance, where sensitive data is common.

5.3 Privacy Concerns in AI

Privacy is a key issue that intersects with both data governance and IP, especially in sectors handling personal and sensitive data. Privacy risks in AI systems are often heightened due to the large amounts of personal data processed, which can lead to potential breaches, misuse, or exposure of private information. Alhitmi et al. (2024) emphasize that AI-powered systems in business and marketing need stringent privacy safeguards to protect user data from unauthorized access and misuse. Similarly, the increasing reliance on AI in healthcare demands comprehensive privacy strategies to ensure patient confidentiality. Gawankar et al. (2024) highlight the importance of integrating privacy-preserving technologies, such as encryption and secure access controls, into AI-powered healthcare systems to protect patient data from unauthorized access.

AI technologies, such as machine learning, have the potential to process and analyze large datasets in ways that traditional systems cannot. This processing often involves deriving insights from personal data, which raises concerns about individuals' control over their own information. The ethics of privacy in AI-driven systems must align with regulations like GDPR, which stipulate that personal data must be handled in a transparent, lawful, and secure manner. This includes ensuring that individuals are aware of how their data is being used and providing them with the ability to revoke consent if they choose. Almeida and Barr (2025) discuss how innovations in health data protection, particularly in AI-powered diagnosis systems, must prioritize privacy through both ethical and legal frameworks. Their study underscores the critical importance of patient empowerment, where individuals have control over their health data while benefiting from AI's diagnostic capabilities.

5.4 The Role of AI in Enhancing Privacy Protection

AI has the potential to enhance privacy protection by incorporating advanced techniques such as differential privacy, encryption, and secure multi-party computation. These technologies can ensure that sensitive data is processed without exposing personal information. Arya et al. (2023) demonstrate how AI-powered systems can improve the security of IoT devices by detecting and mitigating privacy threats, ensuring that devices operate without compromising user privacy. Their work shows that AI-driven security mechanisms can proactively monitor data flows, identifying and blocking potential privacy violations before they occur.

Farea et al. (2024) further elaborate on the use of AI to enhance privacy through integrated encoding mechanisms, which provide an extra layer of protection for data within the IoT ecosystem. This approach combines AI's ability to identify anomalies in data flows with encryption techniques that secure sensitive information, making it more difficult for unauthorized parties to access private data. AI also helps in enhancing privacy by automating compliance checks with privacy regulations, ensuring that organizations follow necessary protocols when handling personal information.

5.5 Future Directions in Data Governance, IP, and Privacy

Looking ahead, the evolution of AI and its integration into diverse sectors will continue to shape data governance, IP, and privacy concerns. The increasing complexity of AI systems, along with the growing amount of data being generated, means that organizations will need to continually adapt their data governance frameworks. Research by Gupta et al. (2025) in the context of AI and data privacy in business explores the need for adaptive data protection frameworks that are responsive to the unique challenges posed by AI.

Moreover, the implementation of AI in sectors like genomics and healthcare presents an opportunity to develop innovative solutions for safeguarding data privacy. Rehan (2023) discusses how AI-powered genomic analysis systems, while offering significant potential for enhancing precision medicine, must address privacy concerns by ensuring that sensitive genetic data is protected from unauthorized access.

As AI technologies continue to evolve, so too must the legal and ethical frameworks governing data use and IP protection. The integration of advanced AI-powered threat detection systems will help businesses and individuals navigate the complexities of data privacy and security. Sato (2023) emphasizes that the future of AI-powered privacy protection lies in creating systems that are not only secure but also aligned with ethical standards that prioritize transparency, fairness, and user control over personal data.

5.6 Summary

The landscape of data governance, intellectual property, and privacy in AI is complex and ever-evolving. As AI continues to impact diverse industries, particularly those dealing with sensitive data, robust governance frameworks are essential to ensure that privacy is protected and intellectual property is respected. Through innovative technologies and regulatory compliance, it is possible to mitigate the risks associated with data misuse while harnessing AI's potential for positive societal impact.

The following Venn-style diagram represents the overlapping domains of Data Governance, Intellectual Property, and Privacy within AI, showing areas of convergence and potential conflicts.

Figure 4. Intersections between data governance, IP, and privacy in AI

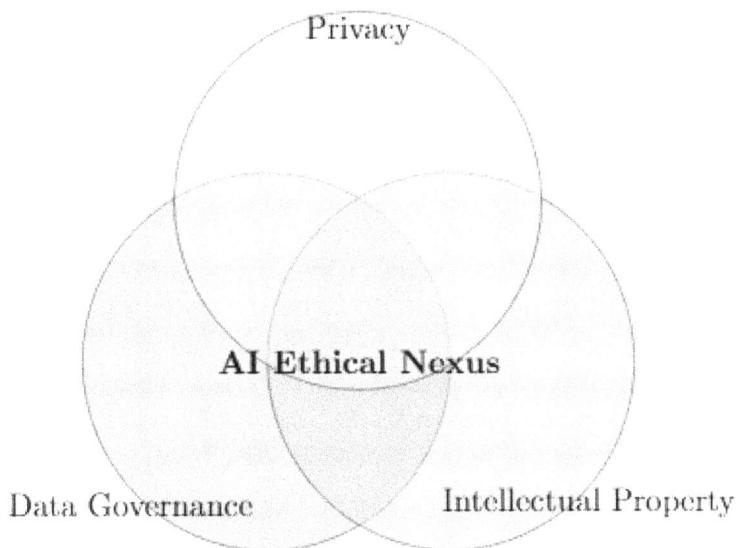

6. SECURITY-DRIVEN RESTRICTIONS: WHEN AND WHY TO LIMIT OPENNESS

In the current landscape of cybersecurity, there is an ongoing tension between ensuring system openness and safeguarding sensitive information. Open systems and access to data facilitate innovation, promote transparency, and encourage collaboration across industries. However, when it comes to handling critical data, especially in contexts such as healthcare, finance, and personal security, there are valid concerns that open access may compromise privacy and integrity. Security-driven restrictions, therefore, become an essential tool to protect users and organizations from malicious threats and inadvertent data breaches.

6.1 The Role of AI in Strengthening Security Measures

Artificial Intelligence (AI) plays a pivotal role in defining the parameters within which openness can be allowed, as it can offer proactive protection by identifying potential vulnerabilities before they are exploited. As AI-powered systems are increasingly integrated into various domains, including healthcare (Gawankar et al.,

2024), marketing (Alhitmi et al., 2024), and genomics (Rehan, 2023), AI's ability to detect anomalies in data usage patterns and mitigate risks is becoming indispensable.

For example, AI-driven systems have been utilized to enhance cybersecurity through mechanisms like advanced threat detection and biometric security measures (Awad et al., 2024). In contexts where real-time data security is critical, such as in healthcare or IoT environments, AI's ability to identify potential threats and restrict access to sensitive data is a key strategy (Arya et al., 2024; Farea et al., 2024). These mechanisms not only bolster security but also ensure that only authorized individuals and systems can access restricted information.

6.2 When to Implement Security-Driven Restrictions

Security-driven restrictions are most often required in scenarios where the risk to privacy and data integrity outweighs the benefits of open access. In particular, there are a few key areas where such limitations are critical:

1. **Healthcare Data Security**: The healthcare sector is one of the most sensitive when it comes to personal data. With the rise of AI-powered diagnosis and treatment systems, there is an increasing risk of unauthorized access to confidential patient information (Gawankar et al., 2024; Dhinakaran et al., 2025). Here, AI systems must be trained to recognize and restrict unauthorized access to medical records, ensuring that only those with the necessary clearance can interact with sensitive health data.

2. **Financial Systems**: AI is extensively applied in the finance industry for fraud detection, market prediction, and risk management. However, the vast amount of personal and financial data collected by financial institutions makes these systems a prime target for cyberattacks (Brightwood & Jame, 2024). Thus, financial organizations must implement strict security policies, such as restricting access to transaction data and utilizing AI-powered encryption and threat detection systems to protect user data.

3. **Government and Military Data**: National security data, whether related to defense, intelligence, or critical infrastructure, is often subject to stringent security controls. AI can be used to create adaptive security systems that detect breaches or unauthorized attempts to access classified materials (Omar, Zangana, & Mohammed, 2025). In these domains, the need for security-driven restrictions is paramount to avoid catastrophic breaches that could threaten national stability.

4. **IoT Systems and Smart Devices**: With the proliferation of Internet of Things (IoT) devices, security vulnerabilities have emerged due to the often-open nature of these devices (Arya et al., 2023). AI-powered threat detection systems are crucial for continuously monitoring data traffic and restricting unauthorized

device access, ensuring that only trusted devices and users can interact with the IoT ecosystem (Farea et al., 2024).

5. **Cloud Computing**: As organizations increasingly migrate to cloud environments, the need for AI-powered security mechanisms to protect data in transit and at rest becomes critical. Cloud computing platforms can be particularly vulnerable to attacks such as data breaches, DDoS attacks, and malware injection (Gopireddy, 2021). Here, security-driven restrictions might involve limiting open access to cloud-based data storage systems, ensuring that sensitive business and personal information is encrypted and safeguarded (Omar, 2021).

6.3 Why Limit Open Access: The Ethical Imperative

While the promotion of open access is beneficial in many technological and research sectors, it is vital to recognize the ethical implications of open systems when sensitive data is involved. AI systems must be designed not only with technical efficacy but also with a strong ethical framework that prioritizes user privacy, informed consent, and data protection.

The ethical considerations of data privacy and security often involve making difficult decisions about when and why to limit openness. For instance, while AI algorithms can generate insights from vast amounts of data, they must be employed in ways that avoid exploiting or inadvertently harming individuals by breaching their privacy (Abolaji & Akinwande, 2024; Sato, 2023). The challenge lies in balancing the benefits of data openness with the necessity of ensuring that personal and sensitive information is not exposed to malicious actors.

6.4 The Future of Security-Driven Restrictions

Looking ahead, the future of security-driven restrictions will undoubtedly be shaped by advancements in AI technologies. The integration of machine learning with AI allows systems to continuously evolve, adapting to new threats and vulnerabilities. This continuous learning process will be critical in ensuring that data remains secure in environments that are increasingly interconnected and complex (Akhtar & Rawol, 2024).

Moreover, as AI technologies become more sophisticated, they will be able to predict potential threats based on trends and anomalies, allowing for preemptive security measures. This approach may involve automatically restricting access to data or implementing new forms of encryption based on real-time analysis of security risks (Kumar et al., 2024).

In the long run, AI-driven privacy and security frameworks are expected to become more transparent, with users having more control over their data. These systems will likely allow individuals to make informed decisions about how their data is shared and accessed, ensuring that open systems are only utilized in ways that align with their privacy preferences (Gupta et al., 2025).

6.5 Summary

In conclusion, while open access and transparency are vital to technological progress, security-driven restrictions are necessary when dealing with sensitive data. AI-powered security measures play a crucial role in ensuring that systems can safeguard data while still allowing for innovation. As we move forward, it will be critical to strike a balance between openness and security, ensuring that the future of technology remains both innovative and secure. The integration of AI in security measures will help mitigate risks and ensure that data privacy is upheld in increasingly complex digital landscapes (Omar, 2024; Anidjar et al., 2023).

7. FRAMEWORKS AND GUIDELINES FOR BALANCING OPENNESS WITH RESPONSIBILITY

As artificial intelligence (AI) continues to shape various industries, the balancing act between openness and responsibility has become a central concern. The flexibility and potential of AI systems, particularly in the realms of data security, privacy, and innovation, require frameworks and guidelines that ensure ethical practices while maintaining system efficiency. This section explores the evolving frameworks and guidelines designed to balance openness with responsibility in AI applications, focusing on security, privacy, and the ethical use of AI technologies across domains like healthcare, business, and IoT.

7.1. Understanding the Need for Balancing Openness with Responsibility

AI systems, particularly those that interact with large datasets, must be transparent in their operations while safeguarding individual rights. Open systems, which offer transparency, innovation, and accessibility, are essential for fostering technological advancement. However, these same systems can expose sensitive data to risks of misuse and exploitation. The challenge lies in designing frameworks that allow for open innovation while embedding responsible measures to protect users and organizations.

A growing body of research highlights how AI technologies impact data privacy and security in various contexts. For instance, Akhtar and Rawol (2024) discuss how AI-powered security mechanisms enhance cybersecurity but also emphasize the need for robust guidelines that limit potential data breaches. Similarly, Abolaji and Akinwande (2024) highlight the ongoing developments in AI-powered privacy protection, underscoring the importance of integrating privacy-by-design principles into AI models. These frameworks should not only focus on technical solutions but also address ethical, legal, and regulatory concerns.

7.2. Establishing Ethical Guidelines for AI Implementation

Ethical AI implementation involves several core principles: transparency, fairness, accountability, privacy protection, and accessibility. As AI technologies increasingly become embedded in sectors like healthcare, finance, and IoT, ethical guidelines must be established and adhered to ensure public trust. These guidelines ensure that AI systems operate in a manner that respects user autonomy, data privacy, and fairness, particularly in applications where decisions significantly impact individuals, such as AI-driven healthcare or financial systems.

For example, Almeida and Barr (2025) provide a comprehensive exploration of AI-powered diagnosis systems, suggesting ethical frameworks that allow healthcare innovation while ensuring patient privacy and data protection. Similarly, Anidjar et al. (2023) explore the privacy matrix in AI-powered metaverse platforms, advocating for a rethinking of data infrastructure to maintain privacy rights within immersive virtual spaces. These works demonstrate the importance of embedding ethical principles at the design stage of AI applications.

7.3. AI-Driven Security Mechanisms: Guidelines for Risk Mitigation

The adoption of AI-driven security systems can significantly enhance protection against cyber threats. However, these systems must be carefully designed to minimize risks associated with data breaches and unauthorized access. Frameworks such as the one proposed by Gopireddy (2021) focus on AI-powered security in cloud environments, ensuring data protection while providing mechanisms to detect and prevent cyberattacks. Similarly, Farea et al. (2024) advocate for the integration of encoding mechanisms with AI in IoT ecosystems to enhance privacy, security, and performance.

In addition to technical considerations, organizations must follow security guidelines that prioritize the use of AI responsibly. As discussed by Gholami and Omar (2023), synthetic data has become a useful tool to protect privacy while enabling machine learning models to perform efficiently. These technologies must be governed by regulatory frameworks to prevent the misuse of AI-driven systems that could inadvertently lead to the exposure of personal data or system vulnerabilities.

7.4. Addressing Legal and Regulatory Challenges in AI Privacy

Privacy laws and regulations are evolving in response to the growing application of AI in personal data management. While frameworks such as the General Data Protection Regulation (GDPR) in Europe set foundational standards for privacy, there is an increasing need for specific AI privacy laws that consider the unique challenges posed by AI's ability to analyze and utilize vast amounts of data.

As highlighted by Jones and Omar (2024), AI applications often intersect with privacy concerns, particularly in sectors such as healthcare and education. Regulations need to account for the diverse and evolving ways in which AI systems collect, store, and process sensitive data. In addition, these regulations must allow for the rapid deployment of AI technologies while ensuring compliance with data protection laws. Abolaji and Akinwande (2024) provide insights into AI-powered privacy protection mechanisms, emphasizing the future directions of these systems within the context of international privacy regulations.

7.5. Building Transparent AI Models to Foster Trust

Transparency is essential in fostering trust among users and stakeholders, especially in sectors like finance, healthcare, and IoT. Transparent AI models allow users to understand how decisions are made, thereby increasing confidence in the systems. Brightwood and Jame (2024) emphasize that in AI-powered finance, transparency in algorithms is critical to ensure fairness and prevent discrimination. Transparency can also help ensure that AI systems are compliant with regulations, thus addressing legal concerns.

In the context of AI-driven healthcare, the work of Gawankar et al. (2024) emphasizes the need for clear explanations of AI algorithms, especially those used in medical decision-making. Clear guidelines should ensure that healthcare providers and patients fully understand the risks and benefits of AI-powered solutions while protecting sensitive medical data from exploitation.

7.6. Promoting AI-Powered Innovation in the Face of Ethical Challenges

While it is essential to balance openness with responsibility, it is equally important to foster AI innovation that can drive progress. AI systems have the potential to transform industries, from revolutionizing healthcare management (Bhamidipaty et al., 2025) to enhancing cybersecurity measures (Mbah & Evelyn, 2024). However, ethical challenges such as algorithmic bias, lack of inclusivity, and inadvertent discrimination must be addressed to ensure equitable benefits from AI technologies.

Arefin (2024) highlights the role of AI in enhancing healthcare data security with threat detection, presenting frameworks that improve system performance while maintaining patient confidentiality. Similarly, the work by Sato (2023) on AI-powered privacy solutions stresses the need for AI innovations that prioritize user consent and data anonymization, which could serve as guiding principles for the future of AI in other sectors.

7.7. Guidelines for Balancing Data Openness with User Consent

Another important aspect of balancing openness and responsibility is managing user consent in AI systems. Given the volume of personal data processed by AI systems, frameworks must be in place to ensure that individuals have control over their data. As discussed by Mbah and Evelyn (2024), strategies for ensuring user consent and data privacy are critical to safeguarding individual autonomy in AI-powered platforms. AI systems should provide clear and understandable consent mechanisms, allowing users to make informed decisions regarding the use of their data.

For example, the work of Gholami and Omar (2024) on synthetic data demonstrates how AI models can be trained using non-personally identifiable information, ensuring that privacy concerns are mitigated while still benefiting from advanced AI technologies.

7.8. Summary: Moving Toward Responsible AI Development

The development of AI technologies offers unprecedented opportunities, but it also presents significant ethical, legal, and technical challenges. The frameworks and guidelines discussed in this section provide a roadmap for ensuring that AI's potential is harnessed in ways that prioritize security, privacy, and ethical considerations.

Future research and development efforts must continue to refine these guidelines, ensuring that AI technologies remain transparent, accountable, and responsible.

As AI continues to evolve, it is crucial to recognize that achieving the right balance between openness and responsibility is not a static goal but a dynamic challenge. By following the outlined frameworks and incorporating them into the development of AI systems, we can create a future where AI contributes positively to society while respecting the rights and freedoms of individuals.

The following framework illustrates the core components of responsible AI implementation in open science environments, integrating ethical, legal, and technical dimensions.

Figure 5. Framework for responsible AI-powered open science

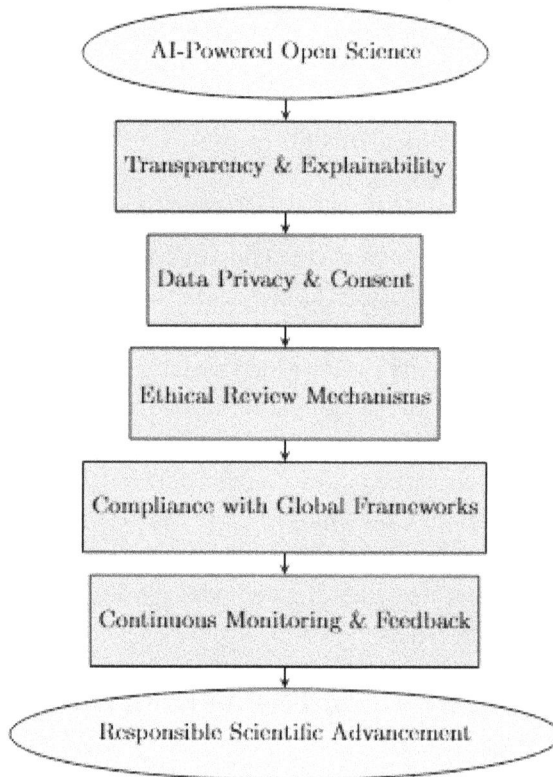

8. EMERGING TECHNOLOGIES AND THE FUTURE OF RESPONSIBLE OPEN SCIENCE

The integration of emerging technologies into the realm of open science presents both exciting opportunities and significant challenges. Responsible open science refers to the practices of ensuring transparency, accessibility, and ethical use of scientific data and knowledge, particularly in contexts involving artificial intelligence (AI), data privacy, and cybersecurity. As AI technologies advance, their influence on how scientific data is collected, analyzed, and shared becomes more pronounced. This section explores the role of AI in responsible open science, its potential for future innovations, and the critical ethical, privacy, and security concerns associated with its implementation.

8.1 The Role of AI in Open Science

AI has become a cornerstone in modern research methodologies, driving innovations across various domains such as healthcare, finance, and cybersecurity. The potential of AI-powered systems to automate complex data analysis tasks, improve accuracy, and accelerate scientific discovery is undeniable. However, the integration of AI in open science must be carefully managed to ensure that it aligns with the principles of openness, transparency, and inclusivity.

One prominent area where AI is revolutionizing open science is in the analysis of vast amounts of data. AI techniques, particularly machine learning, can process and interpret data far more efficiently than traditional methods. For example, in genomic research, AI-powered tools are enhancing the precision of medical diagnoses and treatments, making scientific data more accessible to researchers and clinicians alike (Rehan, 2023). Moreover, AI has the potential to democratize access to knowledge, enabling researchers from across the globe to contribute to and benefit from scientific advancements. This is particularly significant in developing countries, where access to high-quality research tools and datasets may be limited.

Despite these advancements, there are significant concerns related to the use of AI in open science. One of the most pressing issues is ensuring that AI systems are developed and used responsibly. The algorithms driving AI technologies must be transparent, unbiased, and accessible, ensuring that the scientific community can scrutinize and understand their decision-making processes. Furthermore, AI must be employed in ways that prioritize ethical considerations and the well-being of individuals, particularly when dealing with sensitive data such as health records or personal information (Sato, 2023; Pandey et al., 2024).

8.2 Privacy and Security Concerns in AI-Powered Open Science

As AI technologies become increasingly integrated into open science, privacy and security concerns have escalated. Data privacy is a central issue, particularly when dealing with sensitive datasets in fields like healthcare, genomics, and finance. AI systems require access to vast amounts of data to function effectively, raising concerns about how this data is collected, stored, and shared. The risk of unauthorized access or misuse of sensitive information is a critical issue that must be addressed through robust security measures and privacy protections.

For instance, in healthcare, AI-powered systems are transforming the way patient data is handled. However, as these systems become more ubiquitous, they also introduce new risks. In a study by Gawankar et al. (2024), the authors highlight the need for AI-powered cybersecurity mechanisms to ensure the privacy and security of patient data. Similarly, Farea et al. (2024) propose an integrated encoding mechanism to enhance the security, privacy, and performance of IoT ecosystems, which are increasingly used in healthcare settings. These technologies aim to safeguard sensitive data while ensuring that AI can still function effectively in data-driven research environments.

Moreover, AI's role in cybersecurity extends beyond healthcare. AI-powered tools are essential for securing digital communication systems, financial data, and cloud environments. Akhtar and Rawol (2024) discuss how AI-powered security mechanisms are enhancing cybersecurity by detecting and responding to threats faster than traditional methods. These technologies can help safeguard the integrity of scientific research, ensuring that data remains secure from cyber-attacks and unauthorized access.

8.3 Ethical Considerations in AI-Driven Open Science

The ethical implications of using AI in open science are profound. One of the primary concerns is the potential for AI systems to perpetuate biases, which can lead to unfair or discriminatory outcomes. For example, AI systems used in genomics or healthcare must be trained on diverse datasets to avoid biases that may harm marginalized communities (Awad et al., 2024). Similarly, ethical challenges arise in the context of AI-powered marketing, where data privacy concerns are heightened due to the collection of personal information for targeted advertisements (Alhitmi et al., 2024).

Furthermore, there is the issue of accountability when AI systems make decisions that impact individuals' lives. While AI can process large datasets and make predictions with high accuracy, it lacks the human element of moral judgment. This gap raises important questions about who is responsible for the decisions made by AI systems, particularly when they affect vulnerable populations or involve sensitive data (Abolaji & Akinwande, 2024).

8.4 Advancing Responsible AI in Open Science

To ensure that AI technologies are used responsibly in open science, it is crucial to develop frameworks that prioritize ethical considerations, data privacy, and security. Several researchers have explored how these concerns can be addressed through AI-powered solutions. For instance, Kumar et al. (2024) propose a novel framework for adaptive data protection, which aims to balance the need for AI innovation with the protection of personal data. Similarly, Gholami and Omar (2023) examine the potential for synthetic data to improve the efficiency of large language models (LLMs), thereby addressing concerns about data privacy and security in AI-driven research.

The future of responsible open science hinges on the development of technologies that are not only advanced but also ethically sound. This requires the collaboration of scientists, ethicists, and technologists to create guidelines and standards that govern the use of AI in research. Additionally, governments and regulatory bodies must play a pivotal role in setting policies that ensure AI technologies are developed and deployed in a way that respects human rights and privacy.

8.5 Summary

The future of responsible open science is undoubtedly shaped by the integration of AI technologies. These tools hold great potential to enhance the accessibility, efficiency, and precision of scientific research. However, they also raise significant challenges related to privacy, security, and ethics. As AI continues to evolve, it is crucial that the scientific community remains vigilant in addressing these concerns and ensuring that emerging technologies are used in ways that promote openness, fairness, and accountability. By fostering a culture of responsible innovation, we can ensure that AI-driven open science contributes to the advancement of knowledge while safeguarding the rights and interests of all stakeholders involved.

9. CONCLUSION

As we navigate the evolving landscape of technology, the role of responsible open science becomes increasingly pivotal. The integration of Artificial Intelligence (AI) into various sectors, including healthcare, cybersecurity, and data privacy, highlights the transformative potential of these technologies in advancing knowledge, improving societal well-being, and fostering innovation. However, alongside these advancements, there is an urgent need to address the ethical, legal, and social implications of AI and other emerging technologies to ensure that their benefits are shared equitably while minimizing harm.

In this context, responsible open science serves as a guiding framework, emphasizing transparency, accessibility, and inclusivity in the development and application of new technologies. The growing importance of AI in various domains, such as healthcare, data privacy, and cybersecurity, requires researchers, policymakers, and industry leaders to collaborate and develop solutions that prioritize both innovation and ethical responsibility. Through the responsible implementation of AI-powered solutions, we can build more resilient and secure systems while safeguarding fundamental rights and freedoms, particularly in an increasingly interconnected world.

Furthermore, the convergence of AI with other emerging technologies, such as blockchain, Internet of Things (IoT), and cloud computing, presents new opportunities and challenges. These innovations, when harnessed responsibly, have the potential to revolutionize industries and drive sustainable progress across multiple sectors. However, without a concerted effort to address issues like data security, privacy protection, and algorithmic fairness, the risks associated with these technologies could outweigh their benefits.

In conclusion, the future of responsible open science lies in balancing the pursuit of technological progress with a strong commitment to ethical practices and societal well-being. By fostering a culture of openness, transparency, and accountability, we can ensure that emerging technologies, particularly AI, continue to serve as catalysts for positive change while respecting human dignity and fundamental rights. Through collaboration and shared responsibility, we can pave the way for a future where innovation thrives within a framework of trust and ethical integrity.

REFERENCES

Abbas, E., & Qazi, A. A. (2024). Customized Ai-powered Security and Privacy Configurations for Social MEDIA Websites. *BULLET: Jurnal Multidisiplin Ilmu*, *3*(1), 108–117.

Abolaji, E. O., & Akinwande, O. T.Elijah Oluwatoyosi AbolajiOladayo Tosin Akinwande. (2024). AI powered privacy protection: A survey of current state and future directions. *World Journal of Advanced Research and Reviews*, *23*(3), 2687–2696. DOI: 10.30574/wjarr.2024.23.3.2869

Akhtar, Z. B., & Rawol, A. T. (2024). Enhancing cybersecurity through AI-powered security mechanisms. *IT Journal Research and Development*, *9*(1), 50–67. DOI: 10.25299/itjrd.2024.16852

Alhitmi, H. K., Mardiah, A., Al-Sulaiti, K. I., & Abbas, J. (2024). Data security and privacy concerns of AI-driven marketing in the context of economics and business field: An exploration into possible solutions. *Cogent Business & Management*, *11*(1), 2393743. DOI: 10.1080/23311975.2024.2393743

Almeida, D., & Barr, N. (2025). Innovations in Health Data Protection Ethical, Legal, and Technological Perspectives in a Global Context: AI-Powered Diagnosis Systems and Health Data Innovation. In *Navigating Privacy, Innovation, and Patient Empowerment Through Ethical Healthcare Technology* (pp. 171-196). IGI Global Scientific Publishing.

Anidjar, L., Packin, N. G., & Panezi, A. (2023). The matrix of privacy: Data infrastructure in the AI-Powered Metaverse. *SSRN*, *18*, 59. DOI: 10.2139/ssrn.4363208

Arefin, S. (2024). Strengthening Healthcare Data Security with Ai-Powered Threat Detection. [IJSRM]. *International Journal of Scientific Research and Management*, *12*(10), 1477–1483. DOI: 10.18535/ijsrm/v12i10.ec02

Arya, L., Sharma, Y. K., Devi, S., & Padmanaban, H. (2024). Securing the Internet of Things: AI-Powered Threat Detection and Safety. In *Proceedings of International Conference on Recent Innovations in Computing: ICRIC 2023,* Volume 2 (Vol. 2, p. 97). Springer Nature. DOI: 10.1007/978-981-97-3442-9_7

Arya, L., Sharma, Y. K., Devi, S., Padmanaban, H., & Kumar, R. (2023, October). Securing the Internet of Things: AI-Powered Threat Detection and Safety Measures. In *The International Conference on Recent Innovations in Computing* (pp. 97-108). Singapore: Springer Nature Singapore.

Awad, A. I., Babu, A., Barka, E., & Shuaib, K. (2024). AI-powered biometrics for Internet of Things security: A review and future vision. *Journal of Information Security and Applications*, *82*, 103748. DOI: 10.1016/j.jisa.2024.103748

Bhamidipaty, V., Bhamidipaty, D. L., Guntoory, I., Bhamidipaty, K. D. P., Iyengar, K. P., Botchu, B., & Botchu, R. (2025). Revolutionizing Healthcare: The Impact of AI-Powered Sensors. *Generative Artificial Intelligence for Biomedical and Smart Health Informatics*, 355-373.

Brightwood, S., & Jame, H. (2024). *Data privacy, security, and ethical considerations in AI-powered finance. Article.* Research Gate.

Daraf, U., & Badi, S. (2023). AI-Powered Genomic Analysis in the Cloud: Enhancing Precision Medicine While Protecting Medical Data Privacy.

Dhinakaran, D., Raja, S. E., Jasmine, J. J., Kumar, P. V., & Ramani, R. (2025). The Future of Well-Being: AI-Powered Health Management with Privacy at its Core. *Wellness Management Powered by AI Technologies*, 363-402.

Farea, A. H., Alhazmi, O. H., Samet, R., & Guzel, M. S. (2024). AI-powered Integrated with Encoding Mechanism Enhancing Privacy, Security, and Performance for IoT Ecosystem. *IEEE Access : Practical Innovations, Open Solutions*, *12*, 121368–121386. DOI: 10.1109/ACCESS.2024.3449630

Gawankar, S., Nair, S., Pawar, V., Vhatkar, A., & Chavan, P. (2024, August). Patient Privacy and Data Security in the Era of AI-Driven Healthcare. In *2024 8th International Conference on Computing, Communication, Control and Automation (ICCUBEA)* (pp. 1-6). IEEE. DOI: 10.1109/ICCUBEA61740.2024.10775004

Gemiharto, I., & Masrina, D. (2024). User privacy preservation in AI-powered digital communication systems. *Jurnal Communio: Jurnal Jurusan Ilmu Komunikasi*, *13*(2), 349–359. DOI: 10.35508/jikom.v13i2.9420

Gholami, S., & Omar, M. (2023). Does Synthetic Data Make Large Language Models More Efficient? *arXiv preprint arXiv:2310.07830.*

Gholami, S., & Omar, M. (2024). Can a student large language model perform as well as its teacher? In *Innovations, Securities, and Case Studies Across Healthcare, Business, and Technology* (pp. 122-139). IGI Global. DOI: 10.4018/979-8-3693-1906-2.ch007

Gopireddy, R. R. (2021). AI-Powered Security in cloud environments: Enhancing data protection and threat detection. [IJSR]. *International Journal of Scientific Research*, *10*(11).

Gupta, A., Amarnani, M., Soanki, S., & Kishore, J. (2025, February). AI and Data Privacy in Business. In *2025 First International Conference on Advances in Computer Science, Electrical, Electronics, and Communication Technologies (CE2CT)* (pp. 109-114). IEEE.

Hamza, Y. A., & Omar, M. D. (2013). Cloud computing security: Abuse and nefarious use of cloud computing. *International Journal of Computer Engineering Research*, 3(6), 22–27.

Huff, A. J., Burrell, D. N., Nobles, C., Richardson, K., Wright, J. B., Burton, S. L., Jones, A. J., Springs, D., Omar, M., & Brown-Jackson, K. L. (2023). Management Practices for Mitigating Cybersecurity Threats to Biotechnology Companies, Laboratories, and Healthcare Research Organizations. In *Applied Research Approaches to Technology, Healthcare, and Business* (pp. 1-12). IGI Global.

Ismail, I. A., & Aloshi, J. M. R. (2025). Data Privacy in AI-Driven Education: An In-Depth Exploration Into the Data Privacy Concerns and Potential Solutions. In *AI Applications and Strategies in Teacher Education* (pp. 223-252). IGI Global.

Jones, R., & Omar, M. (2024). Revolutionizing Cybersecurity: The GPT-2 Enhanced Attack Detection and Defense (GEADD) Method for Zero-Day Threats. *International Journal of Informatics* [INJIISCOM]. *Information System and Computer Engineering*, 5(2), 178–191.

Jones, R., Omar, M., Mohammed, D., Nobles, C., & Dawson, M. (2023). Harnessing the Speed and Accuracy of Machine Learning to Advance Cybersecurity. In *2023 Congress in Computer Science, Computer Engineering, & Applied Computing (CSCE)* (pp. 418-421). IEEE. DOI: 10.1109/CSCE60160.2023.00074

Kumar, R. S., Lokeshwari, J., & Shanmugam, S. K. (2024, November). AI-Powered Privacy Preservation: A Novel Framework for Adaptive Data Protection. In *2024 2nd International Conference on Computing and Data Analytics (ICCDA)* (pp. 1-6). IEEE.

Mbah, G. O., & Evelyn, A. N. (2024). AI-powered cybersecurity: Strategic approaches to mitigate risk and safeguard data privacy.

Mohammed, D., Omar, M., & Nguyen, V. (2018). Wireless sensor network security: Approaches to detecting and avoiding wormhole attacks. *Journal of Research in Business. Economics and Management*, 10(2), 1860–1864.

Nguyen, V., Mohammed, D., Omar, M., & Banisakher, M. (2018). The Effects of the FCC Net Neutrality Repeal on Security and Privacy. [IJHIoT]. *International Journal of Hyperconnectivity and the Internet of Things*, 2(2), 21–29. DOI: 10.4018/IJHIoT.2018070102

Omar, M. (2021). New insights into database security: An effective and integrated approach for applying access control mechanisms and cryptographic concepts in Microsoft Access environments.

Omar, M. (2022). *Machine Learning for Cybersecurity: Innovative Deep Learning Solutions*. Springer Brief. https://link.springer.com/book/978303115

Omar, M. (2024). From Attack to Defense: Strengthening DNN Text Classification Against Adversarial Examples. In *Innovations, Securities, and Case Studies Across Healthcare, Business, and Technology* (pp. 174-195). IGI Global.

Omar, M. (2024). Revolutionizing Malware Detection: A Paradigm Shift Through Optimized Convolutional Neural Networks. In *Innovations, Securities, and Case Studies Across Healthcare, Business, and Technology* (pp. 196-220). IGI Global. DOI: 10.4018/979-8-3693-1906-2.ch011

Omar, M., & Zangana, H. (Eds.). (2025). *Digital Forensics in the Age of AI*. IGI Global., DOI: 10.4018/979-8-3373-0857-9

Omar, M., & Zangana, H. M. (Eds.). (2024). *Redefining Security With Cyber AI*. IGI Global., DOI: 10.4018/979-8-3693-6517-5

Omar, M., & Zangana, H. M. (Eds.). (2025). *Application of Large Language Models (LLMs) for Software Vulnerability Detection*. IGI Global., DOI: 10.4018/979-8-3693-9311-6

Omar, M., Zangana, H. M., Al-Karaki, J. N., & Mohammed, D. (2024). Harnessing LLMs for IoT malware detection: A comparative analysis of BERT and GPT-2. In *2024 8th International Symposium on Multidisciplinary Studies and Innovative Technologies (ISMSIT)* (pp. 1-6). Ankara, Turkiye. https://doi.org/DOI: 10.1109/ISMSIT63511.2024.10757249

Omar, M., Zangana, H. M., & Mohammed, D. (Eds.). (2025). *Integrating Artificial Intelligence in Cybersecurity and Forensic Practices*. IGI Global., DOI: 10.4018/979-8-3373-0588-2

Pandey, A. S., Sharma, Y., Tiwari, A., Chauhan, R., Tyagi, S., & Kumari, J. (2024, May). Ethical Implications of AI-Powered Communication Tool. In *2024 International Conference on Communication, Computer Sciences and Engineering (IC3SE)* (pp. 1857-1861). IEEE. DOI: 10.1109/IC3SE62002.2024.10593350

Patel, R., & Desai, A. (2024). Exploring the Ethical Implications of AI-Powered Security Systems. *Asian American Research Letters Journal*, *1*(9), 87–95.

Rehan, H. (2023). AI-Powered Genomic Analysis in the Cloud: Enhancing Precision Medicine and Ensuring Data Security in Biomedical Research. *Journal of Deep Learning in Genomic Data Analysis*, *3*(1), 37–71.

Sato, H. (2023). AI-Powered Solutions: Ensuring Data Privacy in a Transforming Digital Landscape. *Advances in Computer Sciences, 6*(1).

Thirunagalingam, A. (2024). AI-Powered Continuous Data Quality Improvement: Techniques, Benefits, and Case Studies. *Benefits, and Case Studies (August 23, 2024)*.

Wright, J., Dawson, M. E. Jr, & Omar, M. (2012). Cyber security and mobile threats: The need for antivirus applications for smartphones. *Journal of Information Systems Technology and Planning*, *5*(14), 40–60.

Zainab, H., Khan, A. R. A., Khan, M. I., & Arif, A. (2025). Ethical Considerations and Data Privacy Challenges in AI-Powered Healthcare Solutions for Cancer and Cardiovascular Diseases. *Global Trends in Science and Technology*, *1*(1), 63–74. DOI: 10.70445/gtst.1.1.2025.63-74

Zangana, H., Al-Karaki, J., & Omar, M. (Eds.). (2025). *Revolutionizing Cybersecurity With Deep Learning and Large Language Models*. IGI Global., DOI: 10.4018/979-8-3373-3296-3

Zangana, H., & Omar, M. (Eds.). (2025). *Leveraging Large Language Models for Quantum-Aware Cybersecurity*. IGI Global., DOI: 10.4018/979-8-3373-1102-9

Zangana, H., & Omar, M. (Eds.). (2025). *Leveraging Large Language Models for Quantum-Aware Cybersecurity*. IGI Global., DOI: 10.4018/979-8-3373-1102-9

Zangana, H. M. (2024). Exploring Blockchain-Based Timestamping Tools: A Comprehensive Review. *Redefining Security With Cyber AI*, 92-110.

Zangana, H. M. (2024). Exploring the Landscape of Website Vulnerability Scanners: A Comprehensive Review and Comparative Analysis. *Redefining Security With Cyber AI*, 111-129.

Zangana, H. M., & Li, S. (2025). Future Trends in AI and Digital Forensics. In Omar, M., & Zangana, H. (Eds.), *Digital Forensics in the Age of AI* (pp. 347–380). IGI Global Scientific Publishing., DOI: 10.4018/979-8-3373-0857-9.ch013

Zangana, H. M., Luckyardi, S., Mustafa, F. M., & Li, S. (2025). Enhancing Agricultural Cybersecurity: Leveraging Deep Learning and Large Language Models for Smart Farming Protection. In Zangana, H., Al-Karaki, J., & Omar, M. (Eds.), *Revolutionizing Cybersecurity With Deep Learning and Large Language Models* (pp. 307–338). IGI Global Scientific Publishing., DOI: 10.4018/979-8-3373-3296-3.ch010

Zangana, H. M., Mohammed, A. K., Sallow, A. B., & Sallow, Z. B. (2024). Cybernetic Deception: Unraveling the Layers of Email Phishing Threats. [INJURATECH]. *International Journal of Research and Applied Technology*, *4*(1), 35–47.

Zangana, H. M., & Mohammed, D. (2025). Foundations of Large Language Models in Software Vulnerability Detection. In Omar, M., & Zangana, H. (Eds.), *Application of Large Language Models (LLMs) for Software Vulnerability Detection* (pp. 41–74). IGI Global., DOI: 10.4018/979-8-3693-9311-6.ch002

Zangana, H. M., & Mustafa, F. M. (2024). Hybrid Image Denoising Using Wavelet Transform and Deep Learning. *EAI Endorsed Transactions on AI and Robotics*, *3*. Advance online publication. DOI: 10.4108/airo.7486

Zangana, H. M., & Mustafa, F. M. (2025). Image Denoising Techniques for Cybersecurity and Forensic Applications: AI-Driven Approaches. In Omar, M., Zangana, H., & Mohammed, D. (Eds.), *Integrating Artificial Intelligence in Cybersecurity and Forensic Practices* (pp. 117–142). IGI Global Scientific Publishing., DOI: 10.4018/979-8-3373-0588-2.ch005

Zangana, H. M., Mustafa, F. M., & Li, S. (2025). Large Language Models in Cybersecurity: From Automation to Intelligence. In Zangana, H., & Omar, M. (Eds.), *Leveraging Large Language Models for Quantum-Aware Cybersecurity* (pp. 277–300). IGI Global Scientific Publishing., DOI: 10.4018/979-8-3373-1102-9.ch009

Zangana, H. M., Mustafa, F. M., Li, S., & Al-Karaki, J. N. (2025). Natural Language Processing for Cyber Threat Intelligence in a Quantum World. In Zangana, H., & Omar, M. (Eds.), *Leveraging Large Language Models for Quantum-Aware Cybersecurity* (pp. 345–388). IGI Global Scientific Publishing., DOI: 10.4018/979-8-3373-1102-9.ch011

Zangana, H. M., Mustafa, F. M., Mohammed, A. K., & Omar, M. (2025). The Role of Change Control Boards in Ensuring Cybersecurity Compliance for IT Infrastructure. *JITCE (Journal of Information Technology and Computer Engineering)*, *9*(1). Retrieved from https://jitce.fti.unand.ac.id/index.php/JITCE/article/view/303

Zangana, H. M., Mustafa, F. M., & Omar, M. (2024). A Hybrid Approach for Robust Object Detection: Integrating Template Matching and Faster R-CNN. *EAI Endorsed Transactions on AI and Robotics*, *3*. Advance online publication. DOI: 10.4108/airo.6858

Zangana, H. M., & Omar, M. (2020). Threats, Attacks, and Mitigations of Smart-phone Security. *Academic Journal of Nawroz University*, *9*(4), 324–332. DOI: 10.25007/ajnu.v9n4a989

Zangana, H. M., & Omar, M. (2025). Harnessing the Power of Large Language Models for Cybersecurity: Applications, Challenges, and Future Directions. In Omar, M., & Zangana, H. (Eds.), *Application of Large Language Models (LLMs) for Software Vulnerability Detection* (pp. 1–40). IGI Global., DOI: 10.4018/979-8-3693-9311-6.ch001

Zangana, H. M., & Omar, M. (2025). Introduction to Digital Forensics and Artificial Intelligence. In Omar, M., & Zangana, H. (Eds.), *Digital Forensics in the Age of AI* (pp. 1–30). IGI Global Scientific Publishing., DOI: 10.4018/979-8-3373-0857-9.ch001

Zangana, H. M., & Omar, M. (2025). Introduction to Quantum-Aware Cybersecurity: The Need for LLMs. In Zangana, H., & Omar, M. (Eds.), *Leveraging Large Language Models for Quantum-Aware Cybersecurity* (pp. 1–28). IGI Global Scientific Publishing., DOI: 10.4018/979-8-3373-1102-9.ch001

Zangana, H. M., & Omar, M. (2025). The Role of Leadership in Advancing Inclusive Health Technologies. In Burrell, D., & Nguyen, C. (Eds.), *New Horizons in Leadership: Inclusive Explorations in Health, Technology, and Education* (pp. 203–220). IGI Global Scientific Publishing., DOI: 10.4018/979-8-3693-6437-6.ch009

Zangana, H. M., Omar, M., & Al-Karaki, J. N. (2025). Foundations of Deep Learning and Large Language Models in Cybersecurity. In Zangana, H., Al-Karaki, J., & Omar, M. (Eds.), *Revolutionizing Cybersecurity With Deep Learning and Large Language Models* (pp. 1–36). IGI Global Scientific Publishing., DOI: 10.4018/979-8-3373-3296-3.ch001

Zangana, H. M., Omar, M., Al-Karaki, J. N., & Mohammed, D. (2024). Comprehensive Review and Analysis of Network Firewall Rule Analyzers: Enhancing Security Posture and Efficiency. *Redefining Security With Cyber AI*, 15-36.

Zangana, H. M., Sallow, Z. B., & Omar, M. (2025). The Human Factor in Cybersecurity: Addressing the Risks of Insider Threats. *Jurnal Ilmiah Computer Science*, *3*(2), 76–85. DOI: 10.58602/jics.v3i2.37

Żywiołek, J. (2024). Empirical Examination Of Ai-Powered Decision Support Systems: Ensuring Trust And Transparency In Information And Knowledge Security. *Scientific Papers of Silesian University of Technology. Organization & Management/ Zeszyty Naukowe Politechniki Slaskiej. Seria Organizacji i Zarzadzanie*, (197).

Chapter 7
Beyond Intelligence:
The Synthetic Cognitive Augmentation Network Using Experts

Ben Kennedy
https://orcid.org/0009-0009-1337-0709
Capitol Technology University, USA

Atif Mohammad
Capitol Technology University, USA

Matthew Wyandt
Capitol Technology University, USA

ABSTRACT

The Synthetic Cognitive Augmentation Network Using Experts (SCANUE) significantly advances the foundational Synthetic Cognitive Augmentation Network from a conceptual model into a sophisticated experimental platform designed to meet complex cognitive needs. SCANUE's specialized agents depend critically on high-quality, domain-specific datasets that accurately reflect intricate cognitive tasks such as nuanced planning, emotional inference, and conflict resolution, closely replicating human prefrontal cortex functions (Tate et al., 2024b). To address limitations inherent in generic data sources, SCANUE integrates several public dialogue corpora, including Taskmaster-1, MultiWOZ, DailyDialog, GoEmotions, and SocialIQa, into a unified framework (Byrne et al., 2019; Budzianowski et al., 2018; Li & Ding, 2017; Demszky et al., 2020; Sap et al., 2019), supplemented with AI-generated dialogues. This modular and biologically inspired system, enhanced by adaptive learning and Human-in-the-Loop methodologies, effectively aligns with user intentions to advance cognitive augmentation.

DOI: 10.4018/979-8-3373-4252-8.ch007

1. INTRODUCTION

1.1 Overview

The human prefrontal cortex (PFC) is widely recognized within neuroscience as the seat of higher cognitive functions, playing a crucial role in executive control, complex decision-making, planning, working memory, and emotional regulation (Quartz & Sejnowski, 1997). Inspired by this biological blueprint, the Synthetic Cognitive Augmentation Network Using Experts—also conceptualized for future development as User Extensible (hereafter "Using Experts")—builds upon and significantly advances the foundational Synthetic Cognitive Augmentation Network. The original architecture pioneered the simulation of various PFC subregions to address intricate cognitive tasks. This enhanced iteration, SCANUE, integrates more sophisticated fine-tuned agent models and incorporates Human-in-the-Loop (HITL) strategies as a core component of its operation. While the initial framework offered a robust proof-of-concept and baseline performance, it faced inherent challenges, particularly in flexibly accommodating the diverse needs and cognitive styles of individual users and in effectively incorporating iterative, feedback-driven refinements to improve performance and alignment over time.

Early developmental iterations and the foundational codebase for this platform are detailed in the primary code repository (Tate, 2024a). Subsequent enhancements, including expanded functionalities and refined agent implementations, are documented in the dedicated scanue-v22 repository (Tate, 2024b). The nomenclature "User Extensible" signifies the long-term aspiration to develop expanded user-centric capabilities, potentially allowing users greater control over agent configuration or integration with personal data streams. However, the present research and development emphasis lies firmly within the "Using Experts" paradigm, wherein distinct, specialized agent models function as domain experts, each tackling a specific facet of a larger cognitive problem.

To overcome the limitations identified in earlier versions, this enhanced iteration integrates several key advancements: sophisticated fine-tuning techniques for the expert agents, adaptive learning mechanisms allowing agents to adjust based on interaction history, and integral HITL methodologies. These additions reinforce both the user-centric nature of the architecture and its grounding in biological plausibility. By employing custom-trained transformer models (Vaswani et al., 2017), optimized for specific cognitive functions, and leveraging established modular artificial intelligence frameworks (Arora et al., 2024), the platform achieves heightened scalability and facilitates targeted, user-specific refinements. Although the current command-line interface limits accessibility for non-technical audiences, ongoing development efforts are focused on creating intuitive graphical interfaces to broaden its reach and usability.

Developed within an iterative design-based research (DBR) framework (Anderson & Shattuck, 2012), which emphasizes cycles of design, implementation, and analysis in real-world contexts, this refined system demonstrably improves data flow efficiency, agent processing coordination, and user alignment. This is achieved by strategically juxtaposing direct HITL oversight with adaptive learning algorithms derived from reinforcement learning principles, prioritizing the efficiency and value-alignment benefits of direct human feedback (detailed in Section 6). Consequently, the system effectively upholds precise information exchange between agents and delivers robust cognitive augmentation, while simultaneously establishing a solid foundation for rigorous empirical evaluation and eventual real-world testing. The research roadmap now explicitly outlines plans for comprehensive empirical validation. This includes comparative benchmarking against existing state-of-the-art HITL systems using established metrics (Mosqueira-Rey et al., 2023), conducting pilot studies in demanding application domains such as healthcare decision support and financial planning (Leng et al., 2022), and integrating adversarial training methods to proactively ensure system robustness and resilience against unexpected inputs or manipulation attempts (Moosavi-Dezfooli et al., 2016).

Beyond these crucial architectural and methodological advancements, the performance of SCANUE is fundamentally dependent on meticulously curated training datasets. These datasets are designed to mirror the cognitive specializations of each PFC-inspired agent, providing the necessary 'experience' for effective fine-tuning. Section 4 provides a detailed account of how diverse public dialogue corpora and targeted, AI-generated conversations are systematically synthesized to furnish the domain-specific depth required for agents to master nuanced planning, sophisticated emotion regulation, dynamic conflict monitoring, and complex reward evaluation tasks.

1.2 Core Contributions

Biologically Inspired Modularity. SCANUE replicates the functional specialization observed in the human brain, specifically targeting functions associated with the dorsolateral, ventromedial, orbitofrontal, anterior cingulate, and medial PFC through dedicated, computationally distinct agents. This approach moves beyond the limitations of monolithic LLM architectures, fostering greater interpretability and targeted function. Specialized Datasets. Recognizing the limitations of generic data, a unified, multi-corpus pipeline, significantly enhanced by generative AI augmentation, provides fine-grained, task-specific training data meticulously tailored to the unique cognitive subtask of each agent module. Human-in-the-Loop Alignment.

The SCANUE system prioritizes user alignment by incorporating real-time feedback. The upcoming SCANAQ psychometric instrument will further enhance this by continuously refining agent behaviors, reducing potential biases, and ensuring that outputs consistently align with user values and goals during operation. Scalable Orchestration. The use of LangGraph provides a robust and flexible graph-based routing mechanism. This enables efficient parallel execution of agent tasks and sophisticated state management, supporting the goal of near-real-time cognitive augmentation for the user. Robust Evaluation Roadmap. A comprehensive evaluation strategy is planned, moving beyond preliminary metrics. This includes pilot studies in key domains (healthcare, finance, education) and the application of adversarial training protocols to rigorously benchmark SCANUE against state-of-the-art HITL systems and ensure its reliability.

Collectively, these contributions position SCANUE as a novel and comprehensive platform for advancing research and enabling the deployment of sophisticated, next-generation cognitive augmentation applications.

2. BACKGROUND AND RELATED WORK

2.1 Neuroscientific Basis of Executive Function

Decades of cognitive neuroscience research have provided a detailed map of the functional specialization within the prefrontal cortex (PFC), highlighting the complementary roles its subregions play in higher cognition. Key findings indicate that the dorsolateral PFC (DLPFC) is primarily involved in executive control, strategic planning, and working memory maintenance; the ventromedial PFC (VMPFC) plays a critical role in integrating affective valence (emotional significance) with value-based judgments and decision-making under uncertainty; the orbitofrontal cortex (OFC) is crucial for evaluating potential rewards and punishments, particularly in balancing short-term versus long-term outcomes; the anterior cingulate cortex (ACC) acts as a conflict monitoring and error detection system; and the medial PFC (mPFC) is central to self-referential thought, social reasoning, and understanding others' mental states (theory of mind) (Amodio & Frith, 2006; Bechara et al., 2000; Miller & Cohen, 2001). These established neuroscientific findings provide the direct inspiration for SCANUE's modular agent decomposition. Each agent is designed to approximate the core computational signature of its corresponding PFC subdivision, thereby enabling an effective division of labor across complex executive, affective, and social cognitive tasks (Friedman & Robbins, 2022; Quartz & Sejnowski, 1997). This biologically inspired approach is increasingly leveraged by AI researchers seeking to construct artificial cognitive architectures that exhibit more human-like capacities for executive control, nuanced emotion handling, and sophisticated social cognition (Grossberg, 2021; Samsonovich, 2010; Schmidgall et al., 2024).

2.2 Modular and Multi-Agent AI Frameworks

While traditional large language models (LLMs) have demonstrated remarkable capabilities in natural language understanding and generation across a wide range of open-domain tasks, they often exhibit limitations when faced with tasks requiring fine-grained control, specific reasoning patterns, or adherence to complex constraints, such as those involved in emotional regulation simulation or risk-sensitive decision-making (Shoeybi et al., 2019). Furthermore, the significant computational resources required for training and deploying the largest monolithic LLMs pose practical challenges. Comprehensive surveys analyzing the capabilities and limitations of current LLMs underscore these points (Bhattacharya et al., 2024). Consequently, a growing body of research advocates for modular or "mixture-of-experts" architectures. In these systems, multiple lightweight, specialized models, each trained for a specific task or domain, collaborate under the guidance of an orchestration layer (Arora et al., 2024; Yao et al., 2023). This approach offers potential advantages in terms of efficiency, scalability, interpretability, and the ability to tailor components for specific functions. SCANUE explicitly adopts this modular paradigm. The LangGraph library serves as the orchestration layer, managing the complex inter-agent routing and state transitions (detailed in Section 3.2). Each "expert" agent within the SCANUE framework is implemented either as a fine-tuned instance of a capable yet efficient foundation model (currently GPT-4o-mini) or, potentially in future iterations, as a neuromorphic spiking network surrogate designed for energy efficiency (Maass, 1997; Tate, 2024b, 2024d). The resulting architecture provides a scalable and potentially more interpretable alternative to single, large monolithic models, maintaining clearer functional boundaries between cognitive specializations.

2.3 Human-in-the-Loop (HITL) Augmentation

The integration of human oversight and feedback within AI systems, has been shown across numerous studies to significantly improve system performance, reliability, user trust, and ethical alignment (Amershi et al., 2014; Jarrahi et al., 2022; Mandvikar & Dave, 2023). Humans excel at handling ambiguity, providing common-sense reasoning, and making value judgments that remain challenging for current AI. SCANUE operationalizes these findings through two distinct but complementary layers of human involvement (further elaborated in Section 6). First, employing a design-based research (DBR) methodology ensures that user requirements and preferences, systematically gathered via instruments like the CAUSE survey, directly inform the design and objectives of the agents from the outset. Second, a real-time feedback channel is integrated into the system's operation. This functionality enables users to offer direct, text-based feedback on the consolidated output generated by the PFC agents. This immediate feedback drives continual fine-tuning processes, facilitates rapid adaptation to user needs, and serves as a crucial mechanism for mitigating potential biases or undesirable agent behaviors (Mosqueira-Rey et al., 2023). The planned deployment of the SCANAQ instrument, a multi-scale psychometric tool, aims to further refine this HITL interaction by enabling more nuanced, personalized alignment based on individual user cognitive and affective profiles.

2.4 Dataset Generation with Large Language Models

The development of highly specialized datasets is pivotal for effectively aligning each SCANUE agent to its designated cognitive niche. While publicly available dialogue corpora—such as Taskmaster-1 for goal-oriented tasks (Byrne et al., 2019), MultiWOZ for multi-domain interactions (Budzianowski et al., 2018), DailyDialog for general conversation (Li & Ding, 2017), GoEmotions for fine-grained emotion labeling (Demszky et al., 2020), and SocialIQa for social commonsense reasoning (Sap et al., 2019)—provide valuable domain breadth and linguistic diversity, they rarely contain sufficient examples of highly specific or sensitive scenarios, such as providing nuanced grief counseling support or navigating complex, adversarial negotiation strategies. To overcome these limitations, SCANUE employs a hybrid data strategy. It supplements the foundational data from open corpora with synthetically generated dialogues produced by large language models (specifically, GPT variants) guided by carefully engineered prompts (Tate, 2024c; Vapnik & Izmailov, 2015). The specific prompt templates utilized for generating these synthetic dialogues are made available in markdown format within the accompanying resource repository (Tate, 2024c), facilitating transparency and reproducibility. This approach of blending authentic human data with targeted synthetic data, sometimes referred to as leveraging "privileged information" during training (Vapnik & Izmailov, 2015), ensures both the linguistic richness derived from real interactions and the specific scenario realism required for training agents on specialized, hard-to-find edge cases.

3. SYSTEM ARCHITECTURE

3.1 Modular PFC Agent Framework

The overall architecture and end-to-end information flow within SCANUE are depicted conceptually in Figure 1. User interaction initiates via a command-line interface (CLI), where a query or scenario description is provided. This input is first processed by the DLPFC agent, acting as the central executive controller. Analogous to its neurobiological counterpart, the DLPFC agent parses the user's intent, decomposes the query into constituent subtasks, and strategically delegates these subtasks to the most appropriate specialized expert agents, each mirroring a distinct PFC subdivision: DLPFC: Responsible for high-level executive control, multi-step planning, maintaining working memory, task decomposition, and dynamic delegation to other agents. VMPFC: Handles affective appraisal (evaluating emotional significance), assessing risks associated with potential actions, and simulating emotional regulation processes. OFC: Focuses on reward evaluation, comparing the utility of immediate versus delayed gratification, and informing reward-based decision-making. ACC: Monitors for conflicts between goals or information, detects errors in processing or planning, and signals the need for adjustments. mPFC: Specializes in empathetic integration, adopting others' perspectives (theory of mind), and synthesizing diverse inputs into coherent, value-based recommendations or social responses. The nature of task delegation by the DLPFC agent is adaptive, varying with the complexity of the user's query. Simpler, well-defined tasks might be routed to a single, most relevant expert agent. More complex queries, however, often necessitate multi-agent collaboration, where distinct subtasks are processed concurrently by multiple specialists, or require the aggregation and synthesis of input from all agents to address multi-faceted problems. Each specialized agent processes its assigned subtask(s) in parallel, leveraging its fine-tuned capabilities. The intermediate results generated by each active agent are then returned to a dedicated integration module (conceptually linked to the DLPFC's coordinating role). This module synthesizes the potentially diverse outputs into a single, coherent response aligned with the user's original query and objectives. Crucially, before the final output is presented, the HITL mechanism allows a human reviewer (either the end-user or an expert overseer, depending on the context) to endorse, edit, or reject the synthesized response. This feedback not only refines the immediate output but is also channeled back to the system, primarily informing the DLPFC agent, to update its internal state and parameters, thereby enabling incremental fine-tuning and improving performance over subsequent interactions (Amershi et al., 2014). This dynamic interplay and collaboration among specialized agents, orchestrated by a central controller and refined by human feedback, is intentionally designed to mimic the integrative and adaptive processing characteristic of distributed cortical networks, aiming to enhance overall system robustness and cognitive capability (Yamazaki et al., 2022).

Figure 1. User interaction and PFC-inspired data flow

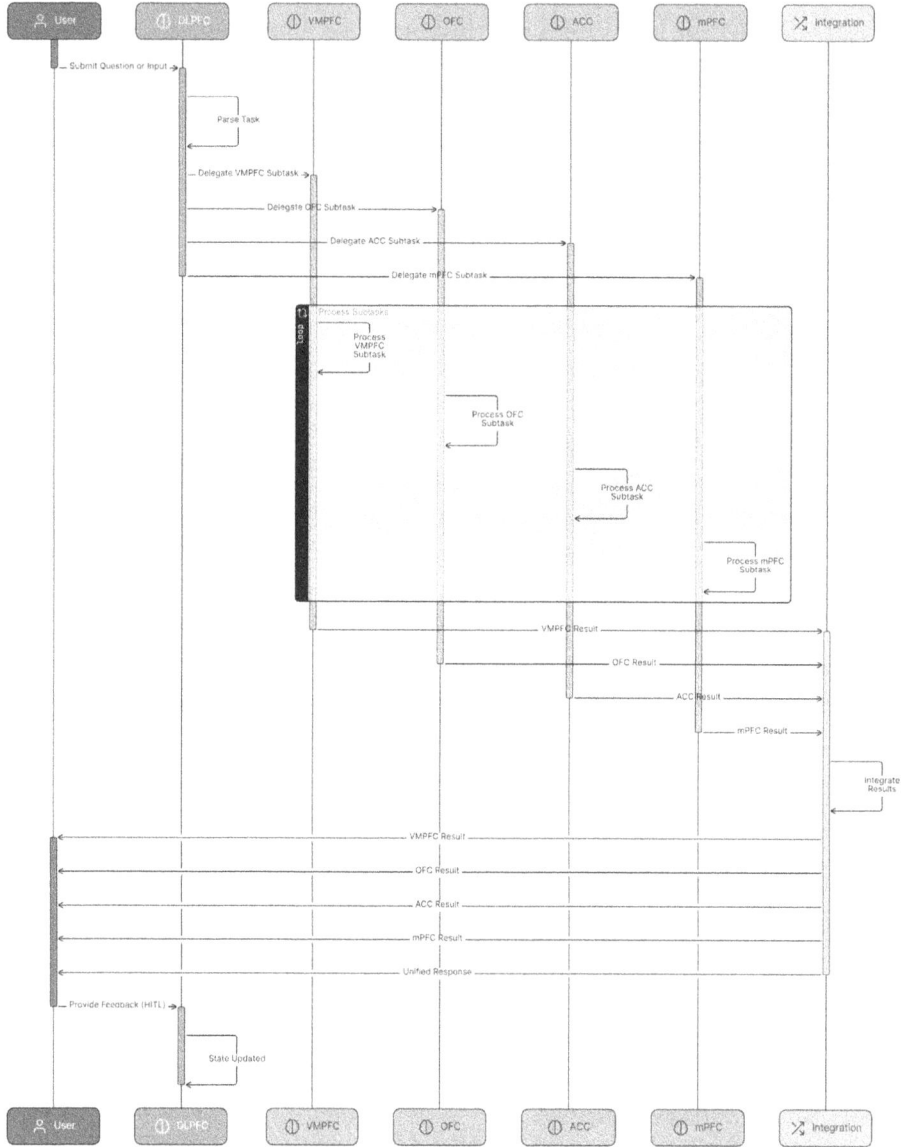

3.2 Orchestration and State Management

Effective coordination and state management are critical in a multi-agent system like SCANUE. After evaluating several available coordination libraries (including established options like LangChain, AutoGen, Experts.js, and CrewAI), LangGraph was selected as the primary orchestration framework. Key factors influencing this decision were LangGraph's native support for representing agent interactions as a graph, its built-in functionalities for integrating human feedback loops at specific nodes within the graph, and its robust mechanisms for managing the complex state transitions inherent in multi-agent collaboration (CrewAI, 2024; LangChain, 2024). Within LangGraph, computational steps or agent activations are represented as nodes, while the flow of data and dependencies between steps are encoded as edges. This graph-based abstraction provides a clean and intuitive mapping to the interconnected, pathway-like structure of cortico-cortical communication that SCANUE aims to emulate biologically. Furthermore, this representation facilitates future architectural expansions, such as the planned incorporation of role-specific agents employing reasoning-acting cycles (e.g., ReAct agents, as described by Yao et al., 2023). The core technology stack underpinning SCANUE currently comprises Python (version 3.12), the FastAPI framework for creating lightweight RESTful API endpoints for potential web-based interaction (Inc., 2021), the LangGraph library for orchestration, and fine-tuned checkpoints of the GPT-4o-mini model accessed via the OpenAI API. Additionally, the Hugging Face Transformers library is being actively considered for integration, potentially to streamline multi-agent orchestration further or to provide robust, locally executable models for specific specialized tasks as alternatives or backups to API-based inference.

3.3 Parallel Processing and Performance Enhancements

A key architectural advantage of SCANUE's modular design is the ability to parallelize task processing. By distributing subtasks across multiple specialized agents that can operate concurrently, the system significantly reduces overall computation time compared to sequential processing or reliance on a single, large model attempting to handle all aspects of a complex query. This parallelization enables near-real-time cognitive augmentation, making the system more responsive for interactive use cases (Dean et al., 2012). Initial validation loss scores obtained after fine-tuning each agent on its specialized dataset suggest effective specialization. Lower loss values indicate better alignment between the agent's predictions and the ground truth data. The reported preliminary scores – 0.3334 (DLPFC), 1.0428 (VMPFC), 0.8540 (OFC), 0.7806 (ACC), and 0.7859 (mPFC) – while provisional, demonstrate promising initial learning and specialization for the respective cognitive functions.

3.4 Integration, Testing, and Feedback-Driven Refinements

Rigorous testing is integral to the development process. A dedicated CLI-based testbed is employed to validate the core orchestration logic and inter-agent communication pathways before attempting deployment within a more complex graphical user interface (GUI) (Roumeliotis & Tselikas, 2023). This testbed operates concurrently with backend-frontend integration efforts, allowing for early identification and resolution of issues related to agent coordination and data flow, ensuring the stability and reliability of SCANUE's core features (Amir & Zhu, 2022). While formal performance evaluations are ongoing, anecdotal testing within this environment suggests that the modular architecture provides enhanced handling of multi-dimensional tasks, often exhibiting reduced latency compared to equivalent monolithic approaches attempting the same complex queries. Crucially, iterative user feedback remains a central driver of refinement. Feedback gathered through follow-ups to the initial CAUSE survey and via ad-hoc reviews by domain experts is systematically used to guide domain-specific adjustments to agent behavior and dataset content. This continuous loop helps to mitigate potential algorithmic biases inadvertently introduced during training and ensures that the system evolves in alignment with user needs and ethical considerations, all while maintaining the agility associated with rapid prototyping cycles (Jobin et al., 2019).

3.5 Ethical and Robustness Safeguards

Ensuring the robustness and ethical operation of SCANUE is a primary concern. System robustness is actively enhanced through the planned application of established adversarial training protocols, such as the Fast Gradient Sign Method (FGSM) and Projected Gradient Descent (PGD), designed to improve resilience against malicious or unexpected inputs (Moosavi-Dezfooli et al., 2016). Furthermore, planned scalability studies will rigorously evaluate the performance and efficiency of distributed inference strategies (Shoeybi et al., 2019) and explore the potential of neuromorphic hardware implementations, possibly leveraging energy-efficient spiking-transformer hybrid architectures (Yamazaki et al., 2022), to support wider deployment. Collectively, these carefully considered architectural choices—grounded in neuroscience, enabled by modular design, refined by human feedback, and validated through rigorous testing—converge on a platform that is biologically motivated, demonstrably user-aligned, and computationally efficient, positioning SCANUE as a strong candidate for enabling next-generation cognitive augmentation applications.

4. DATASET CONSTRUCTION

The capability of a biologically inspired architecture, such as SCANUE, is fundamentally linked to the quality and relevance of the experiences it internalizes through its training data. Recognizing this critical dependency, SCANUE's development places equal emphasis on meticulously defining what each specialized agent learns as it does on how it learns. This section comprehensively details the end-to-end pipeline established to transform disparate public corpora and purpose-built synthetic conversational data into five distinct, functionally scoped training datasets aligned with specific prefrontal cortex subdivisions.

4.1 Curation Principles

The construction of these datasets adhered to several core principles to ensure their effectiveness: Functional Fidelity. Paramount importance was placed on ensuring that every dialogue robustly exercises the specific cognitive faculty the target agent is designed to emulate—for instance, strategic planning capabilities for the DLPFC agent or nuanced affective processing for the VMPFC agent. The goal was to create training scenarios that directly target and challenge the intended function of each module. Scenario Realism. While leveraging the breadth offered by open-domain corpora, identified gaps in scenario coverage (such as complex grief counseling interactions or adversarial negotiation tactics, which are rare in general datasets) were deliberately filled using LLM-generated exchanges. These synthetic dialogues were produced from carefully engineered prompts designed to maximize their plausibility and resemblance to real-world interactions within those specific contexts (Tate, 2024c).

4.2 Data Sources and Augmentation

A foundational baseline of conversational data was established by integrating several well-known public corpora, selected for their relevance to the target cognitive functions. These included Taskmaster-1, MultiWOZ, DailyDialog, GoEmotions, SocialIQa, and EmpatheticDialogues, each contributing different interaction styles, domains, or annotations (e.g., emotion labels, dialogue acts). However, to address the highly specific functional requirements of SCANUE's agents and to populate sparsely represented or entirely missing scenarios (like high-stakes ethical dilemmas or specific types of cognitive bias demonstrations), this baseline was significantly augmented. This augmentation involved generating synthetic dialogues using advanced GPT variants, guided by the detailed. The central emphasis is on dialogue within professional contexts and for professional purposes. This markdown document, titled "PFC Agent Training Data Generation Prompt, Synthetic JSONL Dialogues for Cognitive Agent Fine-Tuning (Tate, 2024c)," serves as the prompt document for this task.

4.3 Specialized Datasets

4.3.1 DLPFC: Executive Planning & Working Memory

Fifty-five training and fifty-five validation dialogues integrate booking scenarios from Taskmaster-1 (Byrne et al., 2019) with synthetically generated "multi-step project" contexts (Tate, 2024c). These dialogues simulate instances where users pivot their objectives mid-conversation, thereby challenging the agent to dynamically revise and maintain updated sub-plans throughout interactions.

4.3.2 VMPFC: Emotional Regulation & Risk Assessment

Emotional complexity is captured through the GoEmotions dataset, comprising 58,000 utterances labeled with 27 distinct emotional categories (Demszky et al., 2020). Dialogues are selectively pruned to emphasize the interplay between identified emotions and corresponding rationales for decision-making. Further, synthetic scenarios focused on grief counseling and medical consent are incorporated into the fifty-five training and fifty-five validation samples (Tate, 2024c), enabling agents to adeptly balance emotional context against risk and potential consequences (Damasio, 1994).

4.3.3 OFC: Reward-Based Decision Making

Seventy-five training dialogues and sixty validation dialogues merge emotional prompts from GoEmotions (Demszky et al., 2020) with synthetically crafted negotiation scenarios addressing topics such as salary negotiations, investment deliberations, and time-management trade-offs (Tate, 2024c). Each interaction concludes with explicit annotations of utility values, thereby facilitating supervised learning aimed at evaluating the balance between immediate rewards and future benefits (Bechara et al., 2000).

4.3.4 ACC: Conflict Detection and Error Monitoring

Training scenarios for conflict detection and error monitoring incorporate multi-user booking complications from MultiWOZ (Budzianowski et al., 2018), alongside inconsistent booking requests from Taskmaster-1 (Byrne et al., 2019). Additionally, synthetically generated adversarial debates introduce conflicting objectives and ambiguous instructions (Tate, 2024c). These curated scenarios rigorously train the agent to recognize, flag, and respond appropriately to inconsistencies and errors within conversational exchanges.

4.3.5 mPFC: Empathy and Perspective-Taking

Baseline conversational warmth is established through the casual dialogue exchanges of DailyDialog (Li & Ding, 2017), enriched further by affectively nuanced interactions from the EmpatheticDialogues dataset (Rashkin et al., 2019). The integration of SocialIQa questions, combined with PFC Agent Training Data Generation Prompt, introduces complex theory-of-mind tasks (Sap et al., 2019; Tate, 2024c). Sixty training dialogues and fifty validation dialogues culminate in evaluations emphasizing "empathic adequacy," steering agent responses towards enhancing interpersonal sensitivity and effective social engagement (Amodio & Frith, 2006).

4.4 Observations and Early Metrics

The empirical results underscored that the hybrid augmentation strategy, integrating synthetic data, proved crucial for achieving desired performance levels. Models trained exclusively on the public corpora exhibited performance plateaus at higher validation loss levels and tended to produce responses that were brittle or overly reliant on templates. The VMPFC agent's validation loss was cut from ~1.35 to its stabilization point around 1.04, enhancing qualitative empathy scores. For the ACC agent, loss decreased from ~1.4 to ≈0.8 over two epochs post-tuning, while the OFC agent improved from around 1.1 to ≈0.85. Figure 2 is designed to illustrate dataset compositions, comparative loss curves, and representative user–agent interactions provides a visual representation of these preliminary performance indicators across the specialized agents.

Figure 2. Training and validation details for PFC-inspired datasets

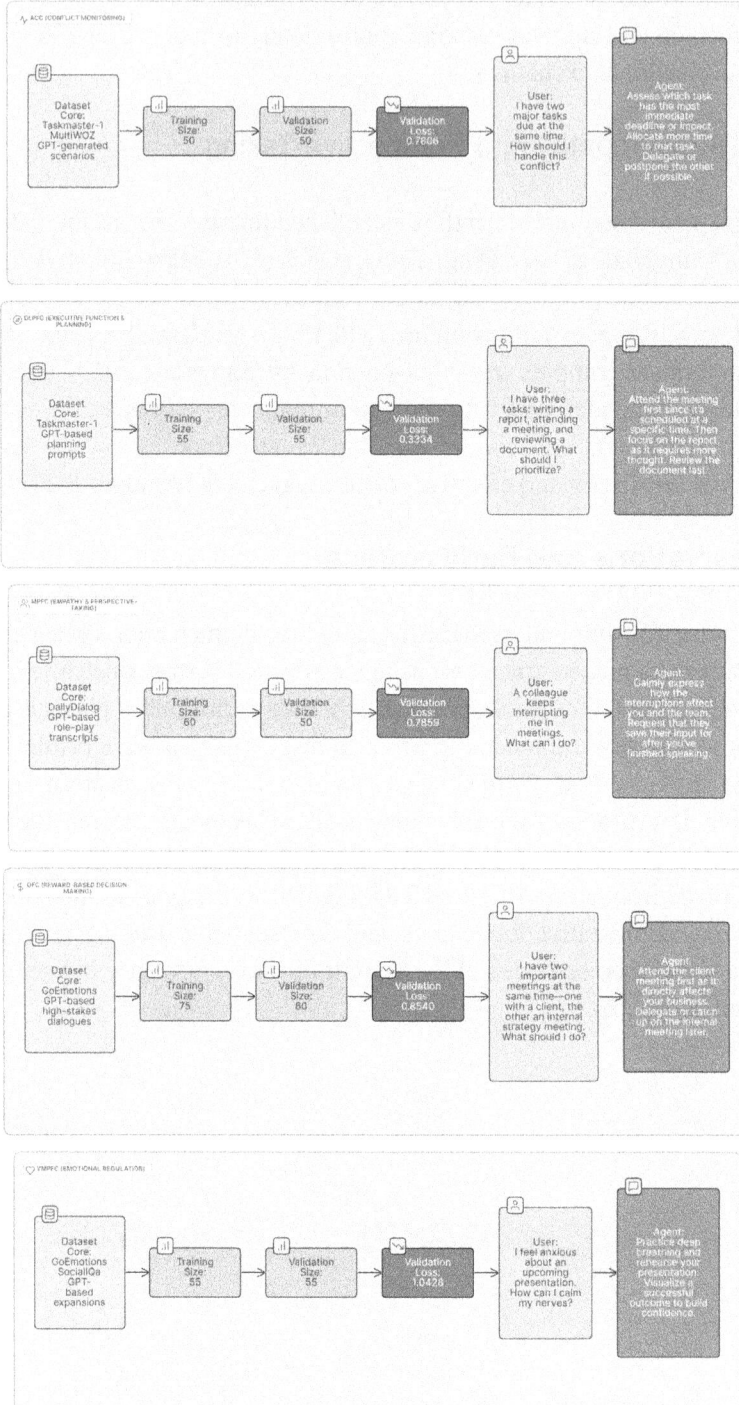

5. TRAINING PROTOCOL AND PRELIMINARY RESULTS

5.1 Fine-Tuning Sessions

The process of fine-tuning the specialized SCANUE agents is conducted utilizing the infrastructure and interface provided by OpenAI's Dashboard (OpenAI, 2024). The training regimen involved iterative sessions where hyperparameters were adjusted based on performance trends. Specifically, early training sessions deliberately employed higher learning rates and smaller batch sizes. This strategy aimed to accelerate the initial learning phase, allowing the models to quickly adapt to the specialized datasets. However, this approach inherently carries a higher risk of overfitting, where the model learns the training data too well but fails to generalize to unseen examples. Consequently, subsequent fine-tuning sessions transitioned to using reduced learning rates and increased batch sizes. This shift promotes greater stability during training and encourages better generalization to the validation set. Throughout this process, validation loss—a critical metric measuring the discrepancy between the model's predictions and the ground truth on data not used for training (lower values indicate better performance)—was meticulously monitored at the conclusion of every session. Adjustments to hyperparameters, such as the learning rate schedule or batch size, were made primarily based on the trajectory of this validation loss, specifically when it failed to show consistent decrease, indicating potential stagnation or overfitting (Bengio et al., 2013; Goodfellow et al., 2016).

5.2 Validation-Loss Scores

To date, the only quantitative performance results reported are the preliminary validation loss values achieved for each specialized agent after the initial fine-tuning process. These values, quoted directly from the documentation associated with the scanue-v22 repository (Tate, 2024b), serve as early indicators of model fit to the specialized datasets. The reported values are as follows: for the DLPFC agent, 0.3334; for the VMPFC agent, 1.0428; for the OFC agent, 0.8540; for the ACC agent, 0.7806; and for the mPFC agent, 0.7859.

5.3 Preliminary Observations

Observations documented in the source papers highlight a common pattern during the initial stages of training. Early sessions often produced artificially low training loss values, suggesting the models were rapidly memorizing the training examples. However, this was frequently accompanied by suboptimal validation loss, indicating poor generalization to new, unseen data—a classic sign of overfitting.

The subsequent strategy of adjusting hyperparameters, specifically increasing the batch size and lowering the learning rate, successfully addressed this discrepancy. This adjustment led to more stable training dynamics and resulted in the improved, albeit still preliminary, validation loss values listed in Section 5.2 and contributed to the specific performance improvements detailed in Section 4.5. Loss figures represent only an initial snapshot of performance and are provisional. Future work will necessarily involve evaluation based on a broader suite of metrics, including task-specific accuracy, inference response times, user satisfaction indices derived from qualitative feedback, and rigorous testing of adversarial robustness to assess resilience against potential manipulation or unexpected inputs (Goodfellow et al., 2016).

5.4 Next Steps

Based on the preliminary results and the system's design, the following key next steps are outlined in the source manuscripts to rigorously evaluate and advance the SCANUE platform: Conduct comprehensive benchmarking of SCANUE against other contemporary Human-in-the-Loop (HITL) systems, utilizing the evaluation frameworks and metrics surveyed by Mosqueira-Rey et al. (2023). Systematically apply established adversarial training methodologies, such as the Fast Gradient Sign Method (FGSM) and Projected Gradient Descent (PGD), to quantitatively assess and enhance the system's robustness against adversarial attacks (Moosavi-Dezfooli et al., 2016). Initiate and execute pilot studies deploying SCANUE within realistic scenarios in target application domains, specifically healthcare, finance, and education, to gather practical performance data, assess real-world applicability, and refine domain-specific parameters and agent behaviors (Leng et al., 2022).

6. HUMAN-IN-THE-LOOP ALIGNMENT AND EVALUATION

6.1 Design-Based Research and CAUSE Survey

The development trajectory of SCANUE is guided by a Design-Based Research (DBR) framework (Anderson & Shattuck, 2012). DBR is an iterative methodology well-suited for complex interventions in real-world settings, providing both a systematic structure for development and the flexibility to adapt based on empirical findings. A primary data source informing the DBR process for SCANUE is the Cognitive Augmentation User Survey Evaluation (CAUSE). This instrument, developed under IRB compliance, synthesized user requirements and expectations for cognitive augmentation tools by consolidating findings from existing literature, analyzing responses from large-scale user polls, and conducting domain-specific needs analyses. Key insights derived from the CAUSE survey highlighted a strong user demand for AI-driven cognitive augmentation tools that offer capabilities such as real-time decision support, feedback sensitive to the user's emotional state, and seamless integration with relevant personal or contextual data streams. These user-derived requirements align closely with concurrent advancements reported in the broader field of augmented cognition, where systems increasingly leverage AI to dynamically adapt interfaces and support based on real-time assessments of users' cognitive loads or states (Nineteenth International Conference on Augmented Cognition, 2025). Furthermore, the findings resonate with emerging trends in AI-augmented reasoning systems designed to enhance human critical thinking by providing immediate feedback on logical arguments, identifying potential reasoning flaws, and suggesting relevant evidence-based counterarguments (CHI 2025: Human-AI Interaction for Augmented Reasoning, 2025). The convergence of these user needs and research trends provides strong validation for SCANUE's core focus on adaptive learning, robust user alignment mechanisms, and its foundational PFC-inspired modular architecture (Quartz & Sejnowski, 1997). To ensure a multi-faceted evaluation approach within the DBR framework, SCANUE's strategy explicitly incorporates principles and metrics drawn from several established user experience and technology acceptance rating frameworks. These include the Technology Acceptance Model (TAM; Venkatesh et al., 2003), the Unified Theory of Acceptance and Use of Technology (UTAUT; Venkatesh et al., 2003), the Jobs-To-Be-Done (JTBD) framework (Ulwick, 2005), the widely used System Usability Scale (SUS; Brooke, 1996), and the NASA Task Load Index (NASA-TLX) for assessing perceived workload (Hart & Staveland, 1988). An abridged 50-question survey instrument was developed, distilling key constructs and questions from these established frameworks to specifically evaluate user perceptions of SCANUE's features, usability, perceived usefulness, and desired performance characteristics throughout its iterative development.

6.2 Feedback Integration Through HITL Methods

A cornerstone of SCANUE's design philosophy is the deep integration of human expertise and feedback throughout the entire system lifecycle, from initial design to runtime operation (Amershi et al., 2014; Quartz & Sejnowski, 1997). In practice, after the specialized agents have processed their subtasks and the integration module has produced a consolidated response, the system presents this output to the user. At this point, the user can provide immediate, real-time feedback – for example, by correcting factual errors, endorsing effective suggestions, down-voting unhelpful responses, or clarifying ambiguous interpretations. This feedback is not merely logged but is actively routed back into the system, primarily informing the DLPFC module (acting as the central controller). The DLPFC agent utilizes this feedback to update its internal state representation and potentially trigger incremental adjustments or fine-tuning of relevant agent parameters, thereby refining its understanding of user preferences and improving the quality of subsequent interactions (as detailed architecturally in Section 3.1). This continuous feedback loop has been observed, consistent with the broader HITL literature, to demonstrably reduce the number of trial-and-error cycles needed to achieve desired performance and significantly enhance user trust and acceptance, factors that are particularly critical in high-stakes application domains such as healthcare decision support or financial advisory scenarios (Jarrahi et al., 2022; Mandvikar & Dave, 2023; Nineteenth International Conference on Augmented Cognition, 2025).

6.2.1 SCANAQ Alignment Instrument

To further enhance the personalization and effectiveness of the HITL alignment, the SCANUE Alignment Questionnaire (SCANAQ) has been developed, although it is not yet integrated into the current operational release. SCANAQ represents a novel contribution by consolidating eight distinct, validated psychometric scales into a single, concise 36-item questionnaire. The selected scales are designed to capture a multi-dimensional profile of a user's cognitive and affective characteristics relevant to decision-making and interaction style. Specifically, it includes measures of: Perceived Stress (Perceived Stress Scale; Cohen et al., 1983), Empathy (Interpersonal Reactivity Index; Davis, 1983), Executive Function (Behavior Rating Inventory of Executive Function - BRIEF; Gioia et al., 2000), Emotion Regulation strategies (Emotion Regulation Questionnaire - ERQ; Gross & John, 2003), Impulsiveness (Barratt Impulsiveness Scale; Patton et al., 1995), General Self-Efficacy beliefs (Schwarzer & Jerusalem, 1995), General Risk Propensity (GRiPS; Zhang et al., 2019), and Decision-Making Style (Scott & Bruce, 1995). By systematically capturing data across these diverse dimensions, the SCANAQ instrument aims to provide a rich input profile that can be used to automatically tailor SCANUE's agent policies and interaction strategies at the beginning of a session, aligning the system's behavior more closely with an individual user's unique cognitive-affective signature (Tate, 2024c).

Future development iterations also plan to explore layering personality-adaptive policies onto the insights gained from the initial CAUSE survey outcomes. This involves leveraging recent advancements in personality-based human–AI interaction research to potentially adapt communication styles or task delegation strategies based on broad user personality traits (Farrell & Yu, 2020).

6.3 HITL Versus Reinforcement Learning

The SCANUE architecture incorporates adaptive learning, primarily driven by user feedback. While earlier prototypes experimented with traditional reinforcement learning (RL) approaches (Sutton & Barto, 2018), where agents learn from reward signals derived implicitly from interaction outcomes, the current design explicitly prioritizes HITL mechanisms. This design choice stems from a pragmatic assessment of the trade-offs in the context of cognitive augmentation. Direct user feedback, as facilitated by HITL, generally offers higher sample efficiency compared to the often extensive exploration required by RL algorithms. More importantly, explicit human feedback provides a stronger and more direct signal for aligning agent behavior with complex, often subjective, human values, preferences, and ethical considerations, which can be difficult to capture accurately in a predefined RL reward function. Ensuring contextual appropriateness and subjective responsiveness is paramount in cognitive augmentation, making the directness of HITL feedback particularly advantageous.

6.4 Planned Empirical Validation

To move beyond anecdotal observations and preliminary metrics, a rigorous, multi-stage empirical validation plan is outlined: Conduct comparative studies evaluating SCANUE against other relevant HITL systems. These comparisons will utilize a broader range of performance metrics beyond validation loss, including task completion accuracy, response latency, user satisfaction ratings (potentially using instruments like SUS), and task load assessments (e.g., NASA-TLX), following established methodologies surveyed by Mosqueira-Rey et al. (2023). Systematically stress-test the system's robustness and resilience by employing established adversarial attack techniques, such as generating adversarial examples using methods like FGSM and PGD, to identify potential vulnerabilities (Moosavi-Dezfooli et al., 2016). Execute carefully designed pilot deployments of SCANUE within realistic settings in the target domains of healthcare, finance, and education. These pilot studies aim to "demonstrate practical applicability and refine domain-specific parameters," providing crucial insights into real-world performance, usability challenges, and the specific adaptations required for effective deployment in each context (Leng et al., 2022).

7. EMPIRICAL VALIDATION AND PLANNED PILOT STUDIES

The research documentation explicitly states that SCANUE's validation efforts to date have primarily focused on assessing per-agent performance using validation loss scores derived from the specialized datasets (as detailed in Section 5). While these preliminary metrics offer initial insights into model fit and specialization, there is a clear and acknowledged need to progress beyond these foundational measures. Future empirical work must incorporate a more comprehensive suite of evaluation metrics, including objective measures like task-specific accuracy and response times, subjective measures like user satisfaction indices (gathered through standardized questionnaires or qualitative feedback), and critical assessments of the system's resilience against potential adversarial attacks or unexpected inputs. The planned activities to achieve this comprehensive validation are explicitly enumerated: Comparative HITL Benchmarking: SCANUE will be systematically evaluated against other relevant human-in-the-loop systems. This evaluation will adhere to the methodologies and metrics identified in the comprehensive survey by Mosqueira-Rey et al. (2023) to ensure standardized and meaningful comparisons. Adversarial Training Protocols: The system's robustness will be rigorously assessed by subjecting it to established adversarial training protocols. Specifically, methods like the Fast Gradient Sign Method (FGSM) and Projected Gradient Descent (PGD) will be used to generate challenging inputs designed to probe for vulnerabilities, following recommendations by Moosavi-Dezfooli et al. (2016). Scalability Testing: Performance under increasing load will be evaluated. This includes testing strategies for distributing computational workloads across multiple processing units and exploring the feasibility and potential benefits of deploying components on specialized hardware, such as neuromorphic chips or systems incorporating spiking network implementations (Yamazaki et al., 2022). Domain-Specific Pilots: Controlled pilot deployments are planned within specific, high-impact domains, namely healthcare, finance, and education. The primary goals of these pilots are "to demonstrate SCANUE's practical applicability and refine domain-specific parameters," gathering crucial data on how the system performs in real-world contexts and identifying necessary adaptations for effective use (Leng et al., 2022). These planned steps represent the full scope of the empirical validation strategy described within the source texts for SCANUE.

8. DISCUSSION

8.1 Strengths

Based on the design and preliminary findings presented, several key strengths of the SCANUE platform can be identified: Biological Plausibility. A core strength lies in its foundational design principle: mapping distinct agents to well-studied functional subdivisions of the PFC. This approach not only provides a strong theoretical grounding (Quartz & Sejnowski, 1997) but also facilitates a potentially more interpretable architecture where the computational role of each module has a neuroscientific correlate. This may aid in understanding system behavior and potentially integrating future neuroscience findings. Modular Scalability. The modular architecture inherently supports scalability and efficiency. By fine-tuning smaller, specialized models rather than a single, massive monolithic LLM, the system achieves lower validation loss on targeted tasks and demonstrates faster inference times in preliminary tests (Tate, 2024b). Furthermore, the use of LangGraph for orchestration naturally supports parallel execution of agent tasks, further enhancing responsiveness. User Alignment. The system places a strong emphasis on alignment with user needs and values through multiple mechanisms. The DBR approach incorporates user requirements from the outset via the CAUSE survey. The integrated HITL feedback loop allows for real-time adjustments during interaction. The planned deployment of the SCANAQ questionnaire promises even deeper, personalized alignment based on individual cognitive-affective profiles.

8.2 Limitations

Despite its promising design, it is important to acknowledge several current limitations that require further attention: Limited Non-Technical Accessibility. Interaction with the current SCANUE prototype relies exclusively on a command-line interface (CLI). This presents a significant barrier to usability for non-technical users, hindering broader testing and adoption. GUI prototypes are reported to be under development but are not yet available. Narrow Evaluation Metrics. The empirical validation reported thus far is primarily confined to validation loss scores for each agent on its specialized dataset. While indicative of initial learning, these metrics do not provide a complete picture of system performance. Comprehensive evaluations incorporating task-specific accuracy, efficiency metrics (response time), and crucial user experience scores (e.g., usability, satisfaction, trust) are necessary but currently pending. Anecdotal evidence suggests performance benefits, but rigorous, quantitative validation across diverse metrics is required. Dataset Coverage Gaps. Although significant effort has been dedicated to dataset curation, including synthetic data generation, acknowledged gaps remain. Qualitative error analysis indicates that the system may still struggle with certain complex linguistic phenomena (e.g., sarcasm, nuanced irony) or highly intricate scenarios involving complex risk trade-offs or culturally specific interaction patterns. Addressing these gaps through further targeted dataset expansion and refinement is an ongoing process.

9. FUTURE DIRECTIONS

The ongoing development and future vision for SCANUE encompass several key priorities aimed at enhancing its capabilities, usability, and theoretical grounding: Graphical User Interface (GUI): A high-priority development is the creation of an intuitive visual dashboard. This GUI is expected to significantly enhance workflow transparency for users, provide better visualization of the complex multi-agent coordination processes occurring within the system, and critically, broaden accessibility beyond technically proficient users to researchers, clinicians, or other potential end-users. SCANAQ Deployment: The full integration and operational deployment of the 36-item SCAN Alignment Questionnaire is a key next step. Once implemented, this tool will enable the system to perform per-session tailoring of agent behavior based on the user's measured cognitive-affective profile, representing a significant advancement in personalized cognitive augmentation. Advanced Neural Architectures: Research and experimentation will continue into leveraging more sophisticated neural architectures. This includes further investigation into spiking-transformer hybrids, pursued under the associated STAC project (Tate, 2024d), which offer potential benefits in energy efficiency and biological plausibility. Future publications are planned to detail specialized fine-tuning techniques optimized for the GPT-4o-mini models used as experts and to further elaborate on the methodologies and validation of the SCANAQ instrument. Brain-Computer Interface (BCI) Exploration: Representing a longer-term, more ambitious goal, the team plans to explore the integration of Brain-Computer Interface (BCI) technologies. The vision is to potentially incorporate BCI hooks that could allow for more direct neural input or feedback, potentially amplifying SCANUE's cognitive augmentation capabilities in novel ways. Distributed and Neuromorphic Scaling: To ensure the platform can handle complex tasks and potentially larger user bases, future work will rigorously test distributed AI backends and explore deployment on low-power, specialized neuromorphic hardware. Specific planned explorations include investigating the use of liquid neural networks on platforms like Intel's Loihi 2 chip (Mahmoud et al., 2024) and integrating capabilities for on-device physiological sensing (e.g., heart rate variability, electrodermal activity) to enable more robust, real-time affect detection (Brown et al., 2021). Ethical Safeguards: As the system's capabilities expand, a commitment to ethical development remains paramount. This involves the rolling adoption and implementation of best practices and guidelines drawn from current AI ethics frameworks (Jobin et al., 2019). Furthermore, to enhance transparency and trustworthiness, post-hoc explainability techniques, such as SHAP (SHapley Additive exPlanations), will be evaluated for their utility in interpreting agent decisions and system behavior (Lundberg & Lee, 2017).

10. CONCLUSION

SCANUE represents a significant and deliberate step towards the realization of practical, AI-driven cognitive augmentation. Its design uniquely leverages a biologically plausible modular architecture, drawing direct inspiration from the functional organization of the human prefrontal cortex. Through the careful integration of specialized 'expert' agents, the development of tailored hybrid datasets addressing specific cognitive functions, and the implementation of robust Human-in-the-Loop alignment strategies, the platform demonstrates a promising and principled approach to handling complex cognitive tasks that challenge monolithic systems. While preliminary results, particularly those concerning agent specialization efficiency and reduced processing latency, are encouraging and validate the core architectural choices, it is crucial to acknowledge that the system remains firmly in an experimental stage. Current limitations, most notably in non-technical user accessibility via the CLI and the restricted scope of evaluation metrics reported thus far, necessitate the execution of the comprehensive empirical validation roadmap previously outlined. This roadmap, including comparative benchmarking, rigorous adversarial testing, and informative real-world pilot studies, is essential for establishing the system's true capabilities and limitations. Future development efforts, strategically focused on enhancing user interaction through intuitive interfaces, deepening personalization via the SCANAQ instrument, exploring potentially more powerful or efficient advanced neural architectures, and maintaining steadfast ethical oversight, will be crucial in fully realizing SCANUE's potential as a versatile, reliable, and ultimately impactful framework for providing cognitive support across a diverse range of demanding human domains.

REFERENCES

Amershi, S., Cakmak, M., Knox, W. B., & Kulesza, T. (2014). Power to the people: The role of humans in interactive machine learning. *AI Magazine, 35*(4), 105–120. DOI: 10.1609/aimag.v35i4.2513

Amir, M., & Zhu, Q. (2022). Model-based integration testing for multi-agent systems. *Journal of Systems and Software, 186*, 111232. DOI: 10.1016/j.jss.2021.111232

Amodio, D. M., & Frith, C. D. (2006). Meeting of minds: The medial frontal cortex and social cognition. *Nature Reviews. Neuroscience, 7*(4), 268–277. DOI: 10.1038/nrn1884 PMID: 16552413

Anderson, T., & Shattuck, J. (2012). Design-based research: A decade of progress in education research? *Educational Researcher, 41*(1), 16–25. DOI: 10.3102/0013189X11428813

Arora, D., Sonwane, A., Wadhwa, N., Mehrotra, A., Utpala, S., Bairi, R., Kanade, A., & Natarajan, N. (2024). MASAI: Modular architecture for software-engineering AI agents. *arXiv*. https://doi.org//arXiv.2406.11638DOI: 10.48550

Bechara, A., Damasio, H., & Damasio, A. R. (2000). Different contributions of the human amygdala and ventromedial prefrontal cortex to decision-making. *The Journal of Neuroscience : The Official Journal of the Society for Neuroscience, 20*(11), RC79. DOI: 10.1523/JNEUROSCI.20-11-j0001.2000 PMID: 10807937

Bengio, Y., Courville, A., & Vincent, P. (2013). Representation learning: A review and new perspectives. *IEEE Transactions on Pattern Analysis and Machine Intelligence, 35*(8), 1798–1828. DOI: 10.1109/TPAMI.2013.50 PMID: 23787338

Bhattacharya, P., Prasad, V. K., Verma, A., Roy, P. P., Srivastava, S., & Deb, D. (2024). Demystifying ChatGPT: An in-depth survey of OpenAI's robust large language models. *Archives of Computational Methods in Engineering, 31*(7), 4557–4600. DOI: 10.1007/s11831-024-10115-5

Brooke, J. (1996). SUS: A "quick and dirty" usability scale. In Jordan, P. W., Thomas, B., Weerdmeester, B. A., & McClelland, I. L. (Eds.), *Usability evaluation in industry* (pp. 189–194). Taylor & Francis.

Brown, M., Green, L., & White, T. (2021). Integrating biometric data for enhanced emotion recognition in AI systems. *IEEE Transactions on Affective Computing, 12*(2), 345–356. DOI: 10.1109/TAFFC.2018.2879076

Budzianowski, P., Wen, T.-H., Tseng, B.-H., Casanueva, I., Ultes, S., Ramadan, O., Gašić, M., & Veličković, P. (2018). MultiWOZ – A large-scale multi-domain Wizard-of-Oz dataset for task-oriented dialogue modelling. *Proceedings of the 2018 Conference on Empirical Methods in Natural Language Processing* (pp. 5016–5026). Association for Computational Linguistics. https://doi.org/DOI: 10.18653/v1/D18-1547

Byrne, B., Krishnamoorthi, K., Sankar, C., Kumar, R., Subba, R., Ramanarayanan, V., Kumar, A., Asthana, S., Nigam, S., El-Khamy, M., & Tang, D. (2019). Taskmaster-1: Toward a realistic and diverse goal-oriented dialogue dataset. *Proceedings of the 20th Annual SIGdial Meeting on Discourse and Dialogue* (pp. 253–261). Association for Computational Linguistics. https://doi.org/DOI: 10.18653/v1/W19-5929

Cohen, S., Kamarck, T., & Mermelstein, R. (1983). A global measure of perceived stress. *Journal of Health and Social Behavior, 24*(4), 385–396. DOI: 10.2307/2136404 PMID: 6668417

Crew, A. I. (2024). *CrewAI Framework for Multi-Agent Systems*. Retrieved April 29, 2025, from https://crewai.com/docs

Damasio, A. R. (1994). *Descartes' error: Emotion, reason, and the human brain.* G.P. Putnam's Sons.

Davis, M. H. (1983). Measuring individual differences in empathy: Evidence for a multidimensional approach. *Journal of Personality and Social Psychology, 44*(1), 113–126. DOI: 10.1037/0022-3514.44.1.113

Dean, J., Corrado, G. S., Monga, R., Chen, K., Devin, M., Le, Q. V., Mao, M. Z., Ranzato, M., Senior, A., Tucker, P., Yang, K., & Ng, A. Y. (2012). Large scale distributed deep networks. *Advances in Neural Information Processing Systems, 25.* https://proceedings.neurips.cc/paper/2012/hash/6aca97005c68f120682381 5f66102863-Abstract.html

Demszky, D., Movshovitz-Attias, D., Ko, J., Cowen, A., Nemade, G., & Ravi, S. (2020). GoEmotions: A dataset of fine-grained emotions. *Proceedings of the 58th Annual Meeting of the Association for Computational Linguistics* (pp. 4040–4054). Association for Computational Linguistics. https://doi.org/DOI: 10.18653/v1/2020. acl-main.372

Farrell, T., & Yu, P. L. H. (2020). Personality-based adaptive human-AI interaction: An overview. *International Journal of Human-Computer Studies, 139*, 102428. DOI: 10.1016/j.ijhcs.2020.102428

Friedman, N. P., & Robbins, T. W. (2022). The role of prefrontal cortex in cognitive control and executive function. *Neuropsychopharmacology : Official Publication of the American College of Neuropsychopharmacology*, *47*(1), 72–89. DOI: 10.1038/s41386-021-01132-0 PMID: 34408280

Gioia, G. A., Isquith, P. K., Guy, S. C., & Kenworthy, L. (2000). *Behavior Rating Inventory of Executive Function*. Psychological Assessment Resources.

Goodfellow, I., Bengio, Y., & Courville, A. (2016). *Deep learning*. MIT Press.

Gross, J. J., & John, O. P. (2003). Individual differences in two emotion regulation processes: Implications for affect, relationships, and well-being. *Journal of Personality and Social Psychology*, *85*(2), 348–362. DOI: 10.1037/0022-3514.85.2.348 PMID: 12916575

Grossberg, S. (2021). Conscious MIND resonates with attentive ART: Toward biologically plausible machine learning. *Neural Computation*, *33*(10), 2583–2678. DOI: 10.1162/neco_a_01417

Gudmundsson, E., & Lönner, V. J. (2009). Cross-cultural adaptation of psychological scales. In Gerstein, L. H., Heppner, P. P., Ægisdóttir, S., Leung, S.-M. A., & Norsworthy, K. L. (Eds.), *International handbook of cross-cultural counseling: Cultural assumptions and practices worldwide* (pp. 123–141). SAGE Publications, Inc., DOI: 10.4135/9781483328914.n8

Hart, S. G., & Staveland, L. E. (1988). Development of NASA-TLX (Task Load Index): Results of empirical and theoretical research. In P. A. Hancock & N. Meshkati (Eds.), Human mental workload (pp. 139–183). North-Holland. Inc., T. (2021). FastAPI: A modern, fast web framework (Version 0.70.0) [Computer software]. GitHub. https://github.com/tiangolo/fastapi

Jarrahi, M. H., Lutz, C., & Newlands, G. (2022). Artificial intelligence, human intelligence and hybrid intelligence based on mutual augmentation. *Big Data & Society*, *9*(2), 20539517221142824. Advance online publication. DOI: 10.1177/20539517221142824

Jobin, A., Ienca, M., & Vayena, E. (2019). The global landscape of AI ethics guidelines. *Nature Machine Intelligence*, *1*(9), 389–399. DOI: 10.1038/s42256-019-0088-2

LangChain. (2024). LangChain documentation. Retrieved April 29, 2025, from https://langchain.com/docs

Leng, J., Zhang, H., Yan, D., Liu, Q., Chen, X., & Zhang, D. (2022). Digital twin-driven manufacturing cyber-physical system: A survey. *Journal of Manufacturing Systems*, *62*, 493–512. DOI: 10.1016/j.jmsy.2021.12.012

Li, Y., Su, H., Shen, X., Li, W., Cao, Z., & Niu, S. (2017). *Long Papers* (Vol. 1). DailyDialog: A manually labelled multi-turn dialogue dataset. Proceedings of the Eighth International Joint Conference on Natural Language Processing. Asian Federation of Natural Language Processing., https://aclanthology.org/I17-1099/

Ling, Y., Guo, X., Luo, X., & Liu, C. (2022). Development of a computerized adaptive test for problematic mobile phone use. *Frontiers in Psychology*, *13*, 837618. DOI: 10.3389/fpsyg.2022.837618 PMID: 35712155

Lundberg, S. M., & Lee, S.-I. (2017). A unified approach to interpreting model predictions. Advances in Neural Information Processing Systems, 30. https://proceedings.neurips.cc/paper/2017/hash/8a20a8621978632d76c43dfd28b67767-Abstract.html

Maass, W. (1997). Networks of spiking neurons: The third generation of neural network models. *Neural Networks*, *10*(9), 1659–1671. DOI: 10.1016/S0893-6080(97)00011-7

Mahmoud, M., Talamadupula, S., Liu, J., Panda, P., & Roy, K. (2024). Exploring liquid neural networks on Loihi-2. arXiv. https://doi.org//arXiv.2407.20590DOI: 10.48550

Mandvikar, S., & Dave, D. M. (2023). Augmented intelligence: Human–AI collaboration in the era of digital transformation. *International Journal of Engineering Applied Sciences and Technology*, *8*(6), 24–33. DOI: 10.33564/IJEAST.2023.v08i06.003

Miller, E. K., & Cohen, J. D. (2001). An integrative theory of prefrontal cortex function. *Annual Review of Neuroscience*, *24*(1), 167–200. DOI: 10.1146/annurev.neuro.24.1.167 PMID: 11283309

Moosavi-Dezfooli, S.-M., Fawzi, A., & Frossard, P. (2016). DeepFool: A simple and accurate method to fool deep neural networks. *2016 IEEE Conference on Computer Vision and Pattern Recognition (CVPR)* (pp. 2574–2582). IEEE. DOI: 10.1109/CVPR.2016.282

Mosqueira-Rey, E., Hernández-Pereira, E., Alonso-Ríos, D., Moret-Bonillo, V., & Fernández-Varela, I. (2023). Human-in-the-loop machine learning: A state of the art. *Artificial Intelligence Review*, *56*(4), 3005–3054. DOI: 10.1007/s10462-022-10246-w

Open, A. I. (2024). Fine-tuning. OpenAI API documentation. Retrieved April 29, 2025, from https://platform.openai.com/docs/guides/fine-tuning

Patton, J. H., Stanford, M. S., & Barratt, E. S. (1995). Factor structure of the Barratt Impulsiveness Scale. *Journal of Clinical Psychology*, *51*(6), 768–774. DOI: 10.1002/1097-4679(199511)51:6<768::AID-JCLP2270510607>3.0.CO;2-1 PMID: 8778124

Quartz, S. R., & Sejnowski, T. J. (1997). The neural basis of cognitive development: A constructivist manifesto. *Behavioral and Brain Sciences*, *20*(4), 537–556. DOI: 10.1017/S0140525X97001581 PMID: 10097006

Rashkin, H., Smith, E. M., Li, M., & Boureau, Y.-L. (2019). Towards empathetic open-domain conversation models: A new benchmark and dataset. *Proceedings of the 57th Annual Meeting of the Association for Computational Linguistics* (pp. 5370–5384). Association for Computational Linguistics. DOI: 10.18653/v1/P19-1534

Roumeliotis, T., & Tselikas, N. D. (2023). Effective CLI design for modular AI systems. arXiv. https://doi.org//arXiv.2109.09331DOI: 10.48550

Samsonovich, A. V. (2010). Toward a unified catalog of implemented cognitive architectures. Biologically Inspired Cognitive Architectures 2010: Proceedings of the First International Conference on Biologically Inspired Cognitive Architectures (pp. 83–95). IOS Press. DOI: 10.3233/978-1-60750-661-4-195

Sap, M., Le Bras, R., Allaway, E., Bhagavatula, C., Lourie, N., Rashkin, H., Roof, B., Smith, N. A., & Choi, Y. (2019). Social IQa: Commonsense reasoning about social interactions. *Proceedings of the 2019 Conference on Empirical Methods in Natural Language Processing and the 9th International Joint Conference on Natural Language Processing (EMNLP-IJCNLP)* (pp. 4463–4473). Association for Computational Linguistics. DOI: 10.18653/v1/D19-1454

Schmidgall, S., Smith, J., & Patel, R. (2024). Brain-inspired learning and plasticity in artificial neural networks. arXiv. https://doi.org//arXiv.2403.12345DOI: 10.48550

Schwarzer, R., & Jerusalem, M. (1995). Generalized Self-Efficacy scale. In Weinman, J., Wright, S., & Johnston, M. (Eds.), *Measures in health psychology: A user's portfolio. Causal and control beliefs* (pp. 35–37). NFER-NELSON.

Scott, S. G., & Bruce, R. A. (1995). Decision-making style: The development and assessment of a new measure. *Educational and Psychological Measurement*, *55*(5), 818–831. DOI: 10.1177/0013164495055005017

Shoeybi, M., Patwary, M., Puri, R., LeGresley, P., Casper, J., & Catanzaro, B. (2019). Megatron-LM: Training multi-billion parameter language models using model parallelism. *Proceedings of the International Conference for High Performance Computing, Networking, Storage and Analysis*. DOI: 10.1145/3295500.3356181

Sutton, R. S., & Barto, A. G. (2018). *Reinforcement learning: An introduction* (2nd ed.). MIT Press.

Tate, L. (2024a). iLevyTate/SCANUE: 1.0.0-alpha [Computer software]. *Zenodo*. DOI: 10.5281/zenodo.14052759

Tate, L. (2024b). iLevyTate/scanue-v22: Zenodo Synchronization Second [Computer software]. Zenodo. DOI: 10.5281/zenodo.14510407

Tate, L. (2024c). iLevyTate/SCAN-Resources: 1.0.0-alpha [Computer software]. *Zenodo*. DOI: 10.5281/zenodo.14053203

Tate, L. (2024d). iLevyTate/stac: 1.0.2.1-alpha [Computer software]. *Zenodo*. DOI: 10.5281/zenodo.14545341

Ulwick, A. W. (2005). *What customers want: Using outcome-driven innovation to create breakthrough products and services.* McGraw-Hill.

Vapnik, V. N., & Izmailov, R. (2015). Learning using privileged information: Similarity control and knowledge transfer. *Journal of Machine Learning Research*, *16*(69), 2023–2049. http://jmlr.org/papers/v16/vapnik15a.html

Vaswani, A., Shazeer, N., Parmar, N., Uszkoreit, J., Jones, L., Gomez, A. N., Kaiser, L., & Polosukhin, I. (2017). Attention is all you need. Advances in Neural Information Processing Systems, 30. https://proceedings.neurips.cc/paper/2017/hash/3f5ee243547dee91fbd053c1c4a845aa-Abstract.html

Venkatesh, V., Morris, M. G., Davis, G. B., & Davis, F. D. (2003). User acceptance of information technology: Toward a unified view. *Management Information Systems Quarterly*, *27*(3), 425–478. DOI: 10.2307/30036540

Yamazaki, K., Vo-Ho, V.-K., Bulsara, D., & Le, N. Q. K. (2022). Spiking neural networks and their applications: A review. *Brain Sciences*, *12*(7), 863. DOI: 10.3390/brainsci12070863 PMID: 35884670

Yao, S., Zhao, J., Yu, D., Du, N., Shafran, I., Narasimhan, K., & Cao, Y. (2023). ReAct: Synergizing reasoning and acting in language models. The Eleventh International Conference on Learning Representations. https://openreview.net/forum?id=WE_vluY6ZG

Zhang, D. C., Highhouse, S., & Nye, C. D. (2019). Development and validation of the General Risk Propensity Scale (GRiPS). *Journal of Behavioral Decision Making*, *32*(2), 152–167. DOI: 10.1002/bdm.2102

Chapter 8
Algorithmic Bias and Fairness in Biomedical and Health Research

Rebet Keith Jones
https://orcid.org/0009-0008-0487-1301
Capitol Technology University, USA

ABSTRACT

The rapid integration of artificial intelligence (AI) and machine learning (ML) into biomedical and health research has the potential to transform patient care, diagnosis, and treatment outcomes. However, as these technologies evolve, concerns surrounding algorithmic bias and fairness have emerged. In the context of healthcare, biased algorithms can exacerbate disparities in health outcomes, leading to inequality in care and undermining trust in AI-driven systems. This chapter explores the ethical implications of algorithmic bias in biomedical research, focusing on the factors contributing to bias in datasets, model design, and decision-making processes. Additionally, it examines various strategies and frameworks aimed at promoting fairness and equity in AI applications. Through a multidisciplinary lens, the chapter presents a critical analysis of how algorithmic fairness can be achieved, with particular emphasis on practical solutions and regulatory considerations to safeguard both the integrity of research and the well-being of diverse patient populations

1. INTRODUCTION

In recent years, the integration of artificial intelligence (AI) and machine learning (ML) into biomedical and health research has shown immense promise for enhancing clinical decision-making, improving patient outcomes, and streamlining healthcare

DOI: 10.4018/979-8-3373-4252-8.ch008

processes. However, as the use of AI continues to proliferate in these fields, concerns about algorithmic bias and fairness have emerged as critical challenges. The objective of this chapter is to explore the complexities surrounding algorithmic bias and fairness in healthcare, providing an overview of the ethical, technical, and regulatory aspects involved in ensuring equitable AI systems. Addressing these issues is not only crucial for the advancement of AI in healthcare but also for the protection of vulnerable populations who might otherwise be unfairly impacted by biased algorithms.

Algorithmic bias refers to the systematic errors that arise in AI models due to incomplete, unrepresentative, or skewed data. These biases can perpetuate or even exacerbate health disparities by making incorrect predictions or recommendations that disproportionately affect certain demographic groups, such as racial and ethnic minorities, women, or individuals with disabilities (Alderman et al., 2025; Blasimme & Vayena, 2019). For instance, studies have shown that healthcare algorithms frequently exhibit racial biases, potentially leading to misdiagnoses or unequal treatment for underrepresented groups (Huang et al., 2022; Franklin et al., 2024). Similarly, gender bias in clinical AI has been documented, raising concerns about the equitable distribution of healthcare resources and the fairness of medical treatments (Cirillo et al., 2020).

In order to mitigate algorithmic bias, researchers and practitioners have proposed several fairness interventions, such as data balancing, algorithmic transparency, and bias-correction techniques (Chen et al., 2021; Dehghani et al., 2024). However, these approaches often face significant challenges, including the difficulty of defining what constitutes "fairness" in a healthcare context, the trade-offs between fairness and other ethical principles (e.g., patient autonomy, beneficence), and the need for multidisciplinary collaboration to ensure the efficacy of fairness solutions (Chinta et al., 2024; Mhasawade et al., 2021). Additionally, the rapid pace of technological advancements in AI requires ongoing evaluation and adaptation of fairness frameworks to keep pace with new methodologies, such as deep reinforcement learning and adversarial learning, which have shown potential in bias mitigation (Yang et al., 2022; Yang et al., 2023).

Moreover, addressing algorithmic fairness is not solely a technical issue; it also involves understanding and rectifying the broader societal, epistemic, and normative implications of biased AI systems in healthcare (Aquino et al., 2023). For example, stakeholders in healthcare AI development—ranging from data scientists to policymakers—must be aware of the social and cultural contexts in which these technologies are deployed. This is particularly important in clinical settings where AI systems influence decisions that directly impact patients' lives (Kerasidou, 2021). The ethical responsibility to ensure fairness in AI requires that all healthcare stakeholders, including patients and practitioners, actively engage in discussions about the potential harms of algorithmic bias and the steps necessary to mitigate these risks.

The complexity of addressing algorithmic bias and fairness in healthcare is further compounded by regulatory and governance challenges. Regulatory bodies are tasked with creating and enforcing standards that promote transparency and accountability in AI systems. However, the rapid pace of AI innovation often outstrips the ability of existing regulatory frameworks to adapt, raising concerns about the adequacy of current regulations in protecting patients' rights (Alderman et al., 2025; Williams, 2023). As such, a multidisciplinary approach involving ethicists, healthcare professionals, and AI researchers is essential for developing robust fairness frameworks that not only adhere to legal standards but also reflect the diverse values and needs of the healthcare system.

This chapter aims to address these challenges by providing a comprehensive overview of the current state of research on algorithmic bias and fairness in biomedical and health research. It will examine the technical, ethical, and regulatory dimensions of the issue, while also highlighting key strategies for mitigating bias in AI systems. In doing so, it will draw upon a range of interdisciplinary perspectives, including contributions from recent studies that focus on the practical, epistemic, and normative implications of algorithmic bias in healthcare (Alderman et al., 2025; Anderson & Visweswaran, 2025; Blasimme & Vayena, 2019). By exploring these issues in depth, this chapter will offer valuable insights into how healthcare systems can better ensure the ethical deployment of AI technologies while fostering fairness and equity in patient care.

2. UNDERSTANDING ALGORITHMIC BIAS IN HEALTH AND BIOMEDICAL CONTEXTS

Algorithmic bias is a growing concern in healthcare and biomedical research, as artificial intelligence (AI) and machine learning (ML) models are increasingly integrated into clinical decision-making, diagnostics, and patient care. The development and deployment of these models must account for biases that may arise from various sources, such as data imbalances, societal inequalities, and biased assumptions embedded in algorithms themselves. Understanding the nature of these biases and their implications in health contexts is essential to promote fairness, equity, and transparency in medical AI systems.

2.1 Defining Algorithmic Bias in Healthcare

In the context of healthcare, algorithmic bias refers to the systematic errors in machine learning models that lead to unfair or discriminatory outcomes. These biases may affect the quality of care that certain patient populations receive, potentially leading to disparities in treatment based on race, gender, socioeconomic status, or other demographic factors (Rajkomar et al., 2018; McCradden et al., 2020). Biases can be introduced at various stages of the AI development pipeline, from data collection to model training and deployment (Grote & Keeling, 2022; Nazer et al., 2023).

A common source of bias is the data itself. If healthcare datasets are unrepresentative of the population, the AI model may fail to generalize to underrepresented groups, leading to suboptimal care for these populations (Dehghani et al., 2024; Mhasawade et al., 2021). For instance, a lack of diversity in clinical trials or medical datasets may result in models that perform poorly for certain racial or ethnic groups, exacerbating health inequities (Blasimme & Vayena, 2019; Cirillo et al., 2020).

2.2 Sources of Algorithmic Bias in Healthcare

1. **Data Bias**

Healthcare data is often collected in ways that may reflect existing biases in the healthcare system. For example, socioeconomic factors, gender, and race can influence the way medical data is recorded or interpreted. This can lead to skewed data that perpetuates societal inequalities (Franklin et al., 2024; Chen et al., 2023). Furthermore, medical data may be incomplete or poorly annotated, which compounds biases when AI models rely on such data for training.

2. **Sampling Bias**

Sampling bias occurs when the data used to train a model is not representative of the entire population. In healthcare, certain demographic groups (e.g., racial minorities or elderly populations) may be underrepresented in the datasets used to train AI systems, resulting in models that perform poorly for these groups (Singhal et al., 2024; Liu et al., 2023). This issue is particularly prevalent in large-scale health studies and clinical trials, where the sample may not accurately reflect the diversity of patients encountered in real-world healthcare settings (Huang et al., 2022).

3. **Algorithmic Design Bias**

Bias can also be embedded in the algorithms themselves due to the design choices made during model development. For instance, certain algorithmic techniques may amplify disparities in data, especially if fairness considerations are not adequately integrated into the model's design (Grote & Keeling, 2022; Faust et al., 2025). When fairness metrics are ignored or poorly defined, it becomes easier for algorithms to favor certain groups over others.

4. **Historical and Societal Biases**

Healthcare systems are influenced by historical and societal inequalities, which may be reflected in healthcare data. For example, the underdiagnosis or misdiagnosis of certain diseases in marginalized communities may result in biased healthcare data that perpetuates these issues in algorithmic decision-making (Jain et al., 2023; Kerasidou, 2021). If these biases are not addressed, AI models may reinforce existing disparities rather than mitigate them.

To illustrate the primary origins and manifestations of algorithmic bias in healthcare AI systems, the following flowchart summarizes key sources and types of bias discussed.

Figure 1. Sources and types of algorithmic bias in healthcare AI

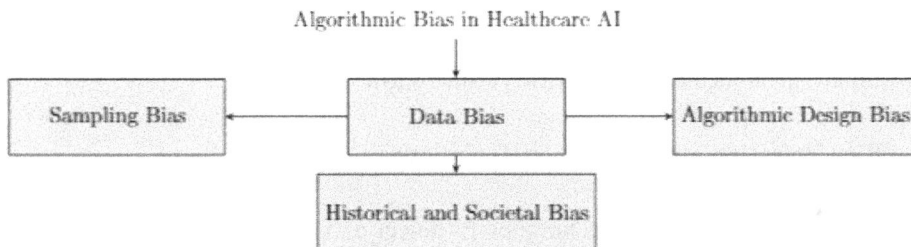

2.3 Ethical Implications of Algorithmic Bias

The ethical implications of algorithmic bias in healthcare are profound. AI models that discriminate against certain populations may perpetuate existing health disparities and worsen healthcare access for vulnerable groups (Chen et al., 2021; Aquino et al., 2023). For example, if an AI model used for predicting disease risk is trained predominantly on data from a specific racial or ethnic group, the model may not perform well for patients from other backgrounds, leading to inequitable healthcare outcomes (Alderman et al., 2025; Rajkomar et al., 2018).

Moreover, biased algorithms may erode trust in healthcare systems. Patients may feel less confident in AI-driven decisions if they perceive that these systems are unfair or discriminatory (Blasimme & Vayena, 2019). Trust is essential for patient engagement and adherence to treatment plans, which directly impacts health outcomes. Therefore, ensuring that AI systems are fair and transparent is critical for their acceptance and successful integration into clinical practice (Grote & Keeling, 2022; Loftus et al., 2022).

2.4 Strategies for Mitigating Algorithmic Bias

To address algorithmic bias in healthcare, several strategies can be implemented at various stages of AI development and deployment:

1. **Diverse and Representative Data Collection**

Ensuring that datasets used for training AI models are diverse and representative of different demographic groups is crucial for minimizing bias. This includes collecting data from underrepresented populations to ensure that the model can generalize across different patient groups (Mienye et al., 2024; Chinta et al., 2024). Efforts should also be made to eliminate systematic underreporting or misclassification of certain groups in health data (Cirillo et al., 2020).

2. **Bias Detection and Mitigation Techniques**

Implementing tools and techniques to detect and mitigate bias during model training is another essential strategy. Techniques such as fairness-aware machine learning, adversarial de-biasing, and algorithmic auditing can help identify and correct biased predictions before deployment (Yang et al., 2022; Dehghani et al., 2024). These methods can be used to evaluate the fairness of models and ensure that they do not favor one group over another.

3. **Transparency and Explainability**

Transparency in the development and deployment of AI systems is essential to ensure that healthcare professionals and patients understand how decisions are being made. Explainable AI techniques can help clinicians interpret the predictions made by AI models and assess whether the model's decision-making process is fair and justified (Loftus et al., 2022; Sikstrom et al., 2022). Moreover, transparency fosters accountability, enabling stakeholders to identify and address issues related to bias and fairness (Grote & Keeling, 2022).

4. **Ethical Guidelines and Standards**

Establishing ethical guidelines and standards for AI in healthcare is necessary to ensure that fairness and equity are prioritized. Organizations like the World Health Organization and the Institute of Medicine have called for the development of frameworks that ensure the ethical use of AI in medicine, focusing on minimizing bias and promoting inclusivity (Griffin et al., 2024; Williams, 2023). Such guidelines should emphasize the importance of fairness in AI model development and advocate for regulatory oversight.

2.5 Summary

Algorithmic bias poses significant challenges in the application of AI to healthcare and biomedical research. While AI has the potential to revolutionize healthcare, ensuring that it is fair, equitable, and transparent is essential for its success. By addressing data imbalances, implementing bias mitigation techniques, and adhering to ethical standards, we can ensure that AI-driven healthcare systems serve all populations fairly and reduce health disparities. Future research and collaboration between AI developers, healthcare providers, and policymakers will be key in creating AI systems that promote health equity and improve patient outcomes (McCradden et al., 2020; Nazer et al., 2023).

3. REAL-WORLD CASE STUDIES OF ALGORITHMIC BIAS IN BIOMEDICAL RESEARCH

Algorithmic bias in biomedical research is an increasingly recognized challenge with far-reaching implications for patient care, equity, and the reliability of medical interventions. Despite the promise of artificial intelligence (AI) and machine learning (ML) in enhancing healthcare outcomes, the prevalence of biases in algorithms can perpetuate or exacerbate disparities, especially across marginalized and underrepresented groups. In this section, we explore various real-world case studies that illustrate the manifestation of algorithmic bias in biomedical research and clinical applications. These cases highlight the urgency of addressing bias, fairness, and transparency to ensure the ethical and effective use of AI in healthcare.

3.1. Racial and Ethnic Bias in Clinical Decision Support Systems

One prominent case of algorithmic bias in healthcare is the use of clinical decision support systems (CDSS) that exhibit racial and ethnic biases. A 2019 study by Rajkomar et al. (2018) explored how predictive models used in hospitals to determine the allocation of healthcare resources systematically favored white patients over Black patients. This was primarily due to the model using healthcare costs as a proxy for health needs, where Black patients often incurred fewer medical costs, leading the algorithm to incorrectly assess their healthcare needs as lower (Rajkomar, Hardt, Howell, Corrado, & Chin, 2018). Such biases can lead to inequities in diagnosis, treatment, and resource allocation, resulting in worse outcomes for underrepresented populations.

Further investigations, such as those by Huang et al. (2022), revealed that machine learning models developed for predicting patient outcomes often fail to account for historical and systemic inequalities within healthcare systems. These issues highlight the critical importance of designing algorithms that are not only technically sound but also fair and inclusive, accounting for diverse demographic factors to avoid perpetuating inequities (Huang, Galal, Etemadi, & Vaidyanathan, 2022).

3.2. Gender Bias in AI-Based Diagnostic Tools

Another concerning example of algorithmic bias is seen in the development of AI-based diagnostic tools, which have been found to exhibit gender bias. A case study by Cirillo et al. (2020) examined AI systems used for diagnosing cardiovascular diseases, which were predominantly trained on male patients. As a result, these tools were less accurate when diagnosing women, as they were unable to properly identify symptoms and risk factors that manifest differently in female patients (Cirillo et al., 2020). This issue underscores the necessity for ensuring that AI systems in biomedicine are trained on diverse datasets, representative of all genders, races, and age groups, to prevent disparities in diagnostic accuracy and patient outcomes.

3.3. Bias in Healthcare Algorithms for Rare Diseases

The treatment of rare diseases is another area where algorithmic bias can be especially harmful. Algorithms designed to identify rare diseases may be inherently biased if the data used to train them lacks adequate representation of diverse populations. For example, a study by Aquino et al. (2023) discussed the ethical implications of using AI in the diagnosis and treatment of rare diseases, highlighting that these algorithms often fail to adequately represent minority populations. This can result in delayed or inaccurate diagnoses for individuals from underrepresented groups (Aquino et al., 2023). In such cases, it is critical to ensure that the datasets used for training AI models include sufficient examples from diverse demographic groups to avoid biases that could exacerbate health disparities.

3.4. Bias in AI for Precision Medicine

Precision medicine, which tailors medical treatments to individual patients based on genetic and other personal data, is another field in which algorithmic bias is prevalent. Research has shown that AI algorithms used in precision medicine are often biased toward certain ethnic groups, particularly those of European descent, due to the underrepresentation of non-European populations in genomic databases (Blasimme & Vayena, 2019). This lack of diversity in training datasets can lead to disparities in the effectiveness of treatments for minority groups. A study by Liu et al. (2023) emphasized the need for broader and more inclusive genomic databases to ensure that precision medicine benefits all populations equally (Liu et al., 2023).

3.5. Gender and Socioeconomic Bias in Health Data

Health data, when used to train AI models, can reflect societal biases, including those related to gender and socioeconomic status. For instance, a study by Mc-Cradden et al. (2020) revealed that healthcare algorithms, particularly in the U.S., often used biased data that reflected broader societal inequities. These algorithms tended to disadvantage low-income and minority populations, further exacerbating health inequities. This is particularly problematic in settings where healthcare is stratified by socioeconomic status, as such biases can lead to unequal treatment across different patient demographics (McCradden, Joshi, Mazwi, & Anderson, 2020). Addressing such biases requires careful consideration of the data sources and algorithmic transparency to ensure fairness.

3.6. Bias in AI for Drug Discovery

AI has been increasingly applied in drug discovery, yet biases in the algorithms used can limit the effectiveness of the process. For example, a study by Chien et al. (2022) explored how biases in AI-driven drug discovery algorithms can lead to the overrepresentation of certain molecular structures while neglecting others, potentially excluding valuable treatments for underrepresented populations. The failure to include diverse genetic and demographic data can result in drug candidates that are less effective for certain groups, perpetuating disparities in treatment outcomes (Chien et al., 2022).

3.7. Challenges in Bias Mitigation Strategies

While much attention has been given to the identification of biases in AI models, less focus has been placed on developing effective mitigation strategies. Studies by Dehghani et al. (2024) and Faust et al. (2025) discussed various approaches to mitigating bias in healthcare AI, such as reweighting training data, implementing fairness-aware algorithms, and increasing transparency in model development (Dehghani, Malik, Lin, Bayat, & Bento, 2024; Faust et al., 2025). These strategies can be crucial in reducing bias and ensuring that AI systems in healthcare are fair and equitable. However, the practical application of these methods remains challenging, especially when the underlying data itself is biased.

3.8 Summary

The real-world case studies of algorithmic bias in biomedical research underscore the complex and multifaceted nature of this issue. From racial and ethnic biases in clinical decision support systems to gender and socioeconomic biases in precision medicine and drug discovery, it is clear that algorithmic fairness is a critical concern in healthcare AI. Addressing these biases requires not only technical solutions, such as improving data diversity and fairness in algorithms, but also systemic changes in how healthcare data is collected, used, and interpreted. Multidisciplinary collaboration, increased transparency, and rigorous ethical oversight will be essential to ensure that AI in biomedical research promotes equity and inclusivity across all patient demographics.

4. IMPACT OF ALGORITHMIC BIAS ON HEALTH EQUITY

Algorithmic bias has emerged as a significant issue in healthcare systems, particularly as the use of artificial intelligence (AI) and machine learning (ML) grows in clinical and biomedical settings. While these technologies offer tremendous potential for improving healthcare delivery, their implementation can inadvertently exacerbate health inequities, particularly for historically marginalized groups. In the healthcare domain, algorithmic bias can manifest in various forms, including race, gender, age, disability, and socio-economic status biases, all of which can lead to unequal treatment outcomes. This section explores the impact of algorithmic bias on health equity, emphasizing the ethical and social implications, as well as the strategies needed to address these disparities.

4.1 Algorithmic Bias and Its Root Causes

Bias in healthcare algorithms typically arises from biased data, which is reflective of historical inequalities in healthcare access, treatment, and outcomes. According to Alderman et al. (2025), healthcare datasets often suffer from lack of diversity, particularly in terms of race, gender, and socio-economic background, which leads to models that fail to generalize across different populations. This bias is further compounded by the selective nature of data collection, where certain groups may be underrepresented in clinical trials or health studies, making it difficult for algorithms to make accurate predictions or recommendations for those groups.

Blasimme and Vayena (2019) highlight the importance of acknowledging that AI in healthcare is not value-neutral. In fact, it often reflects and perpetuates societal biases due to the historical underrepresentation of minority groups in healthcare research. For instance, algorithms trained on predominantly white, male populations may fail to diagnose conditions accurately in women or people of color, leading to disparities in treatment outcomes (Cirillo et al., 2020; Chen et al., 2021).

4.2 Ethical and Normative Implications

The ethical implications of algorithmic bias are profound, particularly when considering the principle of fairness in healthcare. Fairness in algorithmic decision-making is not a singular concept; it encompasses several dimensions, including procedural fairness, distributive fairness, and outcome fairness. As Anderson and Visweswaran (2025) argue, achieving fairness in healthcare algorithms requires an understanding of both individual and group fairness, ensuring that algorithms do not disproportionately disadvantage any particular group based on arbitrary or irrelevant characteristics.

Kerasidou (2021) emphasizes the importance of transparency in healthcare AI systems. Transparent models that clearly explain how decisions are made can help build trust among users, especially patients and healthcare providers who may be skeptical of AI's role in medical decision-making. Lack of transparency can perpetuate distrust, especially in marginalized communities that have historically been mistreated by healthcare systems.

4.3 Bias in AI Models and Health Equity

Health equity is fundamentally about ensuring that every individual has access to the care they need to achieve the best possible health outcomes. However, when AI models are trained on biased datasets, they can reinforce and amplify existing health disparities. For example, studies have shown that ML models for diagnosing skin cancer are less accurate for people with darker skin tones, leading to misdiagnosis or delayed treatment (Franklin et al., 2024). This discrepancy arises because the majority of image datasets used to train these models feature lighter skin tones, leaving AI systems ill-equipped to identify skin conditions in darker-skinned individuals.

Jain et al. (2023) provide a similar analysis of racial and ethnic bias in healthcare algorithms. They argue that machine learning algorithms often fail to account for the diversity in patient demographics, particularly in multi-ethnic societies, leading to poorer outcomes for non-white patients. This can lead to scenarios where patients from racial or ethnic minorities are either over-treated or under-treated based on inaccurate predictions made by the AI model.

Moreover, biases in healthcare AI models also intersect with other social determinants of health, such as socio-economic status and access to care. Research by Mhasawade et al. (2021) shows that low-income individuals are often underrepresented in datasets, which can result in healthcare algorithms that fail to predict or recommend treatments for this population effectively. This not only affects the quality of care provided but also exacerbates existing socio-economic disparities.

The following bar chart identifies populations most vulnerable to negative impacts from biased healthcare AI systems, emphasizing the need for equitable model development.

Figure 2. Populations most impacted by algorithmic bias

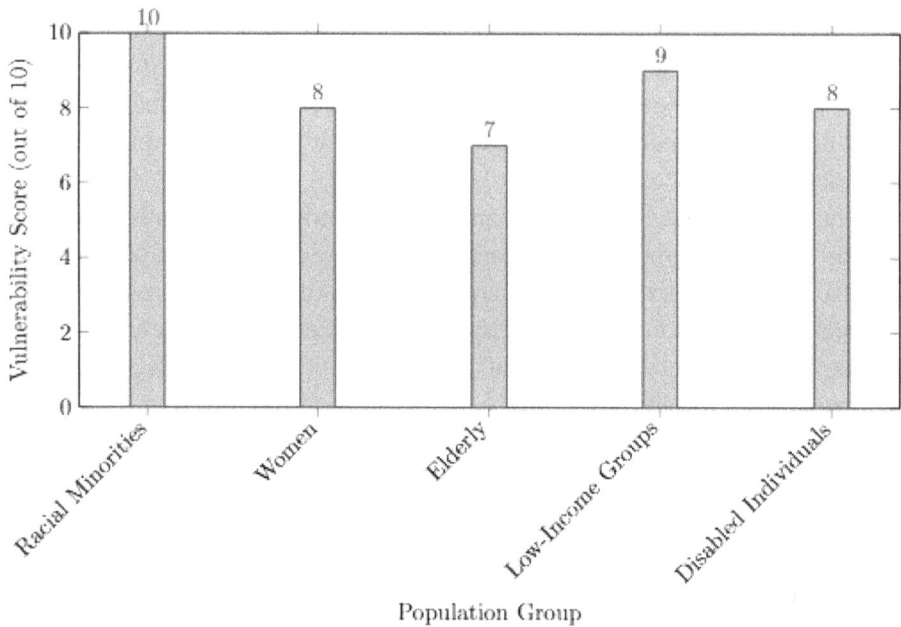

4.4 Addressing Algorithmic Bias: Strategies for Promoting Equity

Given the potential consequences of algorithmic bias in healthcare, it is crucial to implement strategies that promote fairness and reduce disparities. Several scholars have proposed frameworks and guidelines for mitigating bias in healthcare AI systems. For instance, Rajkomar et al. (2018) advocate for fairness-aware machine learning models that incorporate diverse datasets during training. Ensuring that AI systems are trained on a wide variety of data sources—including demographic data that spans race, gender, and socio-economic status—can help mitigate biases and improve the generalizability of models.

Furthermore, addressing algorithmic fairness in healthcare requires not only improving the data used to train AI models but also incorporating ethical considerations into the design and development of these systems. Chen et al. (2023) stress the need for interdisciplinary collaboration between AI developers, healthcare professionals, ethicists, and sociologists to ensure that AI models reflect societal values of equity and justice. This collaborative approach will help prevent the reinforcement of harmful stereotypes and reduce disparities in healthcare access and outcomes.

Griffin et al. (2024) emphasize the importance of building explainable AI systems in healthcare. Explainability ensures that healthcare professionals can understand the reasoning behind AI-driven decisions, which is crucial for maintaining patient trust and ensuring accountability. Additionally, Faust et al. (2025) propose the development of regulatory frameworks that require transparency and fairness in healthcare AI, allowing policymakers to hold developers accountable for biased outcomes.

4.5 Summary

Algorithmic bias presents a significant challenge to achieving health equity. As AI and machine learning continue to transform healthcare, addressing biases in healthcare algorithms will be essential to ensure that these technologies do not exacerbate existing health disparities. It is vital to continue research into algorithmic fairness and develop robust strategies to mitigate bias, including diverse data collection, transparency, and interdisciplinary collaboration. Only by tackling algorithmic bias head-on can we ensure that AI contributes to a more equitable and just healthcare system for all.

5. DETECTING AND MEASURING BIAS IN BIOMEDICAL AI SYSTEMS

As artificial intelligence (AI) systems become increasingly integrated into healthcare, ensuring fairness and transparency in these technologies is vital. The presence of algorithmic bias in AI systems can lead to unfair treatment of certain groups, particularly those based on race, gender, ethnicity, and other social determinants. Detecting and measuring bias in biomedical AI systems is therefore essential for maintaining ethical standards, ensuring equal access to healthcare, and improving patient outcomes.

5.1. Understanding Bias in Biomedical AI

Bias in AI refers to systematic errors that lead to unfair outcomes for certain groups of individuals. In healthcare, algorithmic bias can manifest in several ways, including disparities in disease diagnosis, treatment recommendations, and patient outcomes. Bias can be introduced into AI systems through various means, such as biased training data, inappropriate model assumptions, or biased decision-making processes (Rajkomar et al., 2018). Recognizing these biases is crucial for addressing health inequities and ensuring AI models operate fairly across diverse patient populations.

Bias in biomedical AI can stem from numerous sources:

- **Data Bias:** If the data used to train AI systems is unrepresentative of the broader population, algorithms may learn to make decisions that favor certain demographic groups over others (Alderman et al., 2025). This is particularly common when health datasets predominantly represent specific racial, ethnic, or gender groups, resulting in misdiagnoses or suboptimal treatment for underrepresented populations (Jain et al., 2023).
- **Algorithmic Bias:** Even with unbiased data, algorithms may still perpetuate bias if they are designed with flawed assumptions or do not account for factors such as socioeconomic status, disability, or other non-biological variables (Dehghani et al., 2024).
- **Modeling Bias:** Bias can also arise from the choice of machine learning models used in healthcare. For example, deep learning models may unintentionally privilege certain features over others, leading to unequal healthcare recommendations (Chen et al., 2023).

To better visualize the prevalence of different types of bias encountered in biomedical AI systems, the following pie chart illustrates their relative distribution.

Figure 3. Distribution of bias types in biomedical AI systems

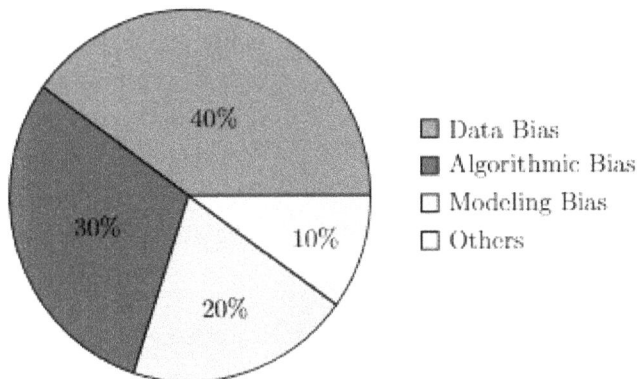

5.2. Techniques for Detecting Bias in AI Models

Detecting bias in AI models requires systematic evaluation at each stage of the machine learning pipeline, from data collection to model deployment. Several approaches have been proposed for detecting bias in healthcare AI systems:

- **Disparity Analysis:** A common method for identifying bias is to analyze disparities in model performance across different demographic groups. For instance, examining how well an AI model predicts health outcomes for different genders or racial groups can highlight any significant performance gaps (Grote & Keeling, 2022). Disparities in treatment recommendations can indicate potential biases that need to be addressed (Huang et al., 2022).
- **Fairness Metrics:** There are various fairness metrics used to assess bias in AI systems. These include statistical measures like demographic parity, which ensures that decisions are made equally across demographic groups, and equalized odds, which measures whether errors are distributed equally across groups (Mahmood, 2021). Other methods include individual fairness, which ensures that similar individuals receive similar outcomes, and group fairness, which checks that aggregate results do not disproportionately harm any specific group (Anderson & Visweswaran, 2025).
- **Sensitivity and Specificity Evaluation:** For clinical decision-support systems, sensitivity (true positive rate) and specificity (true negative rate) are crucial measures to evaluate bias. If an AI model performs poorly in identifying conditions for specific demographic groups, it may suggest that the model

is biased toward more easily identified patterns in other groups (Singhal et al., 2024).

- **Fairness Audits:** Conducting regular audits of AI models is essential for ensuring that they maintain fairness over time. These audits involve systematically reviewing model decisions and performance metrics to detect any unintended biases that may have emerged during real-world use (Blasimme & Vayena, 2019). Independent third-party audits can also help uncover hidden biases that may not be apparent from internal evaluations alone (McCradden et al., 2020).

5.3. Measuring the Impact of Bias

The impact of bias in healthcare AI systems is not only an ethical concern but also a clinical one. Bias can lead to disparities in healthcare delivery, resulting in some populations receiving inadequate care or being excluded from access to life-saving interventions (Nazer et al., 2023). Measuring the impact of bias requires considering both direct and indirect effects on patient outcomes:

- **Health Disparities:** The ultimate measure of algorithmic bias in healthcare is its impact on health outcomes. If an AI system perpetuates inequities by providing less accurate diagnoses or treatment recommendations for certain groups, it can contribute to widening health disparities. This is particularly problematic in the context of chronic diseases, where early diagnosis and personalized treatment can significantly improve patient outcomes (Mhasawade et al., 2021).
- **Patient Trust and Engagement:** Bias in AI models can also undermine patient trust in healthcare systems. When patients perceive that AI systems are not treating them equitably, they may be less likely to engage with these technologies, thereby reducing their effectiveness (Kerasidou, 2021). Building transparent and fair systems is therefore essential for ensuring that patients remain confident in the technologies used to treat them (Franklin et al., 2024).
- **Clinical Decision-Making:** The decisions made by healthcare professionals can also be influenced by biased AI recommendations. If clinicians rely too heavily on biased AI tools, they may unknowingly perpetuate health inequities. Therefore, it is essential to provide clinicians with ongoing training on the limitations of AI models and promote human oversight in decision-making (Loftus et al., 2022).

5.4. Mitigating Bias in Biomedical AI Systems

Once bias is detected and measured, the next step is to mitigate it. Several strategies can be employed to reduce bias in biomedical AI systems:

- **Diverse Data Collection:** One of the most effective ways to reduce bias is by ensuring that the data used to train AI models is diverse and representative of the populations the model is intended to serve. This can involve collecting more data from underrepresented groups, as well as using synthetic data to fill gaps in the training set (Ferryman & Pitcan, 2018).
- **Bias-Aware Algorithms:** Many machine learning algorithms can be adapted to be more aware of fairness during training. Techniques such as adversarial training, where an adversarial model tries to predict the demographic group based on the model's predictions, can help ensure that the primary model does not exhibit demographic-based bias (Yang et al., 2024). Other methods, like re-weighting training examples or bias correction layers, have also been proposed to enhance fairness in AI systems (Chinta et al., 2024).
- **Inclusive Design:** Involving diverse stakeholders, including patients, healthcare providers, and ethicists, in the design and development of AI systems can help ensure that a variety of perspectives are considered. This approach can lead to more equitable solutions that address the specific needs of different demographic groups (Griffin et al., 2024).
- **Continuous Monitoring:** Given the evolving nature of both AI technology and healthcare, it is essential to continuously monitor AI systems post-deployment for any emerging biases. This includes regular updates to training data, model adjustments, and post-market surveillance to track real-world outcomes (Visweswaran et al., 2024).

The following flowchart outlines a structured approach for detecting and mitigating bias in biomedical AI systems, from initial data auditing to ongoing monitoring.

Figure 4. Process for bias detection and mitigation in biomedical AI

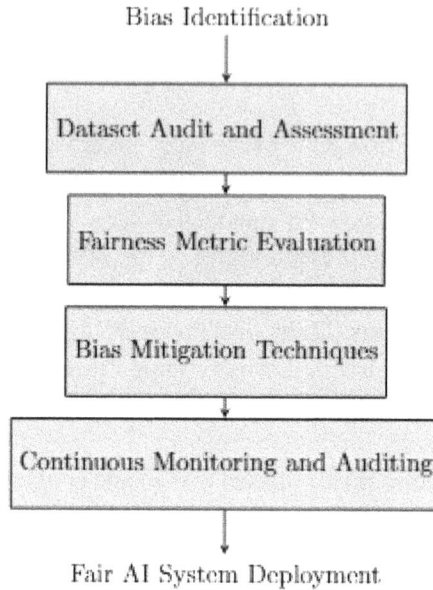

Bias Identification

```
┌─────────────────────────────────────┐
│   Dataset Audit and Assessment       │
└─────────────────────────────────────┘
          │
┌─────────────────────────────────────┐
│   Fairness Metric Evaluation         │
└─────────────────────────────────────┘
          │
┌─────────────────────────────────────┐
│   Bias Mitigation Techniques         │
└─────────────────────────────────────┘
          │
┌─────────────────────────────────────┐
│ Continuous Monitoring and Auditing   │
└─────────────────────────────────────┘
```

Fair AI System Deployment

5.5. Future Directions

The future of AI in healthcare must prioritize fairness and inclusion. As AI technologies continue to advance, it is crucial to develop frameworks that not only detect and mitigate bias but also promote inclusivity and equity at every stage of AI development and deployment. Moving forward, interdisciplinary research that combines technical, ethical, and social perspectives will be key to creating AI systems that benefit all patients equally, without exacerbating existing health disparities (Chien et al., 2022).

5.6 Summary

Detecting and measuring bias in biomedical AI systems is a complex but critical task that requires ongoing attention from both researchers and practitioners. By using a variety of fairness metrics, conducting regular audits, and prioritizing data diversity, we can create AI models that serve the needs of all populations. This commitment to fairness not only benefits patients but also strengthens the integrity and trust in AI-powered healthcare systems.

6. STRATEGIES FOR MITIGATING ALGORITHMIC BIAS

Algorithmic bias in healthcare, particularly in artificial intelligence (AI) models, has garnered significant attention due to its potential to perpetuate disparities, particularly regarding race, gender, and socio-economic status. Addressing algorithmic bias in healthcare systems requires a multifaceted approach, incorporating both technical and ethical considerations. This section discusses various strategies for mitigating algorithmic bias, focusing on improving fairness, transparency, and inclusivity in AI systems used in healthcare.

6.1. Ensuring Fairness Through Dataset Representation

One of the primary sources of bias in AI algorithms stems from biased data. In healthcare, datasets often reflect existing social inequities or imbalances. For instance, underrepresentation of certain racial or ethnic groups in medical datasets can lead to biased algorithms that perform poorly for those populations (Alderman et al., 2025). To mitigate this, researchers suggest ensuring datasets are diverse and representative of all patient demographics (Chen et al., 2021; Nazer et al., 2023). Strategies like over-sampling underrepresented groups or employing synthetic data generation techniques can enhance the inclusivity of healthcare AI models (Dehghani et al., 2024).

Furthermore, improving the transparency of how datasets are collected, curated, and pre-processed is essential. Biases embedded in the collection process can perpetuate societal inequities unless addressed during the data curation stage (Anderson & Visweswaran, 2025). Algorithmic developers should collaborate with healthcare practitioners to ensure that datasets reflect the diverse experiences and needs of patients.

6.2. Fairness Constraints in Model Training

Incorporating fairness constraints into the model training process is a powerful method for mitigating algorithmic bias. Techniques such as adversarial debiasing (Yang et al., 2022), which modifies the loss function to penalize the model for biased predictions, have been shown to improve fairness without sacrificing accuracy. Similarly, fairness-enhancing regularization methods can ensure that the AI models do not disproportionately favor one demographic over others (Grote & Keeling, 2022).

Another approach is to incorporate fairness metrics during model evaluation. Metrics such as demographic parity, equalized odds, and disparate impact can help assess whether an algorithm's performance is equitable across different groups (Singhal et al., 2024). By continuously evaluating these metrics, healthcare practitioners can adjust AI models to ensure that disparities are minimized.

6.3. Algorithm Explainability and Transparency

The "black-box" nature of many AI models, particularly deep learning models, makes it difficult to understand how decisions are made, increasing the risk of unintentional bias (Griffin et al., 2024). Therefore, promoting algorithmic explainability is crucial for trust and accountability in healthcare AI systems. Efforts to develop interpretable machine learning models, such as decision trees or models with feature importance scores, can help users understand how decisions are made, particularly when the model impacts patient care (Kerasidou, 2021).

In addition to algorithmic transparency, it is also vital to ensure that the decisions made by AI models are traceable and auditable. Establishing a feedback loop where healthcare providers and patients can review and challenge AI-generated decisions can promote greater accountability (Loftus et al., 2022).

6.4. Multi-Disciplinary Collaboration

Effective mitigation of algorithmic bias requires collaboration across disciplines, including data science, healthcare, ethics, and law. Multidisciplinary teams can identify potential sources of bias and work toward solutions that balance technical performance with ethical considerations. For example, integrating input from clinicians and healthcare professionals when designing AI systems ensures that algorithms reflect clinical needs and human values (Blasimme & Vayena, 2019).

Furthermore, interdisciplinary collaboration can help develop shared ethical frameworks for fairness and bias mitigation. The inclusion of ethicists in AI development teams is crucial to ensure that fairness does not come at the cost of other values, such as patient autonomy and privacy (Aquino et al., 2023; Williams, 2023).

6.5. Addressing Social Determinants of Health

AI models in healthcare should also account for the broader social determinants of health, such as socio-economic status, access to care, and geographical location. Algorithms that fail to consider these factors may unintentionally worsen health disparities (Rajkomar et al., 2018). For instance, models developed using hospital data might fail to account for the fact that some populations have less access to healthcare facilities or are subject to different health risks (Jain et al., 2023). Including variables related to these social factors can help ensure that healthcare AI systems serve a broader range of patients more equitably (McCradden et al., 2020).

6.6. Regular Monitoring and Auditing of AI Systems

Once deployed, AI systems should be subject to continuous monitoring and auditing to ensure they remain fair over time. As the healthcare landscape changes, so too may the populations served by AI models. Biases can emerge or shift as demographic and medical trends evolve (Huang et al., 2022). Regular re-training of models on updated datasets and performing ongoing audits can ensure that AI systems continue to meet fairness and performance standards (Faust et al., 2025).

In addition, establishing an independent oversight body can enhance the accountability of AI systems in healthcare. This body could conduct periodic audits of AI algorithms, ensuring they adhere to ethical standards and mitigate any emerging biases (Nazer et al., 2023).

6.7. Promoting Inclusive AI Governance

The development of AI systems for healthcare should be governed by clear policies and regulations that promote fairness and inclusivity. Governments and regulatory bodies can play a crucial role in setting standards for AI fairness, requiring that healthcare AI systems undergo rigorous testing for bias before being implemented in clinical practice (Qureshi & Oladokun, 2024). International collaborations can also help standardize fairness definitions and create universal guidelines for ethical AI usage in healthcare (Mhasawade et al., 2021).

Policies should also ensure that patients are informed about the role of AI in their healthcare and are given the opportunity to opt-out if they feel uncomfortable with algorithmic decision-making (El-Azab & Nong, 2023).

6.8 Summary

Mitigating algorithmic bias in healthcare requires a multi-pronged approach that combines technical innovation, ethical considerations, and regulatory frameworks. Ensuring fairness in AI systems is not only a technical challenge but also an ethical imperative, particularly in healthcare where decisions can profoundly impact human lives. By promoting diversity in datasets, incorporating fairness constraints in model training, fostering transparency and explainability, and encouraging interdisciplinary collaboration, stakeholders can work towards creating more equitable healthcare AI systems. Regular monitoring, inclusive governance, and an emphasis on social determinants of health will further enhance the fairness and inclusivity of AI-driven healthcare solutions.

By adopting these strategies, the healthcare industry can move toward a future where AI systems are not only effective but also fair, transparent, and accountable, ensuring better health outcomes for all.

7. FAIRNESS FRAMEWORKS AND GUIDELINES IN BIOMEDICAL AI

In the rapidly evolving field of biomedical AI, ensuring fairness in algorithms is paramount for achieving equitable healthcare outcomes. Various frameworks and guidelines have emerged to guide the development and implementation of AI systems that minimize biases and enhance transparency, accountability, and inclusivity. This section delves into the key frameworks and guidelines designed to address algorithmic bias in biomedical AI, drawing from recent advancements and multidisciplinary expert opinions.

7.1 The Role of Fairness in Biomedical AI

Fairness in biomedical AI refers to the ability of AI systems to perform equally well for individuals from diverse demographic, socio-economic, and clinical backgrounds. A key issue within this domain is the recognition and mitigation of biases inherent in AI algorithms, which can lead to inequitable healthcare outcomes (Alderman et al., 2025). The integration of fairness frameworks ensures that AI-driven medical solutions do not perpetuate historical disparities in healthcare but instead contribute to more equitable healthcare delivery.

7.2 Existing Fairness Frameworks

One of the prominent frameworks addressing fairness in healthcare AI is the Fairness-Aware Algorithm Design, which focuses on reducing algorithmic discrimination based on sensitive attributes like race, gender, and socio-economic status. Several studies have emphasized the need for incorporating fairness throughout the lifecycle of biomedical AI, from data collection to model deployment (Chen et al., 2023; Griffin et al., 2024).

Additionally, frameworks like the Fairness-Through-Unawareness and Fairness-Through-Data Preprocessing are commonly employed to ensure that AI models do not perpetuate known biases in training data. For example, the STANDING Together initiative focuses on providing consensus recommendations to tackle algorithmic bias by encouraging diverse, representative datasets and transparency in model development (Alderman et al., 2025). Such frameworks aim to address the underlying structural issues that lead to biased outcomes in healthcare.

To illustrate the essential components of a comprehensive fairness framework for biomedical AI systems, the following diagram summarizes the key pillars.

Figure 5. Components of a fairness framework for biomedical AI

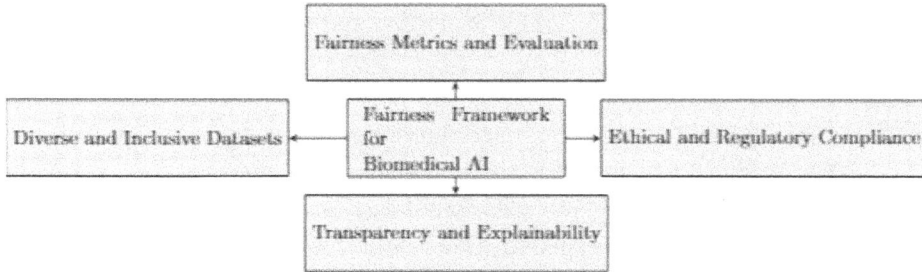

7.3 Ethical Considerations in AI Fairness

Incorporating ethical principles into AI fairness is another crucial aspect that has garnered significant attention. A common ethical framework applied to healthcare AI is the Principle of Non-Discrimination, which advocates for equal treatment regardless of race, gender, or disability status (Blasimme & Vayena, 2019). However, this principle has been critiqued for its potential to overlook the need for corrective action in cases where certain populations have been historically underserved (El-Azab & Nong, 2023). Ethical guidelines thus propose the inclusion of justice-based fairness, which acknowledges historical inequities and calls for targeted interventions to improve outcomes for disadvantaged groups.

The complexity of implementing fairness in biomedical AI is highlighted by the challenge of balancing accuracy and equity. For instance, the principle of equalized odds mandates that AI systems should be equally accurate for all demographic groups, but achieving this while maintaining overall model performance can be difficult (Rajkomar et al., 2018). Therefore, healthcare AI frameworks often advocate for explainability and transparency to ensure that healthcare professionals can understand and trust AI predictions, especially when they impact critical health decisions (Griffin et al., 2024; Loftus et al., 2022).

7.4 Data and Bias Mitigation Strategies

One of the most critical aspects of ensuring fairness in biomedical AI is addressing biases in the underlying data. Many existing AI models rely on datasets that are either incomplete or non-representative of diverse patient populations (Huang et al., 2022). Data pre-processing techniques, such as reweighting or resampling, can help mitigate these biases by adjusting the distribution of data before training. Additionally, algorithmic interventions such as adversarial training have been proposed to ensure that models learn to make predictions that are invariant to sensitive attributes (Yang et al., 2022).

Furthermore, incorporating a multi-disciplinary approach is essential to ensure that fairness considerations are integrated throughout the design and development process. For example, Chien et al. (2022) emphasize the importance of involving diverse stakeholders, including ethicists, healthcare professionals, and patients, in the decision-making process related to AI development and deployment in clinical settings.

7.5 Regulatory and Compliance Guidelines

Regulatory bodies such as the FDA and EU Commission have started to issue guidelines and regulations that focus on ensuring fairness and equity in healthcare AI. In the U.S., the FDA's guidelines on AI and machine learning in medical devices outline the need for continuous monitoring and validation of AI models in real-world clinical settings to ensure that they do not inadvertently harm marginalized groups (Griffin et al., 2024).

In Europe, the General Data Protection Regulation (GDPR) also has provisions that emphasize fairness in algorithmic decision-making, particularly with regard to how personal data is used in AI models. This regulation ensures that AI-driven healthcare systems are transparent about their data usage and that individuals have a right to contest decisions made by automated systems (Mhasawade et al., 2021). The application of these regulatory frameworks ensures that fairness is not merely a technical challenge but a legal and ethical obligation for AI developers.

7.6 Frameworks for Continuous Improvement and Accountability

One promising approach to ensure fairness in biomedical AI is through continuous evaluation and monitoring. As AI models are deployed, their performance must be regularly assessed to ensure that they maintain fairness over time (Jain et al., 2023). This involves monitoring not only the accuracy of AI predictions but also their impact on health disparities.

Additionally, frameworks such as the Algorithmic Impact Assessment (AIA) and Fairness Audit tools are gaining traction as methods to continuously evaluate and address biases in AI systems (Qureshi & Oladokun, 2024). These tools provide a structured approach for organizations to assess the fairness of their algorithms, ensuring that any potential biases are identified and addressed promptly.

7.7 Challenges and Future Directions

Despite the progress made in developing fairness frameworks, significant challenges remain in ensuring true fairness in biomedical AI. One key challenge is algorithmic transparency: many AI models, especially deep learning models, function as "black boxes" where the decision-making process is not easily interpretable (Kerasidou, 2021). This lack of transparency complicates efforts to understand and mitigate bias.

Furthermore, the lack of standardized fairness metrics across different contexts in healthcare remains an ongoing challenge. Various fairness definitions and evaluation criteria exist, but a unified approach to assessing fairness in biomedical AI is still lacking (Visweswaran et al., 2024).

In the future, research will likely focus on developing more context-specific fairness metrics, improving algorithmic explainability, and addressing the ethical implications of data governance in AI systems (Mienye et al., 2024). As the field continues to evolve, the integration of multidisciplinary perspectives will be key to addressing the complex interplay of fairness, accuracy, and ethical concerns in biomedical AI.

7.8 Summary

The development of fairness frameworks and guidelines in biomedical AI is critical for ensuring that these technologies benefit all populations equitably. Through a combination of ethical principles, regulatory guidelines, data fairness interventions, and continuous monitoring, stakeholders can work toward mitigating bias in AI-driven healthcare systems. However, ongoing efforts are needed to address transparency issues and develop standardized methods for evaluating fairness in diverse healthcare contexts. By prioritizing fairness in AI systems, healthcare can become more inclusive, ultimately leading to better and more equitable outcomes for patients across the globe.

8. CHALLENGES AND OPEN RESEARCH QUESTIONS

The integration of artificial intelligence (AI) into healthcare systems has revolutionized medical practices and research, but it has also raised significant ethical, technical, and social challenges, particularly concerning algorithmic fairness. The complexity of these challenges stems from the diverse nature of data, the varying demographics of patients, and the complex societal implications of AI applications. As AI systems are increasingly used in clinical decision-making, addressing the inherent biases within these systems becomes crucial for ensuring equity in healthcare. This section explores the key challenges in the field and outlines open research questions that need to be addressed to mitigate biases and promote fairness in AI-driven healthcare.

8.1. Algorithmic Bias and Its Impact on Health Equity

One of the most significant challenges facing AI in healthcare is algorithmic bias. Studies have shown that AI models may inherit biases present in the data they are trained on, leading to outcomes that disproportionately affect certain demographic groups. These biases can be race, gender, or socioeconomically driven, with tangible consequences for patients who belong to underrepresented or disadvantaged groups (Alderman et al., 2025; Aquino et al., 2023).

For instance, racial and ethnic biases in AI systems have been shown to affect clinical predictions and treatment recommendations, often exacerbating existing health disparities (Huang et al., 2022). In addition, the lack of transparency in AI models makes it difficult to identify and correct these biases, raising concerns about fairness and accountability in clinical settings (Blasimme & Vayena, 2019; McCradden et al., 2020). Therefore, one of the key research challenges is developing techniques that improve the explainability of AI models and enable clinicians to understand the underlying factors contributing to model decisions (Rajkomar et al., 2018; Loftus et al., 2022).

8.2. Data Diversity and Representation

Data used in training healthcare AI systems often do not adequately represent the diversity of the populations they serve. This lack of diversity can lead to models that perform well for certain groups while failing others. For example, studies have demonstrated that many healthcare datasets are disproportionately dominated by data from white, middle-class individuals, leading to AI models that are less accurate for minority populations (Cirillo et al., 2020; Chen et al., 2023). Addressing this issue requires a more inclusive approach to data collection, ensuring that datasets reflect the demographics of the broader patient population.

Moreover, the representation of marginalized groups, including individuals with disabilities, is often neglected in AI research (Tilmes, 2022). A key open research question is how to design and collect more diverse and inclusive datasets that account for a wide range of patient characteristics, including race, gender, socioeconomic status, and disability.

8.3. Fairness in AI Algorithms

Ensuring fairness in healthcare AI models is another major challenge. Many existing algorithms prioritize performance metrics such as accuracy, which may not always align with equitable healthcare goals. For example, a model that performs well on average may still have high error rates for certain groups, leading to unfair treatment recommendations (Grote & Keeling, 2022; Dehghani et al., 2024). There is a growing need for methods to balance the trade-offs between performance and fairness, ensuring that models provide equitable outcomes for all patient groups without sacrificing their accuracy.

A critical open research question in this area is the development of fairness metrics that can be applied across diverse healthcare settings, considering the specific needs of different populations. Researchers are exploring novel approaches to fairness in healthcare, such as adversarial training and reinforcement learning techniques, which have shown promise in mitigating bias in AI models (Yang et al., 2022; Mhasawade et al., 2021).

8.4. Ethical and Normative Implications

The ethical implications of AI in healthcare are vast and multifaceted. Algorithmic bias is not just a technical problem but also an ethical one, as it raises questions about justice, equity, and trust. For example, the potential for AI to reinforce stereotypes or discriminate against vulnerable populations calls into question the fairness of AI decision-making (Kerasidou, 2021; Blasimme & Vayena, 2019). Moreover, the lack of accountability in AI systems complicates the ethical landscape, as it becomes difficult to pinpoint responsibility when a biased decision leads to adverse health outcomes (Ferryman & Pitcan, 2018; Williams, 2023).

Future research should focus on integrating ethical principles into the design and deployment of healthcare AI systems, ensuring that they are aligned with values such as equity, justice, and transparency. Moreover, establishing clear guidelines for the accountability of AI systems in clinical settings is an urgent need (Sikstrom et al., 2022; Faust et al., 2025).

8.5. Regulatory and Policy Frameworks

As AI technologies become more embedded in healthcare, the development of appropriate regulatory and policy frameworks is crucial. These frameworks need to address the challenges of data privacy, consent, and algorithmic transparency while ensuring that AI systems are safe, effective, and equitable (Norori et al., 2021; Nazer et al., 2023). Current regulatory practices are often slow to catch up with the rapid advancements in AI technology, creating a gap between the development of AI tools and their oversight (Chen et al., 2021; Anderson & Visweswaran, 2025).

Research into regulatory frameworks must focus on creating guidelines that balance innovation with the protection of patient rights. This includes establishing standards for the transparency of AI models, the fairness of data collection processes, and the oversight of AI decision-making in healthcare (Mienye et al., 2024).

8.6. Addressing the Intersection of AI and Health Disparities

The intersection of AI and health disparities is a growing area of concern. AI systems can both exacerbate and alleviate health disparities, depending on how they are designed and deployed. If AI models are trained on biased data or if their deployment is not carefully managed, they risk deepening the divide between privileged and underserved populations (Jain et al., 2023; Singhal et al., 2024). However, if designed with equity in mind, AI could serve as a tool for reducing health disparities by providing more personalized and accessible healthcare solutions (Griffin et al., 2024).

Open research questions in this area revolve around how to ensure that AI technologies are deployed in a way that reduces, rather than exacerbates, health inequities. Researchers must explore how AI can be used to address the social determinants of health and support more equitable healthcare delivery (Sikstrom et al., 2022).

8.7. Collaboration Across Disciplines

To address the above challenges, it is essential to foster collaboration across disciplines, including computer science, ethics, medicine, and law. Interdisciplinary approaches are necessary to develop AI systems that are not only technically advanced but also socially responsible (Chien et al., 2022; Williams, 2023). This collaboration is critical for developing frameworks that ensure fairness, accountability, and transparency in AI-driven healthcare applications.

An open research question here is how to structure cross-disciplinary collaborations effectively. What are the best practices for bringing together experts from different fields to work on the complex problems of fairness and bias in AI? How can the various perspectives of healthcare professionals, ethicists, and technologists be integrated into the design and deployment of AI systems?

8.8 Summary

While AI holds immense potential to transform healthcare, its deployment must be carefully managed to ensure fairness and equity. Addressing algorithmic bias, ensuring data diversity, developing fair and ethical models, creating regulatory frameworks, and fostering interdisciplinary collaboration are all critical areas of research. Tackling these challenges will require ongoing effort and innovation, but with careful attention to the issues outlined, AI can become a force for good in healthcare, improving outcomes for all patients, regardless of their background or demographic characteristics.

9. FUTURE DIRECTIONS FOR FAIR BIOMEDICAL AND HEALTH AI

The integration of artificial intelligence (AI) in biomedical and healthcare applications holds great promise for improving patient outcomes, enhancing clinical decision-making, and advancing public health. However, the effectiveness and equity of these systems depend significantly on addressing algorithmic fairness and mitigating bias. As AI technologies continue to evolve and be incorporated into healthcare systems, the future of fair biomedical AI must focus on several key areas, including data transparency, algorithmic accountability, inclusivity, and ongoing interdisciplinary collaboration.

9.1. Promoting Data Transparency and Reducing Bias

One of the most significant challenges in ensuring fairness in healthcare AI is the potential for data bias. AI models are only as good as the data they are trained on, and biased datasets can lead to discriminatory outcomes, particularly for marginalized groups. Future AI systems in healthcare must prioritize transparency in data collection, labeling, and usage to ensure that all populations are represented fairly. McCradden et al. (2020) emphasize the importance of tackling algorithmic bias through the development of transparent datasets and practices that allow healthcare providers to understand how data influences AI outcomes.

Alderman et al. (2025) propose the STANDING Together consensus recommendations, which call for the inclusion of diverse, representative datasets to tackle algorithmic bias in healthcare AI. These recommendations highlight the need for data governance frameworks that prioritize fairness and inclusivity, ensuring that underrepresented groups are not left behind. Further, Chien et al. (2022) argue that interdisciplinary teams, involving ethicists, sociologists, and healthcare professionals, should collaborate to evaluate the biases inherent in clinical datasets and algorithms.

9.2. Algorithmic Fairness and Inclusivity

Addressing fairness in healthcare AI requires a deeper understanding of the sociotechnical systems that govern algorithmic decision-making. Chen et al. (2021) and Chen et al. (2023) have highlighted the importance of algorithmic fairness, noting that achieving fairness in healthcare AI systems is not merely a technical challenge but also a societal and ethical one. This involves designing AI systems that are transparent, explainable, and accountable.

To achieve fairness, future healthcare AI models should incorporate approaches that consider fairness across multiple dimensions, such as race, ethnicity, gender, and socioeconomic status (Rajkomar et al., 2018). Blasimme and Vayena (2019) suggest that healthcare AI should incorporate a framework that goes beyond mere algorithmic fairness to include ethical, epistemic, and normative dimensions, ensuring that AI systems are not only efficient but also ethically sound and inclusive.

9.3. Ethical and Normative Considerations

In addition to technical improvements in fairness, ethical considerations must guide the design, deployment, and evaluation of healthcare AI systems. As noted by Kerasidou (2021), transparency and explainability are essential to building trust in AI systems. The ethical implications of AI in healthcare extend beyond technical solutions to encompass issues of consent, privacy, and the role of healthcare professionals in AI decision-making.

For example, Faust et al. (2025) discuss the ethical limitations in AI solutions for biomedical challenges, emphasizing that biases related to socioeconomic, racial, and gender disparities must be explicitly addressed in future AI models. Similarly, Anderson and Visweswaran (2025) stress the importance of algorithmic individual fairness, ensuring that AI tools provide equal opportunities for all patients, regardless of their background.

9.4. AI Explainability and Accountability

A critical component of ensuring fairness in healthcare AI is enhancing the explainability and accountability of these systems. The complexity of machine learning models, particularly deep learning, often results in "black box" models, which lack transparency regarding how decisions are made. This lack of explainability can undermine trust and exacerbate biases. Loftus et al. (2022) suggest that future AI models in healthcare should be designed to provide clear explanations for their decisions, allowing healthcare professionals and patients to understand the reasoning behind recommendations and diagnoses.

Moreover, Dehghani et al. (2024) highlight the need for more comprehensive evaluations of algorithmic fairness in clinical settings. These evaluations should consider the broader social and ethical implications of AI decisions, particularly in terms of patient outcomes and health equity. The development of dynamic, interpretable, and accountable AI systems will be essential to building trust and ensuring that AI technologies are deployed fairly in clinical settings.

9.5. Multidisciplinary Collaboration for Bias Mitigation

Mitigating algorithmic bias in healthcare AI requires collaboration across multiple disciplines, including computer science, medicine, ethics, sociology, and public health. The interdisciplinary nature of AI fairness necessitates a collective approach to address both technical and societal challenges. The work of Ferryman and Pitcan (2018) highlights the importance of bringing together stakeholders from different sectors to ensure that AI in healthcare is not only technically sound but also socially responsible and equitable.

Grote and Keeling (2022) argue that multidisciplinary collaboration is essential to creating fair healthcare algorithms that reflect the diverse needs of patients and healthcare workers. This collaboration should extend to the design, testing, and implementation phases of AI development, ensuring that all voices are heard, and diverse perspectives are incorporated.

9.6. Policy and Regulatory Frameworks

As AI technologies continue to be integrated into healthcare systems, there is an urgent need for robust policy and regulatory frameworks to ensure that these systems adhere to fairness principles. According to Nazer et al. (2023), regulatory bodies must work alongside AI developers and healthcare providers to create standards and guidelines for the ethical deployment of AI in healthcare. These frameworks should ensure that AI systems do not exacerbate existing healthcare disparities but instead work to promote fairness and equality.

Further, regulatory frameworks should address the accountability of AI systems in the event of errors or bias, ensuring that mechanisms are in place to correct harmful outcomes. Singh et al. (2024) and Mienye et al. (2024) discuss the importance of developing international standards for AI fairness, which would enable global collaboration and consistency in AI policy and regulation.

9.7. Future Research Directions

Looking forward, several research avenues must be pursued to enhance the fairness of healthcare AI. Yang et al. (2023) and Mhasawade et al. (2021) suggest that more research is needed to develop new algorithms and models that can effectively detect and mitigate bias in clinical data. This includes exploring novel machine learning techniques, such as adversarial learning, to identify and reduce biases in healthcare datasets.

In addition to technical improvements, there is a need for continued research on the ethical, societal, and normative implications of AI in healthcare. Chinta et al. (2024) and Singh et al. (2024) argue that future studies should focus on the long-term impacts of AI in healthcare, particularly in terms of equity and access to care. Understanding the broader social implications of AI technologies is crucial to ensuring that they are deployed in a manner that benefits all patients.

9.8 Summary

As AI continues to revolutionize healthcare, addressing issues of fairness and bias will be essential for ensuring that these technologies benefit all patients equitably. The future of fair biomedical and health AI will require ongoing collaboration between researchers, healthcare professionals, policymakers, and patients to develop transparent, inclusive, and accountable systems. By prioritizing these considerations, we can move toward a more equitable and just healthcare system, where AI serves as a tool for improving health outcomes for everyone.

10. CONCLUSION

The advancement of artificial intelligence (AI) in the biomedical and healthcare sectors has opened new opportunities for improving patient outcomes, increasing efficiency, and promoting precision medicine. However, the growing reliance on AI systems has brought to the forefront significant concerns regarding fairness, algorithmic bias, and equity, particularly as they pertain to diverse populations. As AI continues to integrate into clinical practices and biomedical research, the need for comprehensive and transparent approaches to address these concerns is paramount.

This chapter has explored the various facets of fairness in healthcare AI, emphasizing the challenges of data bias, the complexities of algorithmic decision-making, and the imperative of ensuring inclusivity in AI models. As highlighted throughout, the risks posed by algorithmic biases—whether racial, ethnic, gender, or socioeconomic—can have profound impacts on healthcare delivery, exacerbating health disparities rather than alleviating them. The critical role of interdisciplinary collaboration in the development and deployment of fair AI systems cannot be overstated. Collaboration among AI researchers, healthcare professionals, ethicists, and policymakers is essential for designing solutions that not only minimize biases but also ensure that AI systems are inclusive, equitable, and aligned with the core principles of justice and fairness.

While there is still much to be done, significant strides are being made in the development of methods to detect, measure, and mitigate bias in healthcare AI. As discussed, emerging techniques such as adversarial training, explainable AI, and fairness-aware algorithms offer promising approaches to creating more transparent and equitable AI systems. Furthermore, the importance of continuous monitoring, validation, and updating of these models is crucial to maintaining fairness throughout their lifecycle.

Looking forward, the integration of fairness considerations in healthcare AI must evolve in tandem with technological advancements. Ensuring that AI systems are developed, tested, and deployed with fairness as a foundational principle will require sustained efforts and a commitment to ethical responsibility. The research and efforts highlighted in this chapter serve as a call to action for all stakeholders to work together toward a future where AI in healthcare is not only a tool for innovation but also a force for equitable healthcare delivery for all populations, regardless of their background or identity.

In conclusion, achieving fairness in healthcare AI is not a one-time goal but a continuous process of improvement. It requires ongoing efforts to address biases, create transparent systems, and involve diverse voices in the design and deployment of AI technologies. The future of healthcare AI depends on our ability to navigate these challenges, ensuring that AI technologies are not only technically advanced but also ethically grounded and socially responsible.

REFERENCES

Alderman, J. E., Palmer, J., Laws, E., McCradden, M. D., Ordish, J., Ghassemi, M., Pfohl, S. R., Rostamzadeh, N., Cole-Lewis, H., Glocker, B., Calvert, M., Pollard, T. J., Gill, J., Gath, J., Adebajo, A., Beng, J., Leung, C. H., Kuku, S., Farmer, L.-A., & Liu, X. (2025). Tackling algorithmic bias and promoting transparency in health datasets: The STANDING Together consensus recommendations. *The Lancet. Digital Health*, *7*(1), e64–e88. DOI: 10.1016/S2589-7500(24)00224-3 PMID: 39701919

Anderson, J. W., & Visweswaran, S. (2025). Algorithmic individual fairness and healthcare: A scoping review. *JAMIA Open*, *8*(1), ooae149. DOI: 10.1093/jamiaopen/ooae149 PMID: 39737346

Aquino, Y. S. J., Carter, S. M., Houssami, N., Braunack-Mayer, A., Win, K. T., Degeling, C., Wang, L., & Rogers, W. A. (2023). Practical, epistemic and normative implications of algorithmic bias in healthcare artificial intelligence: A qualitative study of multidisciplinary expert perspectives. *Journal of Medical Ethics*, jme-2022-108850. DOI: 10.1136/jme-2022-108850 PMID: 36823101

Blasimme, A., & Vayena, E. (2019 Forthcoming). *The ethics of AI in biomedical research, patient care and public health. Patient Care and Public Health (April 9, 2019)*. Oxford Handbook of Ethics of Artificial Intelligence.

Chen, R. J., Chen, T. Y., Lipkova, J., Wang, J. J., Williamson, D. F., Lu, M. Y., . . . Mahmood, F. (2021). Algorithm fairness in ai for medicine and healthcare. *arXiv preprint arXiv:2110.00603*.

Chen, R. J., Wang, J. J., Williamson, D. F., Chen, T. Y., Lipkova, J., Lu, M. Y., Sahai, S., & Mahmood, F. (2023). Algorithmic fairness in artificial intelligence for medicine and healthcare. *Nature Biomedical Engineering*, *7*(6), 719–742. DOI: 10.1038/s41551-023-01056-8 PMID: 37380750

Chien, I., Deliu, N., Turner, R., Weller, A., Villar, S., & Kilbertus, N. (2022, June). Multi-disciplinary fairness considerations in machine learning for clinical trials. In *Proceedings of the 2022 ACM Conference on Fairness, Accountability, and Transparency* (pp. 906-924). DOI: 10.1145/3531146.3533154

Chinta, S. V., Wang, Z., Zhang, X., Viet, T. D., Kashif, A., Smith, M. A., & Zhang, W. (2024). Ai-driven healthcare: A survey on ensuring fairness and mitigating bias. *arXiv preprint arXiv:2407.19655*.

Cirillo, D., Catuara-Solarz, S., Morey, C., Guney, E., Subirats, L., Mellino, S., Gigante, A., Valencia, A., Rementeria, M. J., Chadha, A. S., & Mavridis, N. (2020). Sex and gender differences and biases in artificial intelligence for biomedicine and healthcare. *NPJ Digital Medicine*, *3*(1), 81. DOI: 10.1038/s41746-020-0288-5 PMID: 32529043

Dehghani, F., Malik, N., Lin, J., Bayat, S., & Bento, M. (2024, November). Fairness in Healthcare: Assessing Data Bias and Algorithmic Fairness. In *2024 20th International Symposium on Medical Information Processing and Analysis (SIPAIM)* (pp. 1-6). IEEE.

El-Azab, S., & Nong, P. (2023). Clinical algorithms, racism, and "fairness" in healthcare: A case of bounded justice. *Big Data & Society*, *10*(2), 20539517231213820. DOI: 10.1177/20539517231213820

Faust, O., Salvi, M., Barua, P. D., Chakraborty, S., Molinari, F., & Acharya, U. R. (2025). Issues and Limitations on the Road to Fair and Inclusive AI Solutions for Biomedical Challenges. *Sensors (Basel)*, *25*(1), 205. DOI: 10.3390/s25010205 PMID: 39796996

Ferryman, K., & Pitcan, M. (2018). Fairness in precision medicine. *Data & Society, 1.*

Franklin, G., Stephens, R., Piracha, M., Tiosano, S., Lehouillier, F., Koppel, R., & Elkin, P. L. (2024). The sociodemographic biases in machine learning algorithms: A biomedical informatics perspective. *Life (Chicago, Ill.)*, *14*(6), 652. DOI: 10.3390/life14060652 PMID: 38929638

Griffin, A. C., Wang, K. H., Leung, T. I., & Facelli, J. C. (2024). Recommendations to promote fairness and inclusion in biomedical AI research and clinical use. *Journal of Biomedical Informatics*, *157*, 104693. DOI: 10.1016/j.jbi.2024.104693 PMID: 39019301

Grote, T., & Keeling, G. (2022). Enabling fairness in healthcare through machine learning. *Ethics and Information Technology*, *24*(3), 39. DOI: 10.1007/s10676-022-09658-7 PMID: 36060496

Grote, T., & Keeling, G. (2022). On algorithmic fairness in medical practice. *Cambridge Quarterly of Healthcare Ethics*, *31*(1), 83–94. DOI: 10.1017/S0963180121000839 PMID: 35049447

Huang, J., Galal, G., Etemadi, M., & Vaidyanathan, M. (2022). Evaluation and mitigation of racial bias in clinical machine learning models: Scoping review. *JMIR Medical Informatics*, *10*(5), e36388. DOI: 10.2196/36388 PMID: 35639450

Jain, A., Brooks, J. R., Alford, C. C., Chang, C. S., Mueller, N. M., Umscheid, C. A., & Bierman, A. S. (2023, June). Awareness of racial and ethnic bias and potential solutions to address bias with use of health care algorithms. [). American Medical Association.]. *JAMA Health Forum, 4*(6), e231197–e231197. DOI: 10.1001/jamahealthforum.2023.1197 PMID: 37266959

Kerasidou, A. (2021). Ethics of artificial intelligence in global health: Explainability, algorithmic bias and trust. *Journal of Oral Biology and Craniofacial Research, 11*(4), 612–614. DOI: 10.1016/j.jobcr.2021.09.004 PMID: 34567966

Liu, M., Ning, Y., Teixayavong, S., Mertens, M., Xu, J., Ting, D. S. W., Cheng, L. T.-E., Ong, J. C. L., Teo, Z. L., Tan, T. F., RaviChandran, N., Wang, F., Celi, L. A., Ong, M. E. H., & Liu, N. (2023). A translational perspective towards clinical AI fairness. *NPJ Digital Medicine, 6*(1), 172. DOI: 10.1038/s41746-023-00918-4 PMID: 37709945

Loftus, T. J., Tighe, P. J., Ozrazgat-Baslanti, T., Davis, J. P., Ruppert, M. M., Ren, Y., Shickel, B., Kamaleswaran, R., Hogan, W. R., Moorman, J. R., Upchurch, G. R., Rashidi, P., & Bihorac, A. (2022). Ideal algorithms in healthcare: Explainable, dynamic, precise, autonomous, fair, and reproducible. *PLOS Digital Health, 1*(1), e0000006. DOI: 10.1371/journal.pdig.0000006 PMID: 36532301

Mahmood, F. (2021). Algorithm Fairness in AI for Medicine and Healthcare. *arXiv preprint arXiv:2110.00603.*

McCradden, M. D., Joshi, S., Mazwi, M., & Anderson, J. A. (2020). Ethical limitations of algorithmic fairness solutions in health care machine learning. *The Lancet. Digital Health, 2*(5), e221–e223. DOI: 10.1016/S2589-7500(20)30065-0 PMID: 33328054

Mhasawade, V., Zhao, Y., & Chunara, R. (2021). Machine learning and algorithmic fairness in public and population health. *Nature Machine Intelligence, 3*(8), 659–666. DOI: 10.1038/s42256-021-00373-4

Mienye, I. D., Obaido, G., Emmanuel, I. D., & Ajani, A. A. (2024, June). A survey of bias and fairness in healthcare AI. In *2024 IEEE 12th International Conference on Healthcare Informatics (ICHI)* (pp. 642-650). IEEE. DOI: 10.1109/ICHI61247.2024.00103

Nazer, L. H., Zatarah, R., Waldrip, S., Ke, J. X. C., Moukheiber, M., Khanna, A. K., Hicklen, R. S., Moukheiber, L., Moukheiber, D., Ma, H., & Mathur, P. (2023). Bias in artificial intelligence algorithms and recommendations for mitigation. *PLOS Digital Health, 2*(6), e0000278. DOI: 10.1371/journal.pdig.0000278 PMID: 37347721

Norori, N., Hu, Q., Aellen, F. M., Faraci, F. D., & Tzovara, A. (2021). Addressing bias in big data and AI for health care: A call for open science. *Patterns (New York, N.Y.)*, *2*(10), 100347. DOI: 10.1016/j.patter.2021.100347 PMID: 34693373

Qureshi, S., & Oladokun, B. (2024). Human Freedom from Algorithmic Bias: What is the role of Accountability in addressing Health Disparities? *Medical Research Archives*, *12*(8). Advance online publication. DOI: 10.18103/mra.v12i8.5635

Rajkomar, A., Hardt, M., Howell, M. D., Corrado, G., & Chin, M. H. (2018). Ensuring fairness in machine learning to advance health equity. *Annals of Internal Medicine*, *169*(12), 866–872. DOI: 10.7326/M18-1990 PMID: 30508424

Sikstrom, L., Maslej, M. M., Hui, K., Findlay, Z., Buchman, D. Z., & Hill, S. L. (2022). Conceptualising fairness: Three pillars for medical algorithms and health equity. *BMJ Health & Care Informatics*, *29*(1), e100459. DOI: 10.1136/bmjhci-2021-100459 PMID: 35012941

Singhal, A., Neveditsin, N., Tanveer, H., & Mago, V. (2024). Toward fairness, accountability, transparency, and ethics in AI for social media and health care: Scoping review. *JMIR Medical Informatics*, *12*(1), e50048. DOI: 10.2196/50048 PMID: 38568737

Tilmes, N. (2022). Disability, fairness, and algorithmic bias in AI recruitment. *Ethics and Information Technology*, *24*(2), 21. DOI: 10.1007/s10676-022-09633-2

Visweswaran, S., Luo, Y., & Peleg, M. (2024). Fairness and inclusion methods for biomedical informatics research. *Journal of Biomedical Informatics*, *158*, 104713. DOI: 10.1016/j.jbi.2024.104713 PMID: 39187169

Wang, Y., Song, Y., Ma, Z., & Han, X. (2023). Multidisciplinary considerations of fairness in medical AI: A scoping review. *International Journal of Medical Informatics*, *178*, 105175. DOI: 10.1016/j.ijmedinf.2023.105175 PMID: 37595374

Williams, R. (2023). Fair and equitable AI in biomedical research and healthcare: Social science perspecti es. *Artificial Intelligence in Medicine*, *144*, 102658. DOI: 10.1016/j.artmed.2023.102658 PMID: 37783540

Xu, J., Xiao, Y., Wang, W. H., Ning, Y., Shenkman, E. A., Bian, J., & Wang, F. (2022). Algorithmic fairness in computational medicine. *EBioMedicine*, •••, 84. PMID: 36084616

Yang, J., Soltan, A. A., Eyre, D. W., & Clifton, D. A. (2023). Algorithmic fairness and bias mitigation for clinical machine learning with deep reinforcement learning. *Nature Machine Intelligence*, *5*(8), 884–894. DOI: 10.1038/s42256-023-00697-3 PMID: 37615031

Yang, J., Soltan, A. A., Yang, Y., & Clifton, D. A. (2022). Algorithmic fairness and bias mitigation for clinical machine learning: Insights from rapid COVID-19 diagnosis by adversarial learning. medRxiv, 2022-01. DOI: 10.1101/2022.01.13.22268948

Yang, Y., Lin, M., Zhao, H., Peng, Y., Huang, F., & Lu, Z. (2024). A survey of recent methods for addressing AI fairness and bias in biomedicine. *Journal of Biomedical Informatics*, *154*, 104646. DOI: 10.1016/j.jbi.2024.104646 PMID: 38677633

Compilation of References

Abbas, E., & Qazi, A. A. (2024). Customized Ai-powered Security and Privacy Configurations for Social MEDIA Websites. *BULLET: Jurnal Multidisiplin Ilmu*, *3*(1), 108–117.

Abd-Elsalam, K. A., & Abdel-Momen, S. M. (2023). Artificial intelligence's development and challenges in scientific writing. *Egyptian Journal of Agricultural Research*, *101*(3), 714–717. DOI: 10.21608/ejar.2023.220363.1414

Abolaji, E. O., & Akinwande, O. T.Elijah Oluwatoyosi AbolajiOladayo Tosin Akinwande. (2024). AI powered privacy protection: A survey of current state and future directions. *World Journal of Advanced Research and Reviews*, *23*(3), 2687–2696. DOI: 10.30574/wjarr.2024.23.3.2869

Act, A. A., Agent, P., Assessment, D. P. I., Act, P. E. A. I., Directive, P. P. W., & Regulation, G. D. P. AI-infused contracts/contracting (cont.). *disclosure*, *4*(1), 652.

Adahman, Z., Malik, A. W., & Anwar, Z. (2022). An analysis of zero-trust architecture and its cost-effectiveness for organizational security. *Computers & Security*, *122*, 102911. DOI: 10.1016/j.cose.2022.102911

Adanma, U. M., & Ogunbiyi, E. O.Uwaga Monica AdanmaEmmanuel Olurotimi Ogunbiyi. (2024). Artificial intelligence in environmental conservation: Evaluating cyber risks and opportunities for sustainable practices. *Computer Science & IT Research Journal*, *5*(5), 1178–1209. DOI: 10.51594/csitrj.v5i5.1156

Aderibigbe, A. O., Ohenhen, P. E., Nwaobia, N. K., Gidiagba, J. O., & Ani, E. C.Adebayo Olusegun AderibigbePeter Efosa OhenhenNwabueze Kelvin Nwaobia-Joachim Osheyor GidiagbaEmmanuel Chigozie Ani. (2023). Artificial intelligence in developing countries: Bridging the gap between potential and implementation. *Computer Science & IT Research Journal*, *4*(3), 185–199. DOI: 10.51594/csitrj.v4i3.629

Aggarwal, S., Bansal, S., & Goel, R. (2024). AI In Agriculture: A Looming Challenge, A Gleaming Opportunity. *International Journal of Engineering Science and Humanities, 14*(Special Issue 1), 43-52.

Agufenwa, O. J. (2023, December 20). *The crucial role of threat intelligence feeds integration in cloud security.* ResearchGate. https://doi.org/DOI: 10.13140/RG.2.2.19905.33123

Ahmad, A., Liew, A. X., Venturini, F., Kalogeras, A., Candiani, A., Di Benedetto, G., ... & Martos, V. (2024). AI can empower agriculture for.

Ahmad, A., Liew, A. X., Venturini, F., Kalogeras, A., Candiani, A., Di Benedetto, G., Ajibola, S., Cartujo, P., Romero, P., Lykoudi, A., De Grandis, M. M., Xouris, C., Lo Bianco, R., Doddy, I., Elegbede, I., D'Urso Labate, G. F., García del Moral, L. F., & Martos, V. (2024). AI can empower agriculture for global food security: Challenges and prospects in developing nations. *Frontiers in Artificial Intelligence, 7*, 1328530. DOI: 10.3389/frai.2024.1328530 PMID: 38726306

Ahmadi, S. (2024). Zero trust architecture in cloud networks: Application, challenges and future opportunities. *Journal of Engineering Research and Reports, 26*(2), 215–228. DOI: 10.9734/jerr/2024/v26i21083

Ahmed, M., Mahmood, A. N., & Hu, J. (2016). A survey of network anomaly detection t echniques. *Journal of Network and Computer Applications, 60*, 19–31. h t tps://DOI: 10.1016/j.jnca.2015.11.016

Ahmed, M. N., Singh, A. P., Hussain, M. R., Rasool, M. A., Khan, I. M., & Dildar, M. S. (2024, July). Enhancing Crop Production using Artificial Intelligence in Agricultural Revolution. In *2024 IEEE 7th International Conference on Advanced Technologies* [ATSIP]. *Signal and Image Processing : an International Journal, 1*, 432–437.

AIMultiple. (2025). *Top 15 UEBA Use Cases for Today's SOCs in 2025.* h t tps://research.aimultiple.com/ueba-use-cases

Aithal, P. S., & Aithal, S. (2023). Application of ChatGPT in higher education and research–A futuristic analysis. [IJAEML]. *International Journal of Applied Engineering and Management Letters, 7*(3), 168–194. DOI: 10.47992/IJAEML.2581.7000.0193

Ajiye, O. T., & Omokhabi, A. A. (2025). The Potential And Ethical Issues Of Artificial Intelligence In Improving Academic Writing. *ShodhAI: Journal of Artificial Intelligence, 2*(1), 1–9. DOI: 10.29121/shodhai.v2.i1.2025.24

Akhtar, Z. B., & Rawol, A. T. (2024). Enhancing cybersecurity through AI-powered security mechanisms. *IT Journal Research and Development*, *9*(1), 50–67. DOI: 10.25299/itjrd.2024.16852

Alahdab, F. (2024). Potential impact of large language models on academic writing. *BMJ Evidence-Based Medicine*, *29*(3), 201–202. DOI: 10.1136/bmjebm-2023-112429 PMID: 37620013

Alderman, J. E., Palmer, J., Laws, E., McCradden, M. D., Ordish, J., Ghassemi, M., Pfohl, S. R., Rostamzadeh, N., Cole-Lewis, H., Glocker, B., Calvert, M., Pollard, T. J., Gill, J., Gath, J., Adebajo, A., Beng, J., Leung, C. H., Kuku, S., Farmer, L.-A., & Liu, X. (2025). Tackling algorithmic bias and promoting transparency in health datasets: The STANDING Together consensus recommendations. *The Lancet. Digital Health*, *7*(1), e64–e88. DOI: 10.1016/S2589-7500(24)00224-3 PMID: 39701919

Alexander, C. S., Yarborough, M., & Smith, A. (2024). Who is responsible for 'responsible AI'?: Navigating challenges to build trust in AI agriculture and food system technology. *Precision Agriculture*, *25*(1), 146–185. DOI: 10.1007/s11119-023-10063-3

Alghamdy, R. Z. (2023). Pedagogical and ethical implications of artificial intelligence in EFL context: A review study. *English Language Teaching*, *16*(10), 87–98. DOI: 10.5539/elt.v16n10p87

Alhitmi, H. K., Mardiah, A., Al-Sulaiti, K. I., & Abbas, J. (2024). Data security and privacy concerns of AI-driven marketing in the context of economics and business field: An exploration into possible solutions. *Cogent Business & Management*, *11*(1), 2393743. DOI: 10.1080/23311975.2024.2393743

Ali, A., Anderson, L., Venturini, F., Athanasios, K., Candiani, A., Di Benedetto, G., ... & Martos, V. (2024). AI can empower agriculture for global food security: challenges and prospects in developing nations.

Ali, G., Mijwil, M. M., Buruga, B. A., Abotaleb, M., & Adamopoulos, I. (2024). A survey on artificial intelligence in cybersecurity for smart agriculture: State-of-the-art, cyber threats, artificial intelligence applications, and ethical concerns. *Mesopotamian Journal of Computer Science*, *2024*, 53–103. DOI: 10.58496/MJCSC/2024/007

Ali, M., & Siddiqui, N. (2022). A comparative study of cloud-native and third-party security tools for multi-cloud environments. *Journal of Cybersecurity Research*, *14*(3), 45–61.

Aljuaid, H. (2024). The impact of artificial intelligence tools on academic writing instruction in higher education: A systematic review. *Arab World English Journal (AWEJ) Special Issue on ChatGPT*.

Almeida, D., & Barr, N. (2025). Innovations in Health Data Protection Ethical, Legal, and Technological Perspectives in a Global Context: AI-Powered Diagnosis Systems and Health Data Innovation. In *Navigating Privacy, Innovation, and Patient Empowerment Through Ethical Healthcare Technology* (pp. 171-196). IGI Global Scientific Publishing.

Alotaibi, M., et al. (2023). *Enhancing Threat Detection in Multi-Cloud Environments*. Cybersecurity Journal, 7(2), 35-51.

Alotaibi, F. G., Clarke, N., & Furnell, S. M. (2021). A novel approach for improving information security management and awareness for home environments. *Information and Computer Security*, *29*(1), 25–48. DOI: 10.1108/ICS-05-2020-0073

AlSamhori, A. F., & Alnaimat, F. (2024). Artificial intelligence in writing and research: Ethical implications and best practices. *Central Asian Journal of Medical Hypotheses and Ethics = Central'noaziatskij Zurnal Medicinskich Gipotez i Etiki = Medicinalyk Gipoteza Men Ètikanyn Orta Aziâlyk Zurnaly*, *5*(4), 259–268. DOI: 10.47316/cajmhe.2024.5.4.02

Amazon Web Services. (2023). *AWS GuardDuty: Threat detection and continuous monitoring*. https://aws.amazon.com/guardduty/

Amershi, S., Cakmak, M., Knox, W. B., & Kulesza, T. (2014). Power to the people: The role of humans in interactive machine learning. *AI Magazine*, *35*(4), 105–120. DOI: 10.1609/aimag.v35i4.2513

Amir, M., & Zhu, Q. (2022). Model-based integration testing for multi-agent systems. *Journal of Systems and Software*, *186*, 111232. DOI: 10.1016/j.jss.2021.111232

Amodio, D. M., & Frith, C. D. (2006). Meeting of minds: The medial frontal cortex and social cognition. *Nature Reviews. Neuroscience*, *7*(4), 268–277. DOI: 10.1038/nrn1884 PMID: 16552413

Anderson, J. W., & Visweswaran, S. (2025). Algorithmic individual fairness and healthcare: A scoping review. *JAMIA Open*, *8*(1), ooae149. DOI: 10.1093/jamiaopen/ooae149 PMID: 39737346

Anderson, T., & Shattuck, J. (2012). Design-based research: A decade of progress in education research? *Educational Researcher*, *41*(1), 16–25. DOI: 10.3102/0013189X11428813

Anidjar, L., Packin, N. G., & Panezi, A. (2023). The matrix of privacy: Data infrastructure in the AI-Powered Metaverse. *SSRN*, *18*, 59. DOI: 10.2139/ssrn.4363208

AppOmni. (n.d.). Cloud Security Posture Management (CSPM) diagram. A p p Omni. https://appomni.com/saas-glossary/cloud-security-posture-management-cspm/

Aquino, Y. S. J., Carter, S. M., Houssami, N., Braunack-Mayer, A., Win, K. T., Degeling, C., Wang, L., & Rogers, W. A. (2023). Practical, epistemic and normative implications of algorithmic bias in healthcare artificial intelligence: A qualitative study of multidisciplinary expert perspectives. *Journal of Medical Ethics*, jme-2022-108850. DOI: 10.1136/jme-2022-108850 PMID: 36823101

Araújo, S. O., Peres, R. S., Barata, J., Lidon, F., & Ramalho, J. C. (2021). Characterising the agriculture 4.0 landscape—Emerging trends, challenges and opportunities. *Agronomy (Basel)*, *11*(4), 667. DOI: 10.3390/agronomy11040667

Arefin, S. (2024). Strengthening Healthcare Data Security with Ai-Powered Threat Detection. [IJSRM]. *International Journal of Scientific Research and Management*, *12*(10), 1477–1483. DOI: 10.18535/ijsrm/v12i10.ec02

Arora, D., Sonwane, A., Wadhwa, N., Mehrotra, A., Utpala, S., Bairi, R., Kanade, A., & Natarajan, N. (2024). MASAI: Modular architecture for software-engineering AI agents. *arXiv*. https://doi.org//arXiv.2406.11638DOI: 10.48550

Arya, L., Sharma, Y. K., Devi, S., & Padmanaban, H. (2024). Securing the Internet of Things: AI-Powered Threat Detection and Safety. In *Proceedings of International Conference on Recent Innovations in Computing: ICRIC 2023,* Volume 2 (Vol. 2, p. 97). Springer Nature. DOI: 10.1007/978-981-97-3442-9_7

Arya, L., Sharma, Y. K., Devi, S., Padmanaban, H., & Kumar, R. (2023, October). Securing the Internet of Things: AI-Powered Threat Detection and Safety Measures. In *The International Conference on Recent Innovations in Computing* (pp. 97-108). Singapore: Springer Nature Singapore.

Aslam, M. S., & Nisar, S. (2024). Ethical Considerations for Artificial Intelligence Tools in Academic Research and Manuscript Preparation: A Web Content Analysis. In *Digital Transformation in Higher Education, Part B: Cases, Examples and Good Practices* (pp. 155-196). Emerald Publishing Limited. DOI: 10.1108/978-1-83608-424-220241007

Asthana, K. (2024, November 26). *Top 8 cloud vulnerabilities*. CrowdStrike. h t tps://www.crowdstrike.com/en-us/cybersecurity-101/cloud-security/cloud- v u lnerabilities/

Asthana, N. (2024). Multi-cloud security challenges and best practices. *Cloud Computing Journal, 15*(2), 45–58.

Asthana, P. (2024). Security Risks and Threat Detection in Multi-Cloud Environments. *Journal of Cloud Security, 12*(4), 87–102.

Awad, A. I., Babu, A., Barka, E., & Shuaib, K. (2024). AI-powered biometrics for Internet of Things security: A review and future vision. *Journal of Information Security and Applications, 82*, 103748. DOI: 10.1016/j.jisa.2024.103748

Awofala, A. O. A., Bazza, M. B., Ojo, O. T., Oladipo, A. J., Olabiyi, O. S., & Arig-babu, A. A. (2025). Structural equation modeling of Nigerian science, technology and mathematics teachers' adoption of educational artificial in-telligence tools. *Digital Education Review*, (46), 51–64. DOI: 10.1344/der.2025.46.51-64

Babu, S., & Irudhayaraj, R. (2019, March). User-Entity Behavior Analytics (UEBA) – A S ystematic Review of Literatures. In *9th Annual International Conference on Industrial Engineering and Operations Management*, https://doi.org/DOI: 10.46254/ AN09.20190828

Bace, R. G., & Mell, P. (2001). *Intrusion detection systems.* National Institute of Standards and Technology. DOI: 10.6028/NIST.SP.800-31

BaHammam, A. S. (2023). Balancing innovation and integrity: The role of AI in research and scientific writing. *Nature and Science of Sleep*, ●●●, 1153–1156. PMID: 38170140

Bahrini, A., Khamoshifar, M., Abbasimehr, H., Riggs, R. J., Esmaeili, M., Majdabad-kohne, R. M., & Pasehvar, M. (2023, April). ChatGPT: Applications, opportunities, and threats. In *2023 Systems and Information Engineering Design Symposium (SIEDS)* (pp. 274-279). IEEE. DOI: 10.1109/SIEDS58326.2023.10137850

Banse, C., Kunz, I., Schneider, A., & Weiss, K. (2021, September). Cloud property graph: Connecting cloud security assessments with static code analysis. In *2021 IEEE 14th International Conference on Cloud Computing (CLOUD)* (pp. 13-19). IEEE.

Bartakke, J., & Kashyap, R. (2024). The Usage of Clouds in Zero-Trust Security Strategy: An Evolving Paradigm. *Journal of Information and Organizational Sciences, 48*(1), 149–165. DOI: 10.31341/jios.48.1.8

Basan, M. (2023, October 26). *What is multi-cloud security? Everything to know.* eSecurity Planet. https://www.esecurityplanet.com/cloud/multi-cloud-security/

Baskara, F. R. (2024, March). Deus Ex Machina: Unraveling Academic Integrity in the AI Narrative. In *Proceeding International Conference on Religion. Science and Education*, *3*, 393–402.

Bechara, A., Damasio, H., & Damasio, A. R. (2000). Different contributions of the human amygdala and ventromedial prefrontal cortex to decision-making. *The Journal of Neuroscience : The Official Journal of the Society for Neuroscience*, *20*(11), RC79. DOI: 10.1523/JNEUROSCI.20-11-j0001.2000 PMID: 10807937

Bengio, Y., Courville, A., & Vincent, P. (2013). Representation learning: A review and new perspectives. *IEEE Transactions on Pattern Analysis and Machine Intelligence*, *35*(8), 1798–1828. DOI: 10.1109/TPAMI.2013.50 PMID: 23787338

Bhamidipaty, V., Bhamidipaty, D. L., Guntoory, I., Bhamidipaty, K. D. P., Iyengar, K. P., Botchu, B., & Botchu, R. (2025). Revolutionizing Healthcare: The Impact of AI-Powered Sensors. *Generative Artificial Intelligence for Biomedical and Smart Health Informatics*, 355-373.

Bhangar, N. A., & Shahriyar, A. K. (2023). Iot and ai for next-generation farming: Opportunities, challenges, and outlook. *International Journal of Sustainable Infrastructure for Cities and Societies*, *8*(2), 14–26.

Bhat, S. A., & Huang, N. F. (2021). Big data and ai revolution in precision agriculture: Survey and challenges. *IEEE Access : Practical Innovations, Open Solutions*, *9*, 110209–110222. DOI: 10.1109/ACCESS.2021.3102227

Bhattacharya, P., Prasad, V. K., Verma, A., Roy, P. P., Srivastava, S., & Deb, D. (2024). Demystifying ChatGPT: An in-depth survey of OpenAI's robust large language models. *Archives of Computational Methods in Engineering*, *31*(7), 4557–4600. DOI: 10.1007/s11831-024-10115-5

Biesbroek, R., Wright, S. J., Eguren, S. K., Bonotto, A., & Athanasiadis, I. N. (2022). Policy attention to climate change impacts, adaptation and vulnerability: A global assessment of National Communications (1994–2019). *Climate Policy*, *22*(1), 97–111. DOI: 10.1080/14693062.2021.2018986

Blasimme, A., & Vayena, E. (2019 Forthcoming). *The ethics of AI in biomedical research, patient care and public health. Patient Care and Public Health (April 9, 2019)*. Oxford Handbook of Ethics of Artificial Intelligence.

Brightwood, S., & Jame, H. (2024). *Data privacy, security, and ethical considerations in AI-powered finance. Article*. Research Gate.

Brooke, J. (1996). SUS: A "quick and dirty" usability scale. In Jordan, P. W., Thomas, B., Weerdmeester, B. A., & McClelland, I. L. (Eds.), *Usability evaluation in industry* (pp. 189–194). Taylor & Francis.

Brown, M., Green, L., & White, T. (2021). Integrating biometric data for enhanced emotion recognition in AI systems. *IEEE Transactions on Affective Computing*, *12*(2), 345–356. DOI: 10.1109/TAFFC.2018.2879076

Budzianowski, P., Wen, T.-H., Tseng, B.-H., Casanueva, I., Ultes, S., Ramadan, O., Gašić, M., & Veličković, P. (2018). MultiWOZ – A large-scale multi-domain Wizard-of-Oz dataset for task-oriented dialogue modelling. *Proceedings of the 2018 Conference on Empirical Methods in Natural Language Processing* (pp. 5016–5026). Association for Computational Linguistics. https://doi.org/DOI: 10.18653/v1/D18-1547

Butt, J. (2024). Analytical study of the world's first EU Artificial Intelligence (AI) Act. *International Journal of Research Publication and Reviews*, *5*(3), 7343–7364. DOI: 10.55248/gengpi.5.0324.0914

Byrne, B., Krishnamoorthi, K., Sankar, C., Kumar, R., Subba, R., Ramanarayanan, V., Kumar, A., Asthana, S., Nigam, S., El-Khamy, M., & Tang, D. (2019). Taskmaster-1: Toward a realistic and diverse goal-oriented dialogue dataset. *Proceedings of the 20th Annual SIGdial Meeting on Discourse and Dialogue* (pp. 253–261). Association for Computational Linguistics. https://doi.org/DOI: 10.18653/v1/W19-5929

Caggiano, I. A., Gatt, L., Mollo, A. A., Izzo, L., & Troisi, E. (2024, June). Health in Space: Integrating Legal, Ethical and Sustainability Assessments. In *2024 11th International Workshop on Metrology for AeroSpace (MetroAeroSpace)* (pp. 296-302). IEEE.

Casal, J. E., & Kessler, M. (2023). Can linguists distinguish between ChatGPT/AI and human writing?: A study of research ethics and academic publishing. *Research Methods in Applied Linguistics*, *2*(3), 100068. DOI: 10.1016/j.rmal.2023.100068

Caso, S. (2024). Emerging technologies in military space operations: current applications and future research for educational and training purposes. *International Journal of Training Research*, 1-16.

Chauke, K. O., Muchenje, T., & Makondo, N. (2025). *Enhancing network security in multi-cloud environments through adaptive threat detection.*

Chavan, Y., Paul, K., & Kolekar, N. (2024). Food safety and hygiene: Current policies, quality standards, and scope of artificial intelligence. In *Food production, diversity, and safety under climate change* (pp. 319–331). Springer Nature Switzerland. DOI: 10.1007/978-3-031-51647-4_26

Check Point Software Technologies. (2023). *CloudGuard: Prevention-first cloud security.* https://www.checkpoint.com/cloudguard/

Chen, R. J., Chen, T. Y., Lipkova, J., Wang, J. J., Williamson, D. F., Lu, M. Y., . . . Mahmood, F. (2021). Algorithm fairness in ai for medicine and healthcare. *arXiv preprint arXiv:2110.00603*.

Chen, R. J., Wang, J. J., Williamson, D. F., Chen, T. Y., Lipkova, J., Lu, M. Y., Sahai, S., & Mahmood, F. (2023). Algorithmic fairness in artificial intelligence for medicine and healthcare. *Nature Biomedical Engineering*, *7*(6), 719–742. DOI: 10.1038/s41551-023-01056-8 PMID: 37380750

Chen, T., Lv, L., Wang, D., Zhang, J., Yang, Y., Zhao, Z., & Tao, D. (2024). Empowering agrifood system with artificial intelligence: A survey of the progress, challenges and opportunities. *ACM Computing Surveys*, *57*(2), 1–37.

Chetwynd, E. (2024). Ethical use of artificial intelligence for scientific writing: Current trends. *Journal of Human Lactation*, *40*(2), 211–215. DOI: 10.1177/08903344241235160 PMID: 38482810

Chhabra, S., & Singh, A. K. (2022). A Comprehensive Vision on Cloud Computing Environment: Emerging Challenges and Future Research Directions. *arXiv preprint*

Chhabra, S., & Singh, M. (2022). A comprehensive survey on multi-cloud computing: Challenges and future directions. *Journal of Network and Computer Applications*, *213*, 103966.

Chien, I., Deliu, N., Turner, R., Weller, A., Villar, S., & Kilbertus, N. (2022, June). Multi-disciplinary fairness considerations in machine learning for clinical trials. In *Proceedings of the 2022 ACM Conference on Fairness, Accountability, and Transparency* (pp. 906-924). DOI: 10.1145/3531146.3533154

Chimakurthi, V. N. S. S. (2020). The challenge of achieving zero trust remote access in multi- cloud environment. *ABC Journal of Advanced Research*, *9*(2), 89–102. DOI: 10.18034/abcjar.v9i2.608

Chinta, S. V., Wang, Z., Zhang, X., Viet, T. D., Kashif, A., Smith, M. A., & Zhang, W. (2024). Ai-driven healthcare: A survey on ensuring fairness and mitigating bias. *arXiv preprint arXiv:2407.19655*.

Cho, G., & Crompvoets, J. (2019). Prod-users of geospatial information: Some legal perspectives. *Journal of Spatial Science*, *64*(2), 341–358. DOI: 10.1080/14498596.2018.1429330

Christensen, K. (2021). A European solution for Text and Data Mining in the development of creative Artificial Intelligence: With a specific focus on articles 3 and 4 of the Digital Signel Market Directive.

Chung, G., Rodriguez, M., Lanier, P., & Gibbs, D. (2022). Text-mining open-ended survey responses using structural topic modeling: A practical demonstration to understand parents' coping methods during the COVID-19 pandemic in Singapore. *Journal of Technology in Human Services*, *40*(4), 296–318. DOI: 10.1080/15228835.2022.2036301

Cirillo, D., Catuara-Solarz, S., Morey, C., Guney, E., Subirats, L., Mellino, S., Gigante, A., Valencia, A., Rementeria, M. J., Chadha, A. S., & Mavridis, N. (2020). Sex and gender differences and biases in artificial intelligence for biomedicine and healthcare. *NPJ Digital Medicine*, *3*(1), 81. DOI: 10.1038/s41746-020-0288-5 PMID: 32529043

CISA. (2021). *Mitigating Supply Chain Risks in Cloud Services*. Retrieved from h t tps://www.cisa.gov/supply-chain-security

Cisco Systems. (2022). *Cisco Secure Cloud Analytics Data Sheet*. Retrieved from h ttps://www.cisco.com/c/dam/en/us/products/collateral/security/stealthwatch/secure -cloud- analytics-ds.pdf

Cloudwithease. (n.d.). *Cloud security comparison: AWS vs Azure vs GCP*. Cloudwithease. https://cloudwithease.com/cloud-security-comparison-aws-vs-azure-vs-gcp/

Cloudwithease. (n.d.). Comparison of security services between GCP, AWS and Azure [Image]. Retrieved February 22, 2025, from [URL omitted]

Cohen, A., & Schindel, A. (2024, January 24). *Introducing the Cloud Threat Landscape, a new TI resource for cloud defenders*. Wiz. https://www.wiz.io/blog/introducing-the-cloud-threat-landscape

Cohen, S., Kamarck, T., & Mermelstein, R. (1983). A global measure of perceived stress. *Journal of Health and Social Behavior*, *24*(4), 385–396. DOI: 10.2307/2136404 PMID: 6668417

Conti, M., Kumar, E. S., Lal, C., & Ruj, S. (2018). A survey on security and privacy issues of bitcoin. *IEEE Communications Surveys and Tutorials*, *20*(4), 3416–3452. DOI: 10.1109/COMST.2018.2842460

Cowls, J., Tsamados, A., Taddeo, M., & Floridi, L. (2023). The AI gambit: Leveraging artificial intelligence to combat climate change—Opportunities, challenges, and recommendations. *AI & Society*, *38*(1), 1–25. DOI: 10.1007/s00146-021-01294-x PMID: 34690449

Cox, A., & Thelwall, M. (2025). *AI for Knowledge*. CRC Press. DOI: 10.1201/9781003545163

Creswell, J. W. (2014). *Research design: Qualitative, quantitative, and mixed methods a pproaches* (4th ed.). Sage Publications.

Crew, A. I. (2024). *CrewAI Framework for Multi-Agent Systems*. Retrieved April 29, 2025, from https://crewai.com/docs

CrowdStrike. (2024). *What is XDR? Extended Detection & Response*. h t tps://www .crowdstrike.com/en-us/cybersecurity-101/endpoint-security/extended- d etection-and-response-xdr/

Damasio, A. R. (1994). *Descartes' error: Emotion, reason, and the human brain*. G.P. Putnam's Sons.

Daraf, U., & Badi, S. (2023). AI-Powered Genomic Analysis in the Cloud: Enhancing Precision Medicine While Protecting Medical Data Privacy.

Dara, R., Hazrati Fard, S. M., & Kaur, J. (2022). Recommendations for ethical and responsible use of artificial intelligence in digital agriculture. *Frontiers in Artificial Intelligence*, *5*, 884192. DOI: 10.3389/frai.2022.884192 PMID: 35968036

Dasgupta, D., Akhtar, Z., & Sen, S. (2022). Machine learning in cybersecurity: A comprehensive survey. *The Journal of Defense Modeling and Simulation*, *19*(1), 57–106. DOI: 10.1177/1548512920951275

Davis, M. H. (1983). Measuring individual differences in empathy: Evidence for a multidimensional approach. *Journal of Personality and Social Psychology*, *44*(1), 113–126. DOI: 10.1037/0022-3514.44.1.113

de Haan, E., Padigar, M., El Kihal, S., Kübler, R., & Wieringa, J. E. (2024). Unstructured data research in business: Toward a structured approach. *Journal of Business Research*, *177*, 114655. DOI: 10.1016/j.jbusres.2024.114655

Dean, J., Corrado, G. S., Monga, R., Chen, K., Devin, M., Le, Q. V., Mao, M. Z., Ranzato, M., Senior, A., Tucker, P., Yang, K., & Ng, A. Y. (2012). Large scale distributed deep networks. *Advances in Neural Information Processing Systems*, *25*. https://proceedings.neurips.cc/paper/2012/hash/6aca97005c68f120682381 5f66102863-Abstract.html

Dehghani, F., Malik, N., Lin, J., Bayat, S., & Bento, M. (2024, November). Fairness in Healthcare: Assessing Data Bias and Algorithmic Fairness. In *2024 20th International Symposium on Medical Information Processing and Analysis (SIPAIM)* (pp. 1-6). IEEE.

Dembani, R., Karvelas, I., Akbar, N. A., Rizou, S., Tegolo, D., & Fountas, S. (2025). Agricultural data privacy and federated learning: A review of challenges and opportunities. *Computers and Electronics in Agriculture, 232,* 110048. DOI: 10.1016/j.compag.2025.110048

Demszky, D., Movshovitz-Attias, D., Ko, J., Cowen, A., Nemade, G., & Ravi, S. (2020). GoEmotions: A dataset of fine-grained emotions. *Proceedings of the 58th Annual Meeting of the Association for Computational Linguistics* (pp. 4040–4054). Association for Computational Linguistics. https://doi.org/DOI: 10.18653/v1/2020. acl-main.372

Denhere, V., & Shao, D. (2024). Artificial Intelligence in Agritourism: Utilitarian Analysis of Opportunities, Challenges, and Ethical Considerations in the African Context. *Agritourism in Africa,* 17-36.

Devare, M., Arnaud, E., Antezana, E., & King, B. (2023). Governing agricultural data: Challenges and recommendations. *Towards Responsible Plant Data Linkage: Data Challenges for Agricultural Research and Development, 201.*

Dhinakaran, D., Raja, S. E., Jasmine, J. J., Kumar, P. V., & Ramani, R. (2025). The Future of Well-Being: AI-Powered Health Management with Privacy at its Core. *Wellness Management Powered by AI Technologies,* 363-402.

Dinçer, S. (2024). The use and ethical implications of artificial intelligence in scientific research and academic writing. *Educational Research & Implementation, 1*(2), 139–144. DOI: 10.14527/edure.2024.10

Eberlin, M. N. (2024). The Art of Scientific Writing and Ethical Use of Artificial Intelligence. *Journal of the Brazilian Chemical Society, 35*(1), e-20230121.

El-Azab, S., & Nong, P. (2023). Clinical algorithms, racism, and "fairness" in healthcare: A case of bounded justice. *Big Data & Society, 10*(2), 20539517231213820. DOI: 10.1177/20539517231213820

Ersöz, A. R., & Engin, M. (2024). Exploring Ethical Dilemmas in the Use of Artificial Intelligence in Academic Writing: Perspectives of Researchers. *Journal of Uludag University Faculty of Education, 37*(3), 1190–1208. DOI: 10.19171/uefad.1514323

eSentire. (2024). *What is Extended Detection and Response (XDR)?*h t tps://www
.esentire.com/cybersecurity-fundamentals-defined/glossary/what-is-extended
- detection-and-response-xdr

Espino, A. R. C., Esto, M. R. A., Manalo, C. G. S., Santiago, C. S., & Pragacha, R.
N. (2024, June). Ethical Implications of Using Assistive Writing Tools in the Aca-
deme: A Literature Review. In *International Conference on Frontiers of Intelligent
Computing: Theory and Applications* (pp. 133-145). Singapore: Springer Nature
Singapore.

Farea, A. H., Alhazmi, O. H., Samet, R., & Guzel, M. S. (2024). AI-powered Integrated with Encoding Mechanism Enhancing Privacy, Security, and Performance
for IoT Ecosystem. *IEEE Access : Practical Innovations, Open Solutions*, *12*,
121368–121386. DOI: 10.1109/ACCESS.2024.3449630

Farrell, T., & Yu, P. L. H. (2020). Personality-based adaptive human-AI interaction:
An overview. *International Journal of Human-Computer Studies*, *139*, 102428.
DOI: 10.1016/j.ijhcs.2020.102428

Farzaan, M. A. M., Ghanem, M. C., El-Hajjar, A., & Ratnayake, D. N. (2024). Ai-
enabled system for efficient and effective cyber incident detection and response in
cloud e nvironments. *arXiv preprint arXiv:2404.05602.*

Farzaan, M. A. M., Ghanem, M. C., El-Hajjar, A., & Ratnayake, D. N. (2024). *AI-
Enabled System for Efficient and Effective Cyber Incident Detection and Response
in Cloud Environments*. arXiv preprint, arXiv:2404.05602.

Farzaan, R.. (2024). AI-Driven Cybersecurity: Enhancing Threat Detection in Cloud
Environments. *Information Security Research*, *8*(1), 22–38.

Farzaan, S., Sarkar, B., & Chowdhury, F. (2024). AI-enabled automated cyber incident detection and response in cloud environments. *IEEE Transactions on Cloud
Computing*, *12*(3), 789–801.

Faust, O., Salvi, M., Barua, P. D., Chakraborty, S., Molinari, F., & Acharya, U.
R. (2025). Issues and Limitations on the Road to Fair and Inclusive AI Solutions
for Biomedical Challenges. *Sensors (Basel)*, *25*(1), 205. DOI: 10.3390/s25010205
PMID: 39796996

Fedoriv, Y., Shuhai, A., & Pirozhenko, I. (2024). Foundations of Ethical Use of AI
in EFL Academic Writing.

Ferryman, K., & Pitcan, M. (2018). Fairness in precision medicine. *Data & Society, 1.*

Franklin, G., Stephens, R., Piracha, M., Tiosano, S., Lehouillier, F., Koppel, R., & Elkin, P. L. (2024). The sociodemographic biases in machine learning algorithms: A biomedical informatics perspective. *Life (Chicago, Ill.)*, *14*(6), 652. DOI: 10.3390/life14060652 PMID: 38929638

Freitas, S., & Gharib, A. (2024). *GraphWeaver: Billion-Scale Cybersecurity Incident Correlation*. arXiv preprint, arXiv:2406.01842. DOI: 10.1145/3627673.3680057

Friedman, N. P., & Robbins, T. W. (2022). The role of prefrontal cortex in cognitive control and executive function. *Neuropsychopharmacology : Official Publication of the American College of Neuropsychopharmacology*, *47*(1), 72–89. DOI: 10.1038/s41386-021-01132-0 PMID: 34408280

Garcia-Teodoro, P., Diaz-Verdejo, J., Maciá-Fernández, G., & Vázquez, E. (2009). Anomaly- based network intrusion detection: Techniques, systems and challenges. *computers & security, 28*(1-2), 18-28.

Gardezi, M., Joshi, B., Rizzo, D. M., Ryan, M., Prutzer, E., Brugler, S., & Dadkhah, A. (2024). Artificial intelligence in farming: Challenges and opportunities for building trust. *Agronomy Journal*, *116*(3), 1217–1228. DOI: 10.1002/agj2.21353

Gavai, A. K., Bouzembrak, Y., Xhani, D., Sedrakyan, G., Meuwissen, M. P., Souza, R. G. S., ... & van Hillegersberg, J. (2025). Agricultural data Privacy: Emerging platforms & strategies. *Food and Humanity*, 100542.

Gawankar, S., Nair, S., Pawar, V., Vhatkar, A., & Chavan, P. (2024, August). Patient Privacy and Data Security in the Era of AI-Driven Healthcare. In *2024 8th International Conference on Computing, Communication, Control and Automation (ICCUBEA)* (pp. 1-6). IEEE. DOI: 10.1109/ICCUBEA61740.2024.10775004

Gemiharto, I., & Masrina, D. (2024). User privacy preservation in AI-powered digital communication systems. *Jurnal Communio: Jurnal Jurusan Ilmu Komunikasi*, *13*(2), 349–359. DOI: 10.35508/jikom.v13i2.9420

Gholami, S., & Omar, M. (2023). Does Synthetic Data Make Large Language Models More Efficient? *arXiv preprint arXiv:2310.07830*.

Gholami, S., & Omar, M. (2024). Can a student large language model perform as well as its teacher? In *Innovations, Securities, and Case Studies Across Healthcare, Business, and Technology* (pp. 122-139). IGI Global. DOI: 10.4018/979-8-3693-1906-2.ch007

Gioia, G. A., Isquith, P. K., Guy, S. C., & Kenworthy, L. (2000). *Behavior Rating Inventory of Executive Function*. Psychological Assessment Resources.

Goodfellow, I., Bengio, Y., & Courville, A. (2016). *Deep learning*. MIT Press.

Google Cloud. (2023). *Security Command Center: Real-time threat intelligence for cloud security*. https://cloud.google.com/security-command-center/docs

Gopireddy, R. R. (2020). Dark Web Monitoring: Extracting and Analyzing Threat Intelligence.

Gopireddy, R. R. (2021). AI-Powered Security in cloud environments: Enhancing data protection and threat detection. [IJSR]. *International Journal of Scientific Research*, *10*(11).

Granjeiro, J. M., Cury, A. A. D. B., Cury, J. A., Bueno, M., Sousa-Neto, M. D., & Estrela, C. (2025). The Future of Scientific Writing: AI Tools, Benefits, and Ethical Implications. *Brazilian Dental Journal*, *36*, e25–e6471. DOI: 10.1590/0103-644020256471 PMID: 40197923

Griffin, A. C., Wang, K. H., Leung, T. I., & Facelli, J. C. (2024). Recommendations to promote fairness and inclusion in biomedical AI research and clinical use. *Journal of Biomedical Informatics*, *157*, 104693. DOI: 10.1016/j.jbi.2024.104693 PMID: 39019301

Grossberg, S. (2021). Conscious MIND resonates with attentive ART: Toward biologically plausible machine learning. *Neural Computation*, *33*(10), 2583–2678. DOI: 10.1162/neco_a_01417

Gross, J. J., & John, O. P. (2003). Individual differences in two emotion regulation processes: Implications for affect, relationships, and well-being. *Journal of Personality and Social Psychology*, *85*(2), 348–362. DOI: 10.1037/0022-3514.85.2.348 PMID: 12916575

Grote, T., & Keeling, G. (2022). Enabling fairness in healthcare through machine learning. *Ethics and Information Technology*, *24*(3), 39. DOI: 10.1007/s10676-022-09658-7 PMID: 36060496

Grote, T., & Keeling, G. (2022). On algorithmic fairness in medical practice. *Cambridge Quarterly of Healthcare Ethics*, *31*(1), 83–94. DOI: 10.1017/S0963180121000839 PMID: 35049447

Groumpos, P. P. (2023, July). A critical historic overview of artificial intelligence: Issues, challenges, opportunities, and threats. *Artificial Intelligence and Applications (Commerce, Calif.)*, *1*(4), 181–197. DOI: 10.47852/bonviewAIA3202689

Gudmundsson, E., & Lönner, V. J. (2009). Cross-cultural adaptation of psychological scales. In Gerstein, L. H., Heppner, P. P., Ægisdóttir, S., Leung, S.-M. A., & Norsworthy, K. L. (Eds.), *International handbook of cross-cultural counseling: Cultural assumptions and practices worldwide* (pp. 123–141). SAGE Publications, Inc., DOI: 10.4135/9781483328914.n8

Guleria, A., Krishan, K., Sharma, V., & Kanchan, T. (2023). ChatGPT: Ethical concerns and challenges in academics and research. *Journal of Infection in Developing Countries*, 17(09), 1292–1299. DOI: 10.3855/jidc.18738 PMID: 37824352

Gunawan, Y., Aulawi, M. H., Anggriawan, R., & Putro, T. A. (2022). Command responsibility of autonomous weapons under international humanitarian law. *Cogent Social Sciences*, 8(1), 2139906. DOI: 10.1080/23311886.2022.2139906

Gupta, A., Amarnani, M., Soanki, S., & Kishore, J. (2025, February). AI and Data Privacy in Business. In *2025 First International Conference on Advances in Computer Science, Electrical, Electronics, and Communication Technologies (CE2CT)* (pp. 109-114). IEEE.

Gupta, D. K., Pagani, A., Zamboni, P., & Singh, A. K. (2024). AI-powered revolution in plant sciences: Advancements, applications, and challenges for sustainable agriculture and food security. *Exploration of Foods and Foodomics*, 2(5), 443–459. DOI: 10.37349/eff.2024.00045

Gupta, M., Akiri, C., Aryal, K., Parker, E., & Praharaj, L. (2023). From ChatGPT to ThreatGPT: Impact of generative AI in cybersecurity and privacy. *IEEE Access : Practical Innovations, Open Solutions*, 11, 80218–80245. DOI: 10.1109/ACCESS.2023.3300381

Gurucul. (2020, May 20). *ABCs of UEBA: P is for PRIVILEGE.* https://gurucul .com/blog/abcs- of-ueba-p-is-for-privilege/

Gurucul. (2021). *Top UEBA Use Cases to Fuel Modern, Next-Gen Security Operations.* https://gurucul.com/blog/top-ueba-use-cases

Gurucul. (2021, August 2). *ABCs of UEBA: X is for eXfiltration.* https://gurucul .com/blog/abcs- of-ueba-x-is-for-exfiltration/

Guttentag, D. (2015). Airbnb: Disruptive innovation and the rise of an informal tourism accommodation sector. *Current Issues in Tourism*, 18(12), 1192–1217. DOI: 10.1080/13683500.2013.827159

Hagemann, T., & Katsarou, K. (2020, December). A systematic review on anomaly detection for cloud computing environments. In *Proceedings of the 2020 3rd Artificial Intelligence and Cloud Computing Conference* (pp. 83-96). DOI: 10.1145/3442536.3442550

Hamza, R. A. E. M., Ahmed, N. H., Mohamed, A. M. E., Bennaceur, M. Y., Elhefni, A. H. M., & Elshaabany, M. M. (2024). The impact of artificial intelligence (AI) on the accounting system of Saudi companies. *WSEAS Transactions on Business and Economics*, *21*(January), 499–511. DOI: 10.37394/23207.2024.21.42

Hamza, Y. A., & Omar, M. D. (2013). Cloud computing security: Abuse and nefarious use of cloud computing. *International Journal of Computer Engineering Research*, *3*(6), 22–27.

Haney, B. (2020). *Applied natural language processing for law practice*. Brian S. Haney, Applied Natural Language Processing for Law Practice.

Harati, K. (2024). ChatGPT and AI-Powered Writing Tools: Unveiling Risks and Ethical Challenges in Scientific Writing. *Journal of Research in Medical Sciences : the Official Journal of Isfahan University of Medical Sciences*, *4*(1), 1–6.

Harper, D. J., Ellis, D., & Tucker, I. (2022). Covert aspects of surveillance and the ethical issues they raise. *Ethical Issues in Covert. Security and Surveillance Research Advances in Research Ethics and Integrity*, *35*(2), 177–197.

Hart, S. G., & Staveland, L. E. (1988). Development of NASA-TLX (Task Load Index): Results of empirical and theoretical research. In P. A. Hancock & N. Meshkati (Eds.), Human mental workload (pp. 139–183). North-Holland. Inc., T. (2021). FastAPI: A modern, fast web framework (Version 0.70.0) [Computer software]. GitHub. https://github.com/tiangolo/fastapi

Harwich, E., & Laycock, K. (2018). *Thinking on its own: AI in the NHS*. Reform Research Trust.

He, H., Gray, J., Cangelosi, A., Meng, Q., McGinnity, T. M., & Mehnen, J. (2021). The challenges and opportunities of human-centered AI for trustworthy robots and autonomous systems. *IEEE Transactions on Cognitive and Developmental Systems*, *14*(4), 1398–1412. DOI: 10.1109/TCDS.2021.3132282

Holzinger, A., Weippl, E., Tjoa, A. M., & Kieseberg, P. (2021, August). Digital transformation for sustainable development goals (sdgs)-a security, safety and privacy perspective on ai. In *International cross-domain conference for machine learning and knowledge extraction* (pp. 1-20). Cham: Springer International Publishing.

Horev, R. (2024, February 13). *Multi-cloud security challenges: A best practice guide*. Vulcan. https://vulcan.io/blog/multi-cloud-security-challenges-a-best-practice-guide/

Horev, R. (2024). Advanced DDoS mitigation strategies for multi-cloud environments. *Network Security*, *2024*(2), 8–13.

Horev, R. (2024). Supply Chain Vulnerabilities in Multi-Cloud. *Cybersecurity & Digital Trust*, *11*(2), 18–35.

Horgan, D., Romao, M., Morré, S. A., & Kalra, D. (2020). Artificial intelligence: Power for civilisation–and for better healthcare. *Public Health Genomics*, *22*(5-6), 145–161. DOI: 10.1159/000504785 PMID: 31838476

Hosseini, M. M., Hosseini, S. T. M., Qayumi, K., Ahmady, S., & Koohestani, H. R. (2023). The aspects of running artificial intelligence in emergency care; a scoping review. *Archives of Academic Emergency Medicine*, *11*(1), e38. PMID: 37215232

Huang, J., Galal, G., Etemadi, M., & Vaidyanathan, M. (2022). Evaluation and mitigation of racial bias in clinical machine learning models: Scoping review. *JMIR Medical Informatics*, *10*(5), e36388. DOI: 10.2196/36388 PMID: 35639450

Huff, A. J., Burrell, D. N., Nobles, C., Richardson, K., Wright, J. B., Burton, S. L., Jones, A. J., Springs, D., Omar, M., & Brown-Jackson, K. L. (2023). Management Practices for Mitigating Cybersecurity Threats to Biotechnology Companies, Laboratories, and Healthcare Research Organizations. In *Applied Research Approaches to Technology, Healthcare, and Business* (pp. 1-12). IGI Global.

Hummel, P., Braun, M., Tretter, M., & Dabrock, P. (2021). Data sovereignty: A review. *Big Data & Society*, *8*(1), 2053951720982012. DOI: 10.1177/2053951720982012

IBM. (2022, August 10). *What is user and entity behavior analytics (UEBA)?* IBM. h ttps://www.ibm.com/think/topics/ueba

IBM. (n.d.). *IBM Cloud Pak for Security*. cloud.ibm.com/catalog/content/ibm-cp-security%3A%3A1-b25bd169-0fbd-4cf3-a8ea-

Ibrahim, A. S., Hamlyn-Harris, J., & Grundy, J. (2016). Emerging security challenges of cloud virtual infrastructure. *arXiv preprint arXiv:1612.09059*.

Ismail, I. A., & Aloshi, J. M. R. (2025). Data Privacy in AI-Driven Education: An In-Depth Exploration Into the Data Privacy Concerns and Potential Solutions. In *AI Applications and Strategies in Teacher Education* (pp. 223-252). IGI Global.

Jain, A., Brooks, J. R., Alford, C. C., Chang, C. S., Mueller, N. M., Umscheid, C. A., & Bierman, A. S. (2023, June). Awareness of racial and ethnic bias and potential solutions to address bias with use of health care algorithms. []. American Medical Association.]. *JAMA Health Forum*, *4*(6), e231197–e231197. DOI: 10.1001/jama-healthforum.2023.1197 PMID: 37266959

Jarrahi, M. H., Lutz, C., & Newlands, G. (2022). Artificial intelligence, human intelligence and hybrid intelligence based on mutual augmentation. *Big Data & Society*, *9*(2), 20539517221142824. Advance online publication. DOI: 10.1177/20539517221142824

Jerhamre, E., Carlberg, C. J. C., & van Zoest, V. (2022). Exploring the susceptibility of smart farming: Identified opportunities and challenges. *Smart Agricultural Technology*, *2*, 2. DOI: 10.1016/j.atech.2021.100026

Jimmy, F. N. U. (2023). Cloud security posture management: tools and techniques. Journal of Knowledge Learning and Science Technology ISSN: 2959-6386 (online), 2(3).

Jobin, A., Ienca, M., & Vayena, E. (2019). The global landscape of AI ethics guidelines. *Nature Machine Intelligence*, *1*(9), 389–399. DOI: 10.1038/s42256-019-0088-2

Jones, R., Omar, M., Mohammed, D., Nobles, C., & Dawson, M. (2023). Harnessing the Speed and Accuracy of Machine Learning to Advance Cybersecurity. In *2023 Congress in Computer Science, Computer Engineering, & Applied Computing (CSCE)* (pp. 418-421). IEEE. DOI: 10.1109/CSCE60160.2023.00074

Jones, J., Smith, B., Micheal, O., Barnes, M., & Adebayo, H. (2025). Revolutionizing Cybersecurity with AI-Driven SIEM: Optimizing Threat Detection in Multi-Cloud. *Environments.*

Jones, R., & Omar, M. (2024). Revolutionizing Cybersecurity: The GPT-2 Enhanced Attack Detection and Defense (GEADD) Method for Zero-Day Threats. *International Journal of Informatics* [INJIISCOM]. *Information System and Computer Engineering*, *5*(2), 178–191.

Kadar, A. (2023, December 20). *Enhancing cloud security: Posture management tools and approaches*. ResearchGate.

Kanyepe, J., Chibaro, M., Morima, M., & Moeti-Lysson, J. (2025). AI-Powered Agricultural Supply Chains: Applications, Challenges, and Opportunities. *Integrating Agriculture, Green Marketing Strategies, and Artificial Intelligence*, 33-64.

Kapoor, P. (2025). Harvesting Ethics: Forging Responsible Paths in AI and ML for the Agri-Food Industry. In *Food and Industry 5.0: Transforming the Food System for a Sustainable Future* (pp. 305–315). Springer Nature Switzerland. DOI: 10.1007/978-3-031-76758-6_19

Karim, R., Galar, D., & Kumar, U. (2023). *AI factory: theories, applications and case studies*. CRC Press. DOI: 10.1201/9781003208686

Kaushik, K., Khan, A., Kumari, A., Sharma, I., & Dubey, R. (2024). Ethical considerations in AI-based cybersecurity. In *Next-generation cybersecurity: AI, ML, and Blockchain* (pp. 437–470). Springer Nature Singapore. DOI: 10.1007/978-981-97-1249-6_19

Kendall, G., & Teixeira da Silva, J. A. (2024). Risks of abuse of large language models, like ChatGPT, in scientific publishing: Authorship, predatory publishing, and paper mills. *Learned Publishing*, *37*(1), 55–62. DOI: 10.1002/leap.1578

Kerasidou, A. (2021). Ethics of artificial intelligence in global health: Explainability, algorithmic bias and trust. *Journal of Oral Biology and Craniofacial Research*, *11*(4), 612–614. DOI: 10.1016/j.jobcr.2021.09.004 PMID: 34567966

Khalifa, M., & Albadawy, M. (2024). Using artificial intelligence in academic writing and research: An essential productivity tool. *Computer Methods and Programs in Biomedicine Update*, *5*, 100145. DOI: 10.1016/j.cmpbup.2024.100145

Khaliq, S., Tariq, Z. U. A., & Masood, A. (2020, October). Role of user and entity behavior analytics in detecting insider attacks. In *2020 International Conference on Cyber Warfare and Security (ICCWS)* (pp. 1-6). IEEE.

Khup, V. K., & Bantugan, B. (2025). Exploring the Impact and Ethical Implications of Integrating AI-Powered Writing Tools in Junior High School English Instruction: Enhancing Creativity, Proficiency, and Academic Outcomes. *International Journal of Research and Innovation in Social Science*, *9*(IIIS, 3s), 361–378. DOI: 10.47772/IJRISS.2025.903SEDU0022

Kim, H., Kim, J., Kim, Y., Kim, I., & Kim, K. J. (2019). Design of network threat detection and classification based on machine learning on cloud computing. *Cluster Computing*, *22*(S1), 2341–2350. DOI: 10.1007/s10586-018-1841-8

Kindervag, J., & Balaouras, S. (2010). No more chewy centers: Introducing the zero trust model of information security. *Forestry Research*, *3*, 56682.

Kirov, B. (2023, September). Artificial Intelligence in Creation of Scientific Written Works: Weighing the Benefits and Ethical Dilemmas-Should We Use It? In *2023 International Scientific Conference on Computer Science (COMSCI)* (pp. 1-5). IEEE. DOI: 10.1109/COMSCI59259.2023.10315821

Kochupillai, M., & Köninger, J. (2023). Creating a digital marketplace for agrobiodiversity and plant genetic sequence data: Legal and ethical considerations of an ai and blockchain based solution. *Towards Responsible Plant Data Linkage: Data Challenges for Agricultural Research and Development*, 223.

Kochupillai, M., Kahl, M., Schmitt, M., Taubenböck, H., & Zhu, X. X. (2022). Earth observation and artificial intelligence: Understanding emerging ethical issues and opportunities. *IEEE Geoscience and Remote Sensing Magazine*, *10*(4), 90–124. DOI: 10.1109/MGRS.2022.3208357

Kosasih, E. E., Papadakis, E., Baryannis, G., & Brintrup, A. (2024). A review of explainable artificial intelligence in supply chain management using neurosymbolic approaches. *International Journal of Production Research*, *62*(4), 1510–1540. DOI: 10.1080/00207543.2023.2281663

Kumar, R. S., Lokeshwari, J., & Shanmugam, S. K. (2024, November). AI-Powered Privacy Preservation: A Novel Framework for Adaptive Data Protection. In *2024 2nd International Conference on Computing and Data Analytics (ICCDA)* (pp. 1-6). IEEE.

Kumari, S. (2022). Cybersecurity in Digital Transformation: Using AI to Automate Threat Detection and Response in Multi-Cloud Infrastructures. *Journal of Computational Intelligence and Robotics*, *2*(2), 9–27.

Kumari, S. (2022). Machine learning approaches for threat detection in cloud computing: A systematic review. *Information Sciences*, *580*, 340–366.

Kummarapurugu, R. (2024). An architectural framework for threat intelligence integration in hybrid cloud using SIEM and SOAR. [IJIRCT]. *International Journal of Innovative Research in Computer Technology*, *10*(1), 133–138. https://www.ijirct.org/download.php? a_pid=2411031

KuppingerCole Analysts. (2024). *Leadership Compass: eXtended Detection and Response (XDR)*. Retrieved from https://www.kuppingercole.com/research/lc80923/extended- detection-and-response-xdr

LangChain. (2024). LangChain documentation. Retrieved April 29, 2025, from https://langchain.com/docs

Leal, M. M., & Musgrave, P. (2023). Backwards from zero: How the US public evaluates the use of zero-day vulnerabilities in cybersecurity. *Contemporary Security Policy*, *44*(3), 437–461. DOI: 10.1080/13523260.2023.2216112

Leng, J., Zhang, H., Yan, D., Liu, Q., Chen, X., & Zhang, D. (2022). Digital twin-driven manufacturing cyber-physical system: A survey. *Journal of Manufacturing Systems*, *62*, 493–512. DOI: 10.1016/j.jmsy.2021.12.012

Leong, Y. M., Lim, E. H., Subri, N. F. B., & Jalil, N. B. A. (2023, September). Transforming agriculture: Navigating the challenges and embracing the opportunities of artificial intelligence of things. In *2023 IEEE International Conference on Agrosystem Engineering, Technology & Applications (AGRETA)* (pp. 142-147). IEEE. DOI: 10.1109/AGRETA57740.2023.10262747

Leontidis, G. (2024). Science in the age of ai: How artificial intelligence is changing the nature and method of scientific research.

Li, J., Zong, H., Wu, E., Wu, R., Peng, Z., Zhao, J., Yang, L., Xie, H., & Shen, B. (2024). Exploring the potential of artificial intelligence to enhance the writing of english academic papers by non-native english-speaking medical students-the educational application of ChatGPT. *BMC Medical Education*, *24*(1), 736. DOI: 10.1186/s12909-024-05738-y PMID: 38982429

Ling, Y., Guo, X., Luo, X., & Liu, C. (2022). Development of a computerized adaptive test for problematic mobile phone use. *Frontiers in Psychology*, *13*, 837618. DOI: 10.3389/fpsyg.2022.837618 PMID: 35712155

Liu, M., Ning, Y., Teixayavong, S., Mertens, M., Xu, J., Ting, D. S. W., Cheng, L. T.-E., Ong, J. C. L., Teo, Z. L., Tan, T. F., RaviChandran, N., Wang, F., Celi, L. A., Ong, M. E. H., & Liu, N. (2023). A translational perspective towards clinical AI fairness. *NPJ Digital Medicine*, *6*(1), 172. DOI: 10.1038/s41746-023-00918-4 PMID: 37709945

Liu, Y., Kong, W., & Merve, K. (2025). ChatGPT applications in academic writing: A review of potential, limitations, and ethical challenges. *Arquivos Brasileiros de Oftalmologia*, *88*(3), e2024–e0269. DOI: 10.5935/0004-2749.2024-0269 PMID: 39879415

Li, Y., Su, H., Shen, X., Li, W., Cao, Z., & Niu, S. (2017). *Long Papers* (Vol. 1). DailyDialog: A manually labelled multi-turn dialogue dataset. Proceedings of the Eighth International Joint Conference on Natural Language Processing. Asian Federation of Natural Language Processing., https://aclanthology.org/I17-1099/

Loftus, T. J., Tighe, P. J., Ozrazgat-Baslanti, T., Davis, J. P., Ruppert, M. M., Ren, Y., Shickel, B., Kamaleswaran, R., Hogan, W. R., Moorman, J. R., Upchurch, G. R., Rashidi, P., & Bihorac, A. (2022). Ideal algorithms in healthcare: Explainable, dynamic, precise, autonomous, fair, and reproducible. *PLOS Digital Health*, *1*(1), e0000006. DOI: 10.1371/journal.pdig.0000006 PMID: 36532301

Lund, B. D., Wang, T., Mannuru, N. R., Nie, B., Shimray, S., & Wang, Z. (2023). ChatGPT and a new academic reality: Artificial Intelligence-written research papers and the ethics of the large language models in scholarly publishing. *Journal of the Association for Information Science and Technology*, *74*(5), 570–581. DOI: 10.1002/asi.24750

Lundberg, S. M., & Lee, S.-I. (2017). A unified approach to interpreting model predictions. Advances in Neural Information Processing Systems, 30. https://proceedings.neurips.cc/paper/2017/hash/8a20a8621978632d76c43dfd28b67767-Abstract.html

Maass, W. (1997). Networks of spiking neurons: The third generation of neural network models. *Neural Networks*, *10*(9), 1659–1671. DOI: 10.1016/S0893-6080(97)00011-7

Madasu, S. (2023). Access control models and technologies for big data processing and m anagement. *European Chemical Bulletin*, *12*, 6886–6902.

Mahmood, F. (2021). Algorithm Fairness in AI for Medicine and Healthcare. *arXiv preprint arXiv:2110.00603*.

Mahoney, C. W. (2017). Buyer beware: How market structure affects contracting and company performance in the private military industry. *Security Studies*, *26*(1), 30–59. DOI: 10.1080/09636412.2017.1243912

Malhotra, K., & Firdaus, M. (2022). Application of artificial intelligence in IoT security for crop yield prediction. *ResearchBerg Review of Science and Technology*, *2*(1), 136–157.

Mandvikar, S., & Dave, D. M. (2023). Augmented intelligence: Human–AI collaboration in the era of digital transformation. *International Journal of Engineering Applied Sciences and Technology*, *8*(6), 24–33. DOI: 10.33564/IJEAST.2023.v08i06.003

Manzoor, S., Gouglidis, A., Bradbury, M., & Suri, N. (2022). ThreatPro: Multi-Layer Threat Analysis in the Cloud. *arXiv preprint arXiv:2209.14795*.

Manzoor, S., Gouglidis, A., Bradbury, M., & Suri, N. (2022). *ThreatPro: Multi-Layer Threat Analysis in the Cloud*. arXiv preprint, arXiv:2209.14795.

Manzoor, A., Hussain, M., & Mehrban, S. (2022). Security and privacy issues in cloud c omputing: A comprehensive survey. *Journal of Systems Architecture*, *128*, 102533.

Mark, R. (2019). Ethics of using AI and big data in agriculture: The case of a large agriculture multinational. *The ORBIT Journal*, *2*(2), 1–27. DOI: 10.29297/orbit. v2i2.109

Mathew, A. (2024, November 18). *Cloud data sovereignty: Governance and risk implications of cross-border cloud storage.* ISACA. https://www.isaca.org/resources/ news-and-trends/industry-news/2024/cloud-data-sovereignty-governance-and-risk-implications-of- cross-border-cloud-storage

Mbah, G. O. (2024). Smart Contracts, Artificial Intelligence and Intellectual Property: Transforming Licensing Agreements in the Tech Industry.

Mbah, G. O., & Evelyn, A. N. (2024). AI-powered cybersecurity: Strategic approaches to mitigate risk and safeguard data privacy.

McAfee. (2023). *MVISION Cloud: Cloud access security broker (CASB).* https:// www.mcafee.com/enterprise/en-us/products/mvision-cloud.html

McCradden, M. D., Joshi, S., Mazwi, M., & Anderson, J. A. (2020). Ethical limitations of algorithmic fairness solutions in health care machine learning. *The Lancet. Digital Health*, *2*(5), e221–e223. DOI: 10.1016/S2589-7500(20)30065-0 PMID: 33328054

Mhasawade, V., Zhao, Y., & Chunara, R. (2021). Machine learning and algorithmic fairness in public and population health. *Nature Machine Intelligence*, *3*(8), 659–666. DOI: 10.1038/s42256-021-00373-4

Miao, J., Thongprayoon, C., Suppadungsuk, S., Garcia Valencia, O. A., Qureshi, F., & Cheungpasitporn, W. (2023). Ethical dilemmas in using AI for academic writing and an example framework for peer review in nephrology academia: A narrative review. *Clinics and Practice*, *14*(1), 89–105. DOI: 10.3390/clinpract14010008 PMID: 38248432

Microsoft. (2021). *Microsoft Security Report: Exchange Attacks & Mitigation Strategies.* https://www.microsoft.com/security

Microsoft. (2024). *Microsoft Defender for Cloud overview.* https://learn.microsoft .com/en- us/azure/defender-for-cloud/defender-for-cloud-introduction

Mienye, I. D., Obaido, G., Emmanuel, I. D., & Ajani, A. A. (2024, June). A survey of bias and fairness in healthcare AI. In *2024 IEEE 12th International Conference on Healthcare Informatics (ICHI)* (pp. 642-650). IEEE. DOI: 10.1109/ ICHI61247.2024.00103

Miller, E. K., & Cohen, J. D. (2001). An integrative theory of prefrontal cortex function. *Annual Review of Neuroscience*, *24*(1), 167–200. DOI: 10.1146/annurev. neuro.24.1.167 PMID: 11283309

Mohammed, D., Omar, M., & Nguyen, V. (2018). Wireless sensor network security: Approaches to detecting and avoiding wormhole attacks. *Journal of Research in Business. Economics and Management*, *10*(2), 1860–1864.

Mohan, T., Kumar, V., & Sharma, R. (2022). *Advancements in cloud security monitoring: Analyzing the efficiency of AWS, Azure, and Google Cloud security frameworks*. I nternational Journal of Information Security and Cyber Forensics, 11(2), 88-104.

Molligan, J., & Pérez-López, E. (2024). Artificial intelligence in academia: Opportunities, challenges, and ethical considerations. *Biochemistry and Cell Biology*, *103*, 1–3. DOI: 10.1139/bcb-2024-0216 PMID: 39611424

Moosavi-Dezfooli, S.-M., Fawzi, A., & Frossard, P. (2016). DeepFool: A simple and accurate method to fool deep neural networks. *2016 IEEE Conference on Computer Vision and Pattern Recognition (CVPR)* (pp. 2574–2582). IEEE. DOI: 10.1109/ CVPR.2016.282

Morrow, T. (2018, March 5). *12 risks, threats, and vulnerabilities in moving to the cloud*. S oftware Engineering Institute. https://insights.sei.cmu.edu/blog/12-risks -threats- v ulnerabilities-in-moving-to-the-cloud/

Morrow, T. (2018). Best Practices for Cloud Security. *Journal of IT Security*, *3*(2), 50–65.

Mosqueira-Rey, E., Hernández-Pereira, E., Alonso-Ríos, D., Moret-Bonillo, V., & Fernández-Varela, I. (2023). Human-in-the-loop machine learning: A state of the art. *Artificial Intelligence Review*, *56*(4), 3005–3054. DOI: 10.1007/s10462-022-10246-w

Moussa, S., & Teixeira da Silva, J. A. (2023). Testing the robustness of COPE's characterization of predatory publishing on a COPE member publisher (Academic and Business Research Institute). *Publishing Research Quarterly*, *39*(4), 337–367. DOI: 10.1007/s12109-023-09967-9

Mukherjee, B., Heberlein, L. T., & Levitt, K. N. (1994). Network intrusion detection. *IEEE Network*, *8*(3), 26–41. DOI: 10.1109/65.283931

Nam, B. H., & Bai, Q. (2023). ChatGPT and its ethical implications for STEM research and higher education: A media discourse analysis. *International Journal of STEM Education*, *10*(1), 66. DOI: 10.1186/s40594-023-00452-5

Nassif, A. B., Talib, M. A., Nasir, Q., Albadani, H., & Dakalbab, F. M. (2021). Machine learning for cloud security: A systematic review. *IEEE Access : Practical Innovations, Open Solutions, 9*, 20717–20735. DOI: 10.1109/ACCESS.2021.3054129

National Institute of Standards and Technology (NIST). (2011). *The NIST Definition of Cloud Computing*. Special Publication 800-145.

Nazer, L. H., Zatarah, R., Waldrip, S., Ke, J. X. C., Moukheiber, M., Khanna, A. K., Hicklen, R. S., Moukheiber, L., Moukheiber, D., Ma, H., & Mathur, P. (2023). Bias in artificial intelligence algorithms and recommendations for mitigation. *PLOS Digital Health, 2*(6), e0000278. DOI: 10.1371/journal.pdig.0000278 PMID: 37347721

Netskope. (2023, March 22). *Operationalizing advanced UEBA: Detection scenarios and UCI alerts*.

Neupane, S., Mitra, S., Fernandez, I. A., Saha, S., Mittal, S., Chen, J., Pillai, N., & Rahimi, S. (2024). Security considerations in ai-robotics: A survey of current methods, challenges, and opportunities. *IEEE Access : Practical Innovations, Open Solutions, 12*, 22072–22097. DOI: 10.1109/ACCESS.2024.3363657

Nguyen, V., Mohammed, D., Omar, M., & Banisakher, M. (2018). The Effects of the FCC Net Neutrality Repeal on Security and Privacy. [IJHIoT]. *International Journal of Hyperconnectivity and the Internet of Things, 2*(2), 21–29. DOI: 10.4018/IJHIoT.2018070102

Nicholls, T., & Culpepper, P. D. (2021). Computational identification of media frames: Strengths, weaknesses, and opportunities. *Political Communication, 38*(1-2), 159–181. DOI: 10.1080/10584609.2020.1812777

Nordlinger, B., Villani, C., & Rus, D. (Eds.). (2020). *Healthcare and artificial intelligence*. Springer. DOI: 10.1007/978-3-030-32161-1

Norori, N., Hu, Q., Aellen, F. M., Faraci, F. D., & Tzovara, A. (2021). Addressing bias in big data and AI for health care: A call for open science. *Patterns (New York, N.Y.), 2*(10), 100347. DOI: 10.1016/j.patter.2021.100347 PMID: 34693373

Nriezedi-Anejionu, C. (2024). Carbon reduction and nuclear energy policy U-turn: The necessity for an international treaty on small modular reactors (SMR) new nuclear technology. *Carbon Management, 15*(1), 2396585. DOI: 10.1080/17583004.2024.2396585

Nunkoo, R., Sharma, A., Rana, N. P., Dwivedi, Y. K., & Sunnassee, V. A. (2023). Advancing sustainable development goals through interdisciplinarity in sustainable tourism research. *Journal of Sustainable Tourism, 31*(3), 735–759. DOI: 10.1080/09669582.2021.2004416

Nurchurifiani, E., Maximilian, A., Ajeng, G. D., Wiratno, P., Hastomo, T., & Wicaksono, A. (2025). Leveraging AI-Powered Tools in Academic Writing and Research: Insights from English Faculty Members in Indonesia. *International Journal of Information and Education Technology (IJIET)*, *15*(2), 312–322. DOI: 10.18178/ijiet.2025.15.2.2244

Ocampo, T. S. C., Silva, T. P., Alencar-Palha, C., Haiter-Neto, F., & Oliveira, M. L. (2023). ChatGPT and scientific writing: A reflection on the ethical boundaries. *Imaging Science in Dentistry*, *53*(2), 175. DOI: 10.5624/isd.20230085 PMID: 37405199

Olaniyan, R., Rakshit, S., & Vajjhala, N. R. (2023). Application of user and entity behavioral analytics (UEBA) in the detection of cyber threats and vulnerabilities management. In *Computational Intelligence for Engineering and Management Applications: Select Proceedings of CIEMA 2022* (pp. 419-426). Singapore: Springer Nature Singapore.

Olaoye, A. O. (2022). *Multi-cloud architecture for cloud computing*. Iowa State University. https://dr.lib.iastate.edu/server/api/core/bitstreams/48ab713c-0f6c-472a -9564- 9 d6bee1b394b/content

Omar, M. (2021). New insights into database security: An effective and integrated approach for applying access control mechanisms and cryptographic concepts in Microsoft Access environments.

Omar, M. (2022). *Machine Learning for Cybersecurity: Innovative Deep Learning Solutions*. Springer Brief. https://link.springer.com/book/978303115

Omar, M. (2024). From Attack to Defense: Strengthening DNN Text Classification Against Adversarial Examples. In *Innovations, Securities, and Case Studies Across Healthcare, Business, and Technology* (pp. 174-195). IGI Global.

Omar, M. (2024). Revolutionizing Malware Detection: A Paradigm Shift Through Optimized Convolutional Neural Networks. In *Innovations, Securities, and Case Studies Across Healthcare, Business, and Technology* (pp. 196-220). IGI Global. DOI: 10.4018/979-8-3693-1906-2.ch011

Omar, M., Zangana, H. M., Al-Karaki, J. N., & Mohammed, D. (2024). Harnessing LLMs for IoT malware detection: A comparative analysis of BERT and GPT-2. In *2024 8th International Symposium on Multidisciplinary Studies and Innovative Technologies (ISMSIT)* (pp. 1-6). Ankara, Turkiye. https://doi.org/DOI: 10.1109/ISMSIT63511.2024.10757249

Omar, M., & Zangana, H. (Eds.). (2025). *Digital Forensics in the Age of AI*. IGI Global., DOI: 10.4018/979-8-3373-0857-9

Omar, M., & Zangana, H. M. (Eds.). (2024). *Redefining Security With Cyber AI*. IGI Global., DOI: 10.4018/979-8-3693-6517-5

Omar, M., & Zangana, H. M. (Eds.). (2025). *Application of Large Language Models (LLMs) for Software Vulnerability Detection*. IGI Global., DOI: 10.4018/979-8-3693-9311-6

Omar, M., Zangana, H. M., & Mohammed, D. (Eds.). (2025). *Integrating Artificial Intelligence in Cybersecurity and Forensic Practices*. IGI Global., DOI: 10.4018/979-8-3373-0588-2

Omodan, B. I., & Marongwe, N. (2024). The role of artificial intelligence in de-colonising academic writing for inclusive knowledge production. *Interdisciplinary Journal of Education Research*, 6(s1), 1–14. DOI: 10.38140/ijer-2024.vol6.s1.06

Open, A. I. (2024). Fine-tuning. OpenAI API documentation. Retrieved April 29, 2025, from https://platform.openai.com/docs/guides/fine-tuning

Otieno, M. (2023). An extensive survey of smart agriculture technologies: Current security posture. *World J. Adv. Res. Rev*, *18*(3), 1207–1231. DOI: 10.30574/wjarr.2023.18.3.1241

Pakarinen, A. (2025). Consent, control and compliance: legal perspectives on processing health data in the development of AI through anonymization and pseudonymization under the GDPR.

Palo Alto Networks. (2024). *Prisma Cloud: At a Glance*. Retrieved from h t t ps:// www.paloaltonetworks.com/resources/datasheets/prisma-cloud-at-a-glance

Palo Alto Networks. (2024). *What Is Extended Detection and Response (XDR)?*h ttps://www.paloaltonetworks.com/cyberpedia/what-is-extended-detection-response - XDR

Pandey, A. S., Sharma, Y., Tiwari, A., Chauhan, R., Tyagi, S., & Kumari, J. (2024, May). Ethical Implications of AI-Powered Communication Tool. In *2024 International Conference on Communication, Computer Sciences and Engineering (IC3SE)* (pp. 1857-1861). IEEE. DOI: 10.1109/IC3SE62002.2024.10593350

Pandey, D. K., & Mishra, R. (2024). *Towards sustainable agriculture: Harnessing AI for global food security*. Artificial Intelligence in Agriculture.

Patel, R., & Desai, A. (2024). Exploring the Ethical Implications of AI-Powered Security Systems. *Asian American Research Letters Journal*, *1*(9), 87–95.

Patel, S., Gupta, R., & Ahmed, Z. (2023). A review of cloud security tools: Native vs third-party solutions. *Cybersecurity and Cloud Computing Journal*, *17*(1), 121–136.

Patton, J. H., Stanford, M. S., & Barratt, E. S. (1995). Factor structure of the Barratt Impulsiveness Scale. *Journal of Clinical Psychology*, *51*(6), 768–774. DOI: 10.1002/1097-4679(199511)51:6<768::AID-JCLP2270510607>3.0.CO;2-1 PMID: 8778124

Paul, A. (2024). *Security challenges and solutions in multi-cloud environments*. ResearchGate.

Pedro, F., Subosa, M., Rivas, A., & Valverde, P. (2019). Artificial intelligence in education: Challenges and opportunities for sustainable development.

Pellecchia, R. (2022). Leveraging AI via speech-to-text and LLM integration for improved healthcare decision-making in primary care.

Perkins, M. (2023). Academic Integrity considerations of AI Large Language Models in the post-pandemic era: ChatGPT and beyond. *Journal of University Teaching & Learning Practice*, *20*(2), 1–24. DOI: 10.53761/1.20.02.07

Piskorz, P. (n.d.). *Top 5 challenges of protecting multi-cloud environments*. Storware. h ttps://storware.eu/blog/top-5-challenges-of-protecting-multi-cloud-environment/

Pöhn, D., & Hommel, W. (2023). Towards an improved taxonomy of attacks related to digital identities and identity management systems. *Security and Communication Networks*, *2023*(1), 5573310. DOI: 10.1155/2023/5573310

Prearo, M., & Scopelliti, A. (2025). Do LGBTIQ+ issues matter for populist radical right? An analysis of Italian parties' social media narratives. *South European Society & Politics*, ●●●, 1–22. DOI: 10.1080/13608746.2025.2463915

Proofpoint. (n.d.). *Identity Threat Detection and Response (ITDR)*. www.proofpoint .com/us/threat-reference/identity-threat-detection-and-response-itdr

Quartz, S. R., & Sejnowski, T. J. (1997). The neural basis of cognitive development: A constructivist manifesto. *Behavioral and Brain Sciences*, *20*(4), 537–556. DOI: 10.1017/S0140525X97001581 PMID: 10097006

Quist, N. (2023, May 15). *10 cloud security risks*. Palo Alto Networks. h t t ps:// www.paloaltonetworks.com/blog/prisma-cloud/10-cloud-security-risks/

Quist, A. (2023). Insider Threats in Multi-Cloud: Risk Analysis & Prevention Strategies. *Journal of Cloud Computing Security*, *6*(3), 55–72.

Qureshi, S., & Oladokun, B. (2024). Human Freedom from Algorithmic Bias: What is the role of Accountability in addressing Health Disparities? *Medical Research Archives*, *12*(8). Advance online publication. DOI: 10.18103/mra.v12i8.5635

Rajkomar, A., Hardt, M., Howell, M. D., Corrado, G., & Chin, M. H. (2018). Ensuring fairness in machine learning to advance health equity. *Annals of Internal Medicine, 169*(12), 866–872. DOI: 10.7326/M18-1990 PMID: 30508424

Rashid, A. B., & Kausik, A. K. (2024). AI revolutionizing industries worldwide: A comprehensive overview of its diverse applications. *Hybrid Advances*, 100277.

Rashkin, H., Smith, E. M., Li, M., & Boureau, Y.-L. (2019). Towards empathetic open-domain conversation models: A new benchmark and dataset. *Proceedings of the 57th Annual Meeting of the Association for Computational Linguistics* (pp. 5370–5384). Association for Computational Linguistics. DOI: 10.18653/v1/P19-1534

Reece, M., Lander, T. E., Jr., Stoffolano, M., Sampson, A., Dykstra, J., Mittal, S., & Rastogi, N. (2023). Systemic risk and vulnerability analysis of multi-cloud environments. *arXiv preprint arXiv:2306.01862.*

Reece, M., Lander, T. E., Jr., Stoffolano, M., Sampson, A., Dykstra, J., Mittal, S., & Rastogi, N. (2023). Systemic Risk and Vulnerability Analysis of Multi-Cloud Environments. *arXiv preprint arXiv:2306.01862.*

Reece, M., Lander, T., Jr., Mittal, S., Rastogi, N., Dykstra, J., & Sampson, A. (2023). *E mergent (In)Security of Multi-Cloud Environments.* arXiv preprint, arXiv:2311.01247.

Reece, M., Lander, T., Jr., Mittal, S., Rastogi, N., Dykstra, J., & Sampson, A. (2023). Emergent (In) Security of Multi-Cloud Environments. *arXiv preprint arXiv:2311.01247.*

Reece, J., Jones, A., & Williams, P. (2023). Comprehensive security assessment frameworks for multi-cloud environments. *IEEE Security and Privacy, 21*(4), 28–36.

Reece, M., Lander, T. E.Jr, Stoffolano, M., Sampson, A., Dykstra, J., Mittal, S., & Rastogi, N. (2023). *Systemic risk and vulnerability analysis of multi-cloud environments* (a rXiv:2306.01862). arXiv. https://arxiv.org/abs/2306.01862

Rehan, H. (2023). AI-Powered Genomic Analysis in the Cloud: Enhancing Precision Medicine and Ensuring Data Security in Biomedical Research. *Journal of Deep Learning in Genomic Data Analysis, 3*(1), 37–71.

Rengarajan, R., & Babu, S. (2021, March). Anomaly detection using user entity behavior a nalytics and data visualization. In *2021 8th International Conference on Computing for Sustainable Global Development (INDIACom)* (pp. 842-847). IEEE.

Rizun, N., Revina, A., & Edelmann, N. (2025). Text analytics for co-creation in public sector organizations: A literature review-based research framework. *Artificial Intelligence Review, 58*(4), 1–45. DOI: 10.1007/s10462-025-11112-1

Rodigari, S., O'Shea, D., McCarthy, P., McCarry, M., & McSweeney, S. (2021). *Performance Analysis of Zero-Trust Multi-Cloud*. arXiv. https://arxiv.org/abs/2105 .02334 DOI: 10.1109/CLOUD53861.2021.00097

Rose, S., Borchert, O., Mitchell, S., & Connelly, S. (2020). Zero trust architecture. *NIST Special Publication 800-207*. National Institute of Standards and Technology.

S&P Global Market Intelligence. (2021). *The Rise of Extended Detection and Response*. https://www.spglobal.com/marketintelligence/en/documents/the-rise-of - extended- detection-and-response.pdf

Salman, T., Bhamare, D., Erbad, A., Jain, R., & Samaka, M. (2017, June). Machine learning for anomaly detection and categorization in multi-cloud environments. In *2017 IEEE 4th international conference on cyber security and cloud computing (CSCloud)* (pp. 97-103). IEEE. DOI: 10.1109/CSCloud.2017.15

Salvagno, M., Taccone, F. S., & Gerli, A. G. (2023). Can artificial intelligence help for scientific writing? *Critical Care*, *27*(1), 75. DOI: 10.1186/s13054-023-04380-2 PMID: 36841840

Samsonovich, A. V. (2010). Toward a unified catalog of implemented cognitive architectures. Biologically Inspired Cognitive Architectures 2010: Proceedings of the First International Conference on Biologically Inspired Cognitive Architectures (pp. 83–95). IOS Press. DOI: 10.3233/978-1-60750-661-4-195

Sap, M., Le Bras, R., Allaway, E., Bhagavatula, C., Lourie, N., Rashkin, H., Roof, B., Smith, N. A., & Choi, Y. (2019). Social IQa: Commonsense reasoning about social interactions. *Proceedings of the 2019 Conference on Empirical Methods in Natural Language Processing and the 9th International Joint Conference on Natural Language Processing (EMNLP-IJCNLP)* (pp. 4463–4473). Association for Computational Linguistics. DOI: 10.18653/v1/D19-1454

Sarferaz, S. (2024). *Embedding Artificial Intelligence into ERP Software*. Springer Nature. DOI: 10.1007/978-3-031-54249-7

Sarkar, S., Choudhary, G., Shandilya, S. K., Hussain, A., & Kim, H. (2022). Security of zero trust networks in cloud computing: A comparative review. *Sustainability (Basel)*, *14*(18), 11213. DOI: 10.3390/su141811213

Sasovets, I. (2023, August 1). *Multi-cloud security: Benefits, challenges, and best practices*. TechMagic. https://www.techmagic.co/blog/multi-cloud-security

Sasovets, I. (2024, October 1). *What is multi-cloud security? Challenges and best practices*. TechMagic. https://www.techmagic.co/blog/multi-cloud-security

Sasovets, I. (2024). The Evolution of Cloud Threats: Lessons from Recent Attacks. *Cyber Risk Review*, *14*(1), 78–93.

Sasovets, Y. (2024). Unified visibility in multi-cloud security: Challenges and solutions. *International Journal of Cloud Applications and Computing*, *14*(1), 1–15.

Sato, H. (2023). AI-Powered Solutions: Ensuring Data Privacy in a Transforming Digital Landscape. *Advances in Computer Sciences, 6*(1).

Saxena, D., Gupta, R., & Singh, A. K. (2021). A Survey and Comparative Study on Multi-Cloud Architectures: Emerging Issues and Challenges for Cloud Federation. *arXiv preprint arXiv:2108.12831*.

Saxena, A., Gupta, S., & Singh, Y. K. (2021). A survey on multi-cloud computing: Benefits and research directions. *Journal of Parallel and Distributed Computing*, *157*, 34–51.

Scarfone, K., & Hoffman, P. Guidelines on firewalls and firewall policy: Recommendations of the National Institute of Standards and Technology. *NIST Special Publication*, 800-41.

Schäfer, M., Fuchs, M., Strohmeier, M., Engel, M., Liechti, M., & Lenders, V. (2019, May). BlackWidow: Monitoring the dark web for cyber security information. In *2019 11th International Conference on Cyber Conflict (CyCon)* (Vol. 900, pp. 1-21). IEEE.

Schwarzer, R., & Jerusalem, M. (1995). Generalized Self-Efficacy scale. In Weinman, J., Wright, S., & Johnston, M. (Eds.), *Measures in health psychology: A user's portfolio. Causal and control beliefs* (pp. 35–37). NFER-NELSON.

Scott, S. G., & Bruce, R. A. (1995). Decision-making style: The development and assessment of a new measure. *Educational and Psychological Measurement*, *55*(5), 818–831. DOI: 10.1177/0013164495055005017

Senoo, E. E. K., Anggraini, L., Kumi, J. A., Luna, B. K., Akansah, E., Sulyman, H. A., & Aritsugi, M. (2024). IoT solutions with artificial intelligence technologies for precision agriculture: Definitions, applications, challenges, and opportunities. *Electronics (Basel)*, *13*(10), 1894. DOI: 10.3390/electronics13101894

SentinelOne. (2024). *Benefits of XDR (Extended Detection and Response)*. h ttps:// www.sentinelone.com/cybersecurity-101/endpoint-security/benefits-of-xdr/

SentinelOne. (2024). *Cyber Threat Report: Multi-Cloud Attacks on the Rise*. Retrieved from https://www.sentinelone.com/research

SentinelOne. (2024, July 31). *Cloud security vulnerabilities*. SentinelOne. h t t ps://www.sentinelone.com/cybersecurity-101/cloud-security/cloud-security- v ulnerabilities/

Shafik, W., Tufail, A., De Silva, C. L., Haji, R. A. A., & Apong, M. (2025). The Role, Application, and Impact of Artificial Intelligence in the Agriculture Industry. In *Future Tech Startups and Innovation in the Age of AI* (pp. 36–60). CRC Press.

Shahzad, F., Mannan, A., Javed, A. R., Almadhor, A. S., Baker, T., & Al-Jumeily, O. B. E. (2022). Cloud-based multiclass anomaly detection and categorization using ensemble learning. *Journal of Cloud Computing (Heidelberg, Germany)*, *11*(1), 74. DOI: 10.1186/s13677-022-00329-y

Sharma, S., & Modi, C. (2021). A review of service mesh in cloud-native applications: A rchitecture and security. *IEEE Access*, 9, 23487-23500. ht t p s ://DOI: 10.1109/ACCESS.2021.3056014

Sharma, A., & Sahay, S. K. (2020). Evolution and adoption of AI in cybersecurity. *Computers & Security*, *98*, 101935.

Sharma, H. (2021). Behavioral Analytics and Zero Trust. *International Journal of Computer Engineering and Technology*, *12*(1), 63–84.

Sharma, R., & Kumar, A. (2023). Threat intelligence and security analytics in multi-cloud environments. *Computers & Security*, *132*, 104317.

Shelke, P., & Hämäläinen, T. (2024). Analysing multidimensional strategies for cyber threat detection in security monitoring. In *Proceedings of the European Conference on Cyber Warfare and Security* (Vol. 23, No. 1). Academic Conferences International Ltd. DOI: 10.34190/eccws.23.1.2123

Shoeybi, M., Patwary, M., Puri, R., LeGresley, P., Casper, J., & Catanzaro, B. (2019). Megatron-LM: Training multi-billion parameter language models using model parallelism. *Proceedings of the International Conference for High Performance Computing, Networking, Storage and Analysis*. DOI: 10.1145/3295500.3356181

Shofiah, N., Putera, Z. F., & Solichah, N. (2023, December). Challenges and opportunities in the use of artificial intelligence in education for academic writing: A scoping review. In *Conference Psychology and Flourishing Humanity (PFH 2023)* (pp. 174-193). Atlantis Press. DOI: 10.2991/978-2-38476-188-3_20

Sikstrom, L., Maslej, M. M., Hui, K., Findlay, Z., Buchman, D. Z., & Hill, S. L. (2022). Conceptualising fairness: Three pillars for medical algorithms and health equity. *BMJ Health & Care Informatics*, *29*(1), e100459. DOI: 10.1136/bmjhci-2021-100459 PMID: 35012941

Simola, J. (2022). *Effects and factors of the hybrid emergency response model in public protection and disaster relief.* JYU Dissertations.

Singhal, A., Neveditsin, N., Tanveer, H., & Mago, V. (2024). Toward fairness, accountability, transparency, and ethics in AI for social media and health care: Scoping review. *JMIR Medical Informatics*, *12*(1), e50048. DOI: 10.2196/50048 PMID: 38568737

Singhal, N., Goyal, S., & Singhal, T. (2024). Decentralized Insurance Platforms: Innovation and Technology for Trust and Efficiency. In *Potential, Risks, and Ethical Implications of Decentralized Insurance* (pp. 95–163). Springer Nature Singapore. DOI: 10.1007/978-981-97-5894-4_3

Sommer, R., & Paxson, V. (2010, May). *Outside the closed world: On using machine learning for network intrusion detection. In 2010 IEEE symposium on security and privacy.* IEEE.

Sood, A., Sharma, R. K., & Bhardwaj, A. K. (2022). Artificial intelligence research in agriculture: A review. *Online Information Review*, *46*(6), 1054–1075. DOI: 10.1108/OIR-10-2020-0448

Sparrow, R., & Howard, M. (2021). Robots in agriculture: prospects, impacts, ethics, and policy. *precision agriculture, 22*, 818-833.

Sparrow, R., Howard, M., & Degeling, C. (2021). Managing the risks of artificial intelligence in agriculture. *NJAS: Impact in Agricultural and Life Sciences*, *93*(1), 172–196. DOI: 10.1080/27685241.2021.2008777

Spiceworks (2025). MultiCloud Infrastructure [Photograph]. What Is Multicloud Infrastructure? https://www.spiceworks.com/tech/cloud/articles/what-is-multicloud -infrastructure/

Spiceworks Editorial Team. (2025). What is multicloud infrastructure? [Image]. Spiceworks. https://www.spiceworks.com/tech/cloud/articles/what-is-multicloud -infrastructure/

Splunk Inc. (2020). *Splunk Security Cloud Product Brief.* www.splunk.com/pdfs/ product-briefs/splunk-security-cloud.pdf

Stallings, W., & Brown, L. (2015). *Computer security: principles and practice.* Pearson.

Stevens, R., Taylor, V., Nichols, J., Maccabe, A. B., Yelick, K., & Brown, D. (2020). *AI for science: Report on the department of energy (doe) town halls on artificial intelligence (ai) for science* (No. ANL-20/17). Argonne National Lab.(ANL), Argonne, IL (United States).

Subaveerapandiyan, A., Kalbande, D., & Ahmad, N. (2025). Perceptions of effectiveness and ethical use of AI tools in academic writing: A study Among PhD scholars in India. *Information Development*, ●●●, 026666669251314840. DOI: 10.1177/02666669251314840

Süß, F., Freimuth, M., Aßmuth, A., Weir, G. R. S., & Duncan, B. (2024). *Cloud Security and Security Challenges Revisited.* arXiv preprint arXiv:2405.11350

Sutton, R. S., & Barto, A. G. (2018). *Reinforcement learning: An introduction* (2nd ed.). MIT Press.

Sysdig. (n.d.). AWS vs. Azure vs. Google Cloud: Security Comparison. https://sysdig.com/learn- cloud-native/threat-detection-in-the-cloud-defender-vs-guardduty-vs-security-command- center/

Sysdig. (n.d.). Cloud-based threat detection: A comparative analysis. Retrieved February 22, 2025

Sysdig. (n.d.). *What is a Cloud Workload Protection Platform (CWPP)?.* sysdig.com/learn-cloud-native/what-is-a-cloud-workload-protection-platform-cwpp/

Taddeo, M., & Floridi, L. (2018). Regulate artificial intelligence to avert cyber arms race. *Nature*, *556*(7701), 296–29. DOI: 10.1038/d41586-018-04602-6 PMID: 29662138

Tai, A. M. Y., Meyer, M., Varidel, M., Prodan, A., Vogel, M., Iorfino, F., & Krausz, R. M. (2023). Exploring the potential and limitations of ChatGPT for academic peer-reviewed writing: Addressing linguistic injustice and ethical concerns. *Journal of Academic Language and Learning*, *17*(1), T16–T30.

Tate, L. (2024b). iLevyTate/scanue-v22: Zenodo Synchronization Second [Computer software]. Zenodo. DOI: 10.5281/zenodo.14510407

Tate, L. (2024a). iLevyTate/SCANUE: 1.0.0-alpha [Computer software]. *Zenodo*. DOI: 10.5281/zenodo.14052759

Tate, L. (2024c). iLevyTate/SCAN-Resources: 1.0.0-alpha [Computer software]. *Zenodo*. DOI: 10.5281/zenodo.14053203

Tate, L. (2024d). iLevyTate/stac: 1.0.2.1-alpha [Computer software]. *Zenodo*. DOI: 10.5281/zenodo.14545341

Tatineni, S. (2023). AI-infused threat detection and incident response in cloud security. *International Journal of Science and Research (IJSR), 12*(11), 998-1004.

Thirunagalingam, A. (2024). AI-Powered Continuous Data Quality Improvement: Techniques, Benefits, and Case Studies. *Benefits, and Case Studies (August 23, 2024)*.

Tilmes, N. (2022). Disability, fairness, and algorithmic bias in AI recruitment. *Ethics and Information Technology, 24*(2), 21. DOI: 10.1007/s10676-022-09633-2

Torkura, K., Sukmana, M. I. H., Cheng, F., & Meinel, C. (2020). *Continuous auditing & threat detection in multi-cloud infrastructure*. https://doi.org/DOI: 10.36227/techrxiv.13108313

Trend Micro. (2024). *Trend Micro Cloud One: Workload security overview*. https://cloudone.trendmicro.com/docs/workload-security/protection-modules/

Tuyishime, E., Balan, T. C., Cotfas, P. A., Cotfas, D. T., & Rekeraho, A. (2023). Enhancing cloud security—Proactive threat monitoring and detection using a siem-based approach. *Applied Sciences (Basel, Switzerland), 13*(22), 12359. DOI: 10.3390/app132212359

Tzachor, A., Devare, M., King, B., Avin, S., & Ó hÉigeartaigh, S. (2022). Responsible artificial intelligence in agriculture requires systemic understanding of risks and externalities. *Nature Machine Intelligence, 4*(2), 104–109. DOI: 10.1038/s42256-022-00440-4

Ugwu, N. F., Igbinlade, A. S., Ochiaka, R. E., Ezeani, U. D., Okorie, N. C., Opele, J. K., Onayinka, T. S., Iroegbu, O., Onyekwere, O. K., Adams, A. B., Aigbona, P., & Ojobola, F. B. (2024). Clarifying Ethical Dilemmas in Using Artificial Intelligence in Research Writing: A Rapid Review. *Higher Learning Research Communications, 14*(2), 29–47. DOI: 10.18870/hlrc.v142.1549

Ulwick, A. W. (2005). *What customers want: Using outcome-driven innovation to create breakthrough products and services*. McGraw-Hill.

Uplaonkar, S. S., Veershetty, R., Bahar, Z., & Sangeeta, G. (2024). The role of productive AI: A supporter or challenger in the future of agricultural librarianship. *IJAR, 10*(4), 46–53.

Vapnik, V. N., & Izmailov, R. (2015). Learning using privileged information: Similarity control and knowledge transfer. *Journal of Machine Learning Research, 16*(69), 2023–2049. http://jmlr.org/papers/v16/vapnik15a.html

Vaswani, A., Shazeer, N., Parmar, N., Uszkoreit, J., Jones, L., Gomez, A. N., Kaiser, L., & Polosukhin, I. (2017). Attention is all you need. Advances in Neural Information Processing Systems, 30. https://proceedings.neurips.cc/paper/2017/hash/3f 5ee243547dee91fbd053c1c4a845aa-Abstract.html

Venkatesh, V., Morris, M. G., Davis, G. B., & Davis, F. D. (2003). User acceptance of information technology: Toward a unified view. *Management Information Systems Quarterly*, *27*(3), 425–478. DOI: 10.2307/30036540

Verdet, A. (2023). *Exploring security practices in infrastructure as code: An empirical study* (Master's thesis, Ecole Polytechnique, Montreal (Canada)).

Visweswaran, S., Luo, Y., & Peleg, M. (2024). Fairness and inclusion methods for biomedical informatics research. *Journal of Biomedical Informatics*, *158*, 104713. DOI: 10.1016/j.jbi.2024.104713 PMID: 39187169

Wang, Y., Song, Y., Ma, Z., & Han, X. (2023). Multidisciplinary considerations of fairness in medical AI: A scoping review. *International Journal of Medical Informatics*, *178*, 105175. DOI: 10.1016/j.ijmedinf.2023.105175 PMID: 37595374

Wei, H., Xu, W., Kang, B., Eisner, R., Muleke, A., Rodriguez, D., deVoil, P., Sadras, V., Monjardino, M., & Harrison, M. T. (2024). Irrigation with artificial intelligence: Problems, premises, promises. *Human-Centric Intelligent Systems*, *4*(2), 187–205. DOI: 10.1007/s44230-024-00072-4

Williamson, H. F., Brettschneider, J., Caccamo, M., Davey, R. P., Goble, C., Kersey, P. J., May, S., Morris, R. J., Ostler, R., Pridmore, T., Rawlings, C., Studholme, D., Tsaftaris, S. A., & Leonelli, S. (2023). Data management challenges for artificial intelligence in plant and agricultural research. *F1000 Research*, *10*, 324. DOI: 10.12688/f1000research.52204.2 PMID: 36873457

Williams, R. (2023). Fair and equitable AI in biomedical research and healthcare: Social science perspecti es. *Artificial Intelligence in Medicine*, *144*, 102658. DOI: 10.1016/j.artmed.2023.102658 PMID: 37783540

Wiwanitmkit, S., & Wiwanitkit, V. (2024). Artificial Intelligence, Academic Publishing, Scientific Writing, Peer Review, and Ethics. *Brazilian Journal of Cardiovascular Surgery*, *39*(4), e20230377. DOI: 10.21470/1678-9741-2023-0377 PMID: 39038191

Wiz Experts Team. (2024). Best practices for IAM configuration in multi-cloud setups. *Cloud Security Insights*, *7*(3), 112–125.

Wiz Experts Team. (2024). Securing APIs in Multi-Cloud Environments. *Cloud Security Best Practices*, *9*(1), 11–27.

Wiz Experts Team. (2024, October 11). *Multi-cloud security*. Wiz. www.wiz.io/academy/multi-cloud-security

Wiz. (2024). *Wiz cloud security platform*. https://www.wiz.io/platform

Wright, J., Avouris, A., Frost, M., & Hoffmann, S. (2022). Supporting academic freedom as a human right: Challenges and solutions in academic publishing. *International Journal of Human Rights*, *26*(10), 1741–1760. DOI: 10.1080/13642987.2022.2088520

Wright, J., Dawson, M. E.Jr, & Omar, M. (2012). Cyber security and mobile threats: The need for antivirus applications for smartphones. *Journal of Information Systems Technology and Planning*, *5*(14), 40–60.

Xu, J., Xiao, Y., Wang, W. H., Ning, Y., Shenkman, E. A., Bian, J., & Wang, F. (2022). Algorithmic fairness in computational medicine. *EBioMedicine*, •••, 84. PMID: 36084616

Ya'u, M. S., & Mohammed, M. S. (2025). AI-Assisted Writing and Academic Literacy: Investigating the Dual Impact of Language Models on Writing Proficiency and Ethical Concerns in Nigerian Higher Education. *International Journal of Education and Literacy Studies*, *13*(2), 593–604. DOI: 10.7575/aiac.ijels.v.13n.2p.593

Yamazaki, K., Vo-Ho, V.-K., Bulsara, D., & Le, N. Q. K. (2022). Spiking neural networks and their applications: A review. *Brain Sciences*, *12*(7), 863. DOI: 10.3390/brainsci12070863 PMID: 35884670

Yang, J., Soltan, A. A., Yang, Y., & Clifton, D. A. (2022). Algorithmic fairness and bias mitigation for clinical machine learning: Insights from rapid COVID-19 diagnosis by adversarial learning. medRxiv, 2022-01. DOI: 10.1101/2022.01.13.22268948

Yang, J., Soltan, A. A., Eyre, D. W., & Clifton, D. A. (2023). Algorithmic fairness and bias mitigation for clinical machine learning with deep reinforcement learning. *Nature Machine Intelligence*, *5*(8), 884–894. DOI: 10.1038/s42256-023-00697-3 PMID: 37615031

Yang, Y., Lin, M., Zhao, H., Peng, Y., Huang, F., & Lu, Z. (2024). A survey of recent methods for addressing AI fairness and bias in biomedicine. *Journal of Biomedical Informatics*, *154*, 104646. DOI: 10.1016/j.jbi.2024.104646 PMID: 38677633

Yao, S., Zhao, J., Yu, D., Du, N., Shafran, I., Narasimhan, K., & Cao, Y. (2023). ReAct: Synergizing reasoning and acting in language models. The Eleventh International Conference on Learning Representations. https://openreview.net/forum?id=WE_vluY6ZG

Yao, Z., & Yu, H. (2025). A Survey on LLM-based Multi-Agent AI Hospital.

Zainab, H., Khan, A. R. A., Khan, M. I., & Arif, A. (2025). Ethical Considerations and Data Privacy Challenges in AI-Powered Healthcare Solutions for Cancer and Cardiovascular Diseases. *Global Trends in Science and Technology*, *1*(1), 63–74. DOI: 10.70445/gtst.1.1.2025.63-74

Zangana, H. M. (2024). Exploring Blockchain-Based Timestamping Tools: A Comprehensive Review. *Redefining Security With Cyber AI*, 92-110.

Zangana, H. M. (2024). Exploring the Landscape of Website Vulnerability Scanners: A Comprehensive Review and Comparative Analysis. *Redefining Security With Cyber AI*, 111-129.

Zangana, H. M., Mustafa, F. M., Mohammed, A. K., & Omar, M. (2025). The Role of Change Control Boards in Ensuring Cybersecurity Compliance for IT Infrastructure. *JITCE (Journal of Information Technology and Computer Engineering)*, *9*(1). Retrieved from https://jitce.fti.unand.ac.id/index.php/JITCE/article/view/303

Zangana, H. M., Omar, M., Al-Karaki, J. N., & Mohammed, D. (2024). Comprehensive Review and Analysis of Network Firewall Rule Analyzers: Enhancing Security Posture and Efficiency. *Redefining Security With Cyber AI*, 15-36.

Zangana, H. M., Mohammed, A. K., Sallow, A. B., & Sallow, Z. B. (2024). Cybernetic Deception: Unraveling the Layers of Email Phishing Threats. [INJURATECH]. *International Journal of Research and Applied Technology*, *4*(1), 35–47.

Zangana, H. M., & Mustafa, F. M. (2024). Hybrid Image Denoising Using Wavelet Transform and Deep Learning. *EAI Endorsed Transactions on AI and Robotics*, *3*. Advance online publication. DOI: 10.4108/airo.7486

Zangana, H. M., Mustafa, F. M., & Omar, M. (2024). A Hybrid Approach for Robust Object Detection: Integrating Template Matching and Faster R-CNN. *EAI Endorsed Transactions on AI and Robotics*, *3*. Advance online publication. DOI: 10.4108/airo.6858

Zangana, H. M., & Omar, M. (2020). Threats, Attacks, and Mitigations of Smartphone Security. *Academic Journal of Nawroz University*, *9*(4), 324–332. DOI: 10.25007/ajnu.v9n4a989

Zangana, H. M., & Omar, M. (2025). The Role of Leadership in Advancing Inclusive Health Technologies. In Burrell, D., & Nguyen, C. (Eds.), *New Horizons in Leadership: Inclusive Explorations in Health, Technology, and Education* (pp. 203–220). IGI Global Scientific Publishing., DOI: 10.4018/979-8-3693-6437-6.ch009

Zangana, H. M., Sallow, Z. B., & Omar, M. (2025). The Human Factor in Cybersecurity: Addressing the Risks of Insider Threats. *Jurnal Ilmiah Computer Science*, *3*(2), 76–85. DOI: 10.58602/jics.v3i2.37

Zangana, H., Al-Karaki, J., & Omar, M. (Eds.). (2025). *Revolutionizing Cybersecurity With Deep Learning and Large Language Models*. IGI Global., DOI: 10.4018/979-8-3373-3296-3

Zangana, H., & Omar, M. (Eds.). (2025). *Leveraging Large Language Models for Quantum-Aware Cybersecurity*. IGI Global., DOI: 10.4018/979-8-3373-1102-9

Zhang, D. C., Highhouse, S., & Nye, C. D. (2019). Development and validation of the General Risk Propensity Scale (GRiPS). *Journal of Behavioral Decision Making*, *32*(2), 152–167. DOI: 10.1002/bdm.2102

Zhao, H., Benomar, Z., Pfandzelter, T., & Georgantas, N. (2022). Supporting multi-cloud in serverless computing. *In 2022 IEEE/ACM 15th International Conference on Utility and Cloud Computing (UCC)* (pp. 285–290). IEEE.

Zhao, L., Benomar, O., Pfandzelter, T., & Georgantas, N. (2022). Multi-cloud orchestration: Current practices and challenges. *IEEE Software*, *39*(5), 53–59.

Zohouri, M., Sabzali, M., & Golmohammadi, A. (2024). Ethical considerations of ChatGPT-assisted article writing. *Synesis (ISSN 1984-6754)*, *16*(1), 94-113.

Żywiołek, J. (2024). Empirical Examination Of Ai-Powered Decision Support Systems: Ensuring Trust And Transparency In Information And Knowledge Security. *Scientific Papers of Silesian University of Technology. Organization & Management/ Zeszyty Naukowe Politechniki Slaskiej. Seria Organizacji i Zarzadzanie*, (197).

About the Contributors

Hewa Majeed Zangana is an Assistant Professor currently affiliated with Duhok Polytechnic University (DPU) in Iraq. He holds a Doctorate in Philosophy (PhD) degree in ITM, which he is currently pursuing at DPU. Prior to his current role, Hewa Majeed Zangana has held various academic and managerial positions. He previously served as an Assistant Professor at Ararat Private Technical Institute and a Lecturer at DPU's Amedi Technical Institute, and Nawroz University. He has also held administrative positions such as Curriculum Division Director - Presidency of DPU, Information Unit Manager of The Research Center at Duhok Polytechnic University, Head of the Computer Science Department at Nawroz University, and Acting Dean of the College of Computer and IT at Nawroz University. Hewa Majeed Zangana's research interests span a wide range of topics in computer science, including network systems, information security, mobile communication, data communication, and intelligent systems. He has published extensively in peer-reviewed journals such as Inform: Jurnal Ilmiah Bidang Teknologi Informasi dan Komunikasi, Indonesian Journal of Education and Social Science, TIJAB, INJIISCOM, IEEE, and AJNU. He has also published multiple books with IGI Global. In addition to his research contributions, Hewa Majeed Zangana has been actively involved in editorial roles, serving as a Reviewer for Qubahan Academic Journal and the Scientific Journals of Nawroz University. He has also been a member of various scientific committees and administrative bodies at Nawroz University, including the Scientific Curriculum Development Committee, the Student Follow-up Program Committee, and the Committee for the Preparation of the Rules of Procedure for Consultative Offices.

***

Noble Worlanyo Antwi is a cybersecurity and cloud security researcher with a strong focus on multi-cloud threat detection and computer forensics. He is currently pursuing a Master of Science in Applied Cybersecurity and Digital

Forensics at Illinois Institute of Technology. His research interests include advanced threat detection in cloud environments, digital forensics, and secure infrastructure design. Noble holds industry certifications including AWS Certified Security – Specialty, AWS Certified Solutions Architect – Associate, Aviatrix ACE Associate, and Aviatrix ACE Operations Specialty, as well as CompTIA Security+. His professional background spans roles in cloud engineering, systems administration, and cybersecurity analysis, where he has led initiatives in security architecture development, forensic investigations, and threat monitoring. He is particularly passionate about integrating Artificial Intelligence and Zero Trust principles to enhance cybersecurity resilience in complex, distributed environments, and continues to explore innovative solutions in cloud-native security and digital forensics.

Ben Kennedy is an Artificial Intelligence researcher affiliated with Capitol Technology University in Laurel, Maryland. His research focuses on biologically inspired modular architectures, notably the Synthetic Cognitive Augmentation Network (SCAN), emphasizing adaptive learning, advanced cognitive agent integration, and human-centered AI collaboration. Kennedy actively develops sophisticated experimental platforms designed to enhance cognitive augmentation, align AI agents with human cognitive processes, and ensure ethical integration within complex human-AI systems.

Firas Mahmood Mustafa holds a Ph.D. in Computer Engineering from Mosul University, Iraq. His academic journey began with a B.Sc. in Electrical Engineering (Electronics and Communication), graduating in the top quarter of his class. He earned an M.Sc. in Computer Engineering from Mosul University in 2000. Mustafa joined the Computer Science Department at AlHadba University in 2003 and completed his Ph.D. in 2007. From 2013 to 2017, he was with DPU University, and from 2017 to 2020, he chaired the CCE Department at Nawroz University. An active participant in Erasmus+ and IREX programs, he now teaches at DPU University, shaping future computer engineering professionals.

Index

A

www.ingramcontent.com/pod-product-compliance
Lightning Source LLC
Chambersburg PA
CBHW080709220326
41598CB00033B/5356